Waste Input-Output Analysis

ECO-EFFICIENCY IN INDUSTRY AND SCIENCE

VOLUME 26

For other titles published in this series, go to
www.springer.com/series/5887

Waste Input-Output Analysis

Concepts and Application to Industrial Ecology

Shinichiro Nakamura and Yasushi Kondo

 Springer

Shinichiro Nakamura
Graduate School of Economics
Waseda University
1-6-1 Nishi-waseda
Shinjuku-ku, Tokyo, 169-8050
Japan
nakashin@waseda.jp

Yasushi Kondo
Graduate School of Economics
Waseda University
1-6-1 Nishi-waseda
Shinjuku-ku, Tokyo, 169-8050
Japan
ykondo@waseda.jp

ISBN: 978-1-4020-9901-4   e-ISBN: 978-1-4020-9902-1

Library of Congress Control Number: 2008943987

Printed on acid-free paper

9 8 7 6 5 4 3 2 1

springer.com

Preface

Industrial ecology (IE) is a rapidly growing scientific discipline that is concerned with the sustainability of industrial systems under explicit consideration of its interdependence with natural systems. In recent years, there has been an ever-increasing awareness about the applicability of Input-Output Analysis (IOA) to IE, in particular to LCA (life cycle assessment) and MFA (material flow analysis). This is witnessed in the growing number of papers at ISIE (International Society for Industrial Ecology) conferences, which use IOA, and also by the installment of subject editors on IOA in the International Journal of Life Cycle Assessment. It can be said that IE has become a major field of application for IOA. The broadening of users of IOA from various backgrounds implies a need for a self-contained textbook on IOA that can meet the needs of students and practitioners without compromising on basic concepts and the latest developments. This book was written with the aim of filling this need, and is primarily addressed to students and practitioners of IE.

As the title suggests, the core contents of the book have grown out of our research in IOA of waste management issues over the last decade. We have been fascinated by the versatile nature of IOA with regard to various technical issues of waste management in particular, and to IE in general. For us (both economists by training), IOA has turned out to be extremely useful in establishing productive communication with scientists and engineers interested in IE.

Many people have helped us in writing this book. Special thanks go to Kohji Takase and Shigemi Kagawa for helpful comments on major parts of the draft, and to Anthony Newell for his kind help in checking the English. We have also benefited from helpful correspondence with Manfred Lenzen, Keisuke Nansai, Helga Weisz, Erik Diezenbacher, Jan Minx, Makiko Tsukui, Kiyoshi Fujikawa, Noboru Yoshida, Ettore Settanni, and Yasuaki Iwasaki. We would like to thank Chen Lin, too, for his assistance in preparing TeX files.

Tokyo
November 2008

Shinichiro Nakamura
Yasushi Kondo

v

Contents

List of Figures

List of Tables

Chapter 1
Introduction

Abstract This chapter gives the aim of the book, and an outline of its contents as well. With its explicit consideration of the physical flows of waste and the activity of waste management (including the recycling of waste materials), the Waste Input-Output model (WIO) has made it possible to take into account the entire life cycle phases of a product within the framework of (hybrid) IOA. This is a book on the concepts and application of WIO for students and practitioners of IE. In order to make it a self-contained textbook on IOA, the first part is devoted to the standard IOA, while the second part is reserved for issues which are closely related to WIO including the environmental extensions of IOA.

1.1 The Aim of the Book

With regard to application to issues related to product life cycles, the standard IOA suffers from the weakness that it does not consider the physical aspects of the end-of-life phase, that is, the physical flows of waste and the activity of waste management. While the standard IOA has been extended to environmental IO models by Leontief [6] and Duchin [2], among others, their applicability to issues of waste and waste management has been of a rather limited nature: they assume the presence of a one-to-one correspondence between wastes and waste treatments, which does not reflect the realities of waste management. To cope with this problem, we developed a Waste Input-Output model (WIO) [7,8] that explicitly considers the physical flows of waste and the activity of waste management, including the recycling of waste materials.

With its closure of the loop of a product life cycle, WIO has made it possible to take into account the entire life cycle phases of a product within the framework of (hybrid) IOA. The major aim of this book is to make WIO accessible to students and practitioners of IE. In-depth discussion on the concepts and applications of WIO constitutes a distinguishing feature of this book.

S. Nakamura, Y. Kondo, *Waste Input-Output Analysis*, Eco-Efficiency in Industry and Science 26, © Springer Science+Business Media B.V. 2009

1.2 Outline of the Book's Content

Because WIO involves a conceptual extension of IOA, we find it beneficial to the readers, especially to those with little knowledge or experience of IOA, to provide opportunities to become familiar with the most basic concepts of IOA. We therefore divide the book into two parts, devoting the first part to the standard IOA, while the second part is reserved for the environmental extensions of IOA that are relevant to WIO.

Figure 1.1 shows the outline of the book, and the interdependence among its components, that is, chapters. The large arrows connecting the boxes referring to chapters indicate the logical connections. For instance, Chapter 5 uses the concepts discussed in Chapter 3, while Chapter 3 uses the concepts discussed in Chapter 2. On the other hand, Chapter 4 is only loosely related to the other chapters: there is no large arrow going into or coming from this chapter. The arrows in broken lines refer to individual items in the respective chapters on which the subjects being pointed

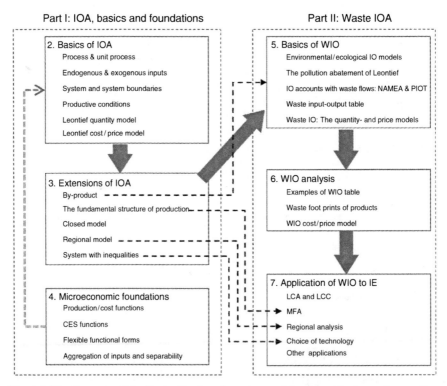

Fig. 1.1 The Structure of the Book: The Interdependence of Sections. The Large Arrows Connecting the Boxes Referring to Chapters Indicate the Logical Connections. The Arrows in Broken Lines Refer to Individual Items in the Respective Chapters on Which the Subjects Being Pointed to Are Based.

to are based. For instance, "MFA" in Chapter 7 makes use of the concept of "the fundamental structure of production" in Chapter 3. The exception to the above rule is the arrow in a broken line connecting the item "by-product" to Chapter 5, which indicates that the entire chapter is based on the concept of "by-product" as discussed in Chapter 3.

Readers with little or no prior knowledge of IOA are recommended to start from Chapter 2, which begins with the simplest case of a one-sector model. While this case may appear odd with the multi-sectoral feature of IOA, we find this case useful to help readers become familiar with the basic concepts of IOA without being unduly bothered by mathematics: simple arithmetic suffices. Readers with little knowledge of IE will also find the chapter useful because it introduces the basic concepts of IE such as a process, a unit process, a system, a system boundary, and a functional unit. The simplest model is subsequently extended to the general case of n sectors based on matrix algebra, the basics of which are also briefly explained to the extent required for the practical use of IOA.

Chapter 3 deals with the extensions of IOA that are necessary for practical applications, namely regional (spatial) extensions and the inclusion of by-products, as well as other topics of interests for IE. Regional extensions are necessary to account for the plain fact that economies are connected to each other through the transboundary flows of goods and services, just as the sectors of an economy are connected to each other through the flow of goods and services. The US–Japan international IO tables are shown as a real example of a multi-regional IO table.

Waste and emissions, which are objects of great concern in IE, are typical examples of by-products. Two definitions of by-product, by-product I and by-product II, are introduced, which play a fundamental role in the discussion on WIO in Chapter 5. Other topics to be dealt with include the closing of the model, the replacement of equalities by inequalities in IOA, the fundamental structure of production, and the Ghosh model.

The closing of the IO model refers to making endogenous the final demand sectors, which are exogenous in the standard IOA. The introduction of inequalities refers to allowing for the presence of resource constraints, and also opens the way for considering the possibility of choice among alternative processes, an issue which will also be discussed in Chapter 7 within the framework of WIO. The fundamental structure of production refers to the identification of basic patterns in the IO table that can be obtained by rearranging the ordering of sectors. This concept will also be used in Chapter 7 in the derivation of a new methodology of WIO-based MFA.

Chapter 4 is devoted to providing some basics of economics that are closely related to IOA. IOA has its origins in economics, and is deeply rooted in microeconomics in particular. Basic knowledge of the microeconomic foundations of IOA will therefore be useful for a better understanding of its underlying features. This will particularly be the case when one is interested in generalizing the basic framework of IOA to accommodate topics of interest that may not be adequately dealt with within the existing framework. First, some basic concepts of great relevance to IOA in microeconomics, such as the concept of production- and cost functions, will be discussed. When it comes to practical application, general concepts need to be specified in order to be quantitatively implemented using real data. Issues related to the specification of technology in microeconomics constitute the second topic of this chapter. Some remarks on CGE (the Computable General Equilibrium model) with regard to its specification of technology will also be made. Finally, as a summary of the whole discussion, the microeconomic characteristics of IOA will be derived. Readers with little interest in economics can skip this chapter without significant loss.

Chapter 5 introduces the basic concepts of WIO tables and the WIO model. WIO is a variant of environmental Input-Output (EIO) models, and focuses on issues of waste management (waste treatment and recycling). Familiarity with major EIO models is helpful in understanding the main features of WIO. This chapter starts with a brief review of EIO models from rather broad areas, which include economy–environment (ecology) IO tables and the models of Daly [1], Isard [4], and Hannon [3], IO-based energy analysis, and IO models of emissions. It then focuses on EIO models with a close affinity to WIO that deal with pollution (waste) and its abatement (treatment), including the seminal work of Leontief [6], and its extension by Duchin, as well as the development of input-output tables incorporating waste flows (the Dutch NAMEA [5] and the German PIOT [12]). The relevance of EIO models to waste management issues is discussed, and the need for its extension is pointed out. The concept of WIO is then introduced as a concrete solution to meet this need.

Chapter 6 is concerned with the application of the WIO model to real data, and the derivation of the WIO cost-price model as well. First, as a real example of a WIO table, a WIO table for a small city in northern Japan is shown, characterized by ambitious management programs of municipal solid waste (MSW) with a remarkably high rate of recycling. Secondly, large scale WIO tables for Japan with 50 to about 400 endogenous sectors with dozens of waste types are introduced, and the results of applications based on the WIO quantity model are shown. Also dealt with in this chapter is the issue of adjusting for the imports of waste in WIO, which were not touched upon in Chapter 5.

A distinguishing feature of a waste treatment process consists in its dependence on the chemical properties of the incoming waste. The third topic of this chapter deals with the consideration of this "dynamic nature of a waste treatment process" in WIO by use of engineering models. Fourth, the cost and price counterpart of the WIO quantity model, the WIO price model [9], is derived. Numerical examples are given for illustration.

Finally, Chapter 7 deals with the application of WIO to LCA, environmental life cycle costing (LCC) [12], and MFA. Closing the loop of a product life cycle calls for introducing the use phase into WIO, which remained unconsidered in the previous chapters. Alternative ways are shown for introducing the use phase into WIO. The closing of a product life cycle results in the WIO-LCA model, which can be used for life cycle inventory analysis. Introducing the use phase into the WIO price model discussed in Chapter 6 gives the cost counterpart of WIO-LCA, the WIO-LCC model, which can be used for an environmental LCC. Numerical examples are given to illustrate these methods, followed by examples of application to real data on air-conditioners and washing machines.

We then proceed to a new methodology of MFA, WIO-MFA, which is capable of considering the physical flows of any number of materials simultaneously at the level of detail determined by the underlying IO table [10, 11]. In particular, this methodology provides a simple and economical way to convert a monetary input-output table to a physical input-output table, the independent compilation of which can be quite costly. This methodology makes use of information on the material composition of products, which can be estimated by exploiting the triangular nature of the matrix of input coefficients discussed in Chapter 3. Some results of application to the flow of metals in Japan are shown.

Other topics to be dealt with in Chapter 7 refer to the extensions to a regional model and to a linear programming problem (LP). The regional extension addresses the importance of regional aspects in waste management owing to the fact that the locations of waste generation may considerably differ from the locations where they can be recycled. Results of its implementation with reference to Japanese regional IO tables and to the IO table for Tokyo are shown. The extension to an LP addresses the issue associated with the possibility of the choice among alternative technologies referring to both production and waste treatment. A brief survey of other applications of WIO in diverse fields of Industrial Ecology closes this final chapter, and thus the book.

1.3 Putting IOA in Practice by Excel

Excel provides an easy and readily accessible way of applying IOA to real data. In particular, all the calculations in this book, except for the nonlinear WIO model based on a system engineering model, can be carried out by Excel. For those who are not familiar with the use of Excel for calculations involving arrays, which apply to IOA, we have prepared some examples where Excel is used for several topics discussed in the text. At the end of Chapter 2, the reader will find examples where Excel is used for calculating the matrix of input coefficients and the Leontief inverse matrix, for solving the Leontief quantity-and-price model, and for consolidating an input-output table to one with a smaller number of sectors.

The second set of examples with Excel appears at the end of Chapter 3. They refer to the Leontief quantity model in the presence of competitive imports, and the use of Solver to solve linear programming problems that occur in connection with the allocation of scarce resources and the choice of technology from among alternatives.

References

1. Daly, H. (1968). On economics as a life science. *Journal of Political Economy, 76*, 392–406.
2. Duchin, F. (1990). The conversion of biological materials and waste to useful products. *Structural Change and Economic Dynamics, 1*, 243–261.

3. Hannon, B. (1973). The structure of ecosystems. *Journal of Theoretical Biology, 41*, 535–546.

4. Isard, W., Bassett, K., Choguill, C., Furtado, J., Izumita, R., Kissin, J., Romanoff, E., Seyfarth, R., & Tatlock, R. (1968). On the linkage of socio-economic and ecologic systems. *Papers in Regional Science, 21*, 79–99.

5. Keuning, S. J., van Dalen, J., & de Haan, M. (1999). The Netherlands' NAMEA; presentation, usage and future extensions. *Structural Change and Economic Dynamics, 10*, 15–37.

6. Leontief, W. (1970). Environmental repercussions and the economic structure: An input-output approach. *Review of Economics and Statistics, 52*, 262–271.

7. Nakamura, S. (1999). Input-output analysis of waste cycles. In First International Symposium on Environmentally Conscious Design and Inverse Manufacturing, Proceedings. Los Alamitos, CA: IEEE Computer Society.

8. Nakamura, S. & Kondo, Y. (2002). Input-output analysis of waste management. *Journal of Industrial Ecology, 6*(1), 39–63.

9. Nakamura, S. & Kondo, Y. (2006). A waste input output life-cycle cost analysis of the recycling of end-of-life electrical home appliances. *Ecological Economics, 57*, 494–506.

10. Nakamura, S. & Nakajima, K. (2005). Waste input-output material flow analysis of metals in the Japanese economy. *Materials Transactions, 46*, 2550–2553.

11. Nakamura, S., Nakajima, K., Kondo, Y., & Nagasaka, T. (2006). The waste input-output approach to materials flow analysis concepts and application to base metals. *Journal of Industrial Ecology, 11*(4), 50–63.

12. Stahmer, C. Kuhn, M., & Braun, N. (1998). Physical input-output tables for Germany, 1990. Eurostat Working Papers 2/1998/B/1, European Commission, Luxembourg.

Part I
Input-Output Analysis

Chapter 2
Basics of Input-Output Analysis

Abstract This chapter deals with the basic concepts of input-output analysis (IOA) such as the representation of a production process in terms of the flow of inputs and outputs, the input and output coefficients, the Leontief inverse coefficients, the productive conditions, the Leontief quantity- and price models, as well as input-output tables. First we consider the simplest case of an economy consisting of a single producing sector. While this case may appear odd with the multi-sectoral feature of IOA, we find this case useful to help readers become familiar with the basic concepts of IOA without being unduly bothered by mathematics: simple arithmetic suffices. The multi-sectoral feature of IOA is then considered for the case of an economy with two producing sectors. The Leontief quantity- and price models are also derived, which involve systems of simultaneous equations. Rewriting these systems of simultaneous equations in terms of matrix algebra yields the basic results of IOA that can be applied to the general case of n producing sectors.

2.1 The One-Sector Model

For a simple economy that consists of one production sector, rice, the basic concepts of input-output analysis (henceforth, IOA) are introduced. Consideration of this simple case makes it possible to convey to the readers the major concepts of IOA such as input coefficients, Leontief inverse coefficients, the Leontief quantity- and price models, issues of substitution among alternative inputs, and environmental applications, based on simple calculation involving scalars alone without resorting to matrix algebra.

2.1.1 Input and Output in a Productive Economy

Quantitative representation of production and consumption processes is at the heart of IOA as well as industrial ecology (IE). A process is represented by the flow

S. Nakamura, Y. Kondo, *Waste Input-Output Analysis*, Eco-Efficiency in Industry and Science 26, © Springer Science+Business Media B.V. 2009

of inputs and outputs. Combination of processes with mutually interrelated flows of inputs and outputs gives a system. Based on the definition of the system under consideration, inputs can further be divided into endogenous and exogenous inputs. A process can be reduced to a unit process when the amounts of inputs and outputs are proportional to the level of production or activity. For a system composed of given unit processes, its productiveness is determined by the features of the unit processes.

2.1.1.1 Representation of a Process

It is usual to define a (production or consumption) process as a black box, in the sense that the reaction or transformation processes within the box are not taken into account, and one is concerned only with the inputs to and the outputs from the process ([2], p.38). Following this convention, a production process is quantitatively represented by the amount of inputs and outputs entering and leaving the process (Figure 2.1).

A sector is characterized by the operation/activity of a given process. In the current case of a simple economy, the rice production sector is characterized by the operation of the rice production process. Another important part of the economy is the household sector characterized by the consumption activity.

For the sake of simple illustration, suppose in the production process of rice, that rice seeds, water, land, and labor occur as the inputs, and rice as well as waste (residues), such as leaves and stalks, occur as the outputs (we abstain from counting as an item of output the water leaving the process, including that which has evaporated, although it may be necessary in a Material Flow Analysis (MFA) study, where one pays great attention to the conservation of mass between inputs and outputs) (Figure 2.2). The input of land refers to the use of a given size and type of land or the use of services provided by that land over a certain period of time, and is measured in square meters. In a similar vein, the input of labor refers to the use of services supplied by a given number and type of workers over a certain period of time, and is measured by hours of work. In the following discussion, each type of inputs and outputs is assumed to be homogeneous; in particular, it is assumed that there is no qualitative difference in land and labor.

Fig. 2.1 The Black Box Representation of a Production Process with Three Inputs and Two Outputs: The Circles Refers to Inputs and Outputs, While the Box Refers to a Process. The Arrow Indicates the Direction of the Relevant Flow.

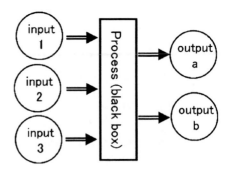

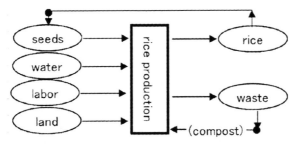

Fig. 2.2 The Production Process of Rice. A Part of Production Waste Reenters the Process as an Ingredient of Compost.

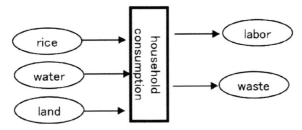

Fig. 2.3 Process Representation of Household Consumption.

The use of agricultural tools is neglected (except for primitive ones which can be produced without any specialization), because by assumption there are no sectors that specialize in the production of agricultural tools or in the production of materials such as metals from which the tools would be made. For the same reason, the input of fertilizer is also neglected because of the absence of a sector specializing in its production. However, it is assumed that after harvesting rice its residues are left in the field to decay and become compost that helps maintain the fertility of the field.

In a similar fashion, household consumption can be represented as a process where rice, water, and land (for dwelling) enter as inputs, and labor and waste (consumption waste, waste water, and human waste) occur as outputs (Figure 2.3). For the sake of simplicity, the use of utensils and shelter (to the extent they need specialized supplying sectors) is also neglected. Waste from the household sector can also be returned to the production process as an ingredient of compost.

2.1.1.2 Exogenous Inputs and Endogenous Inputs

An economy can be represented as a system where processes are mutually connected by the web of the flow of inputs and outputs among them. Integration of the production (Figure 2.2) and consumption (Figure 2.3) processes through its flow of inputs and outputs results in Figure 2.4, which gives a system view of this simple economy.

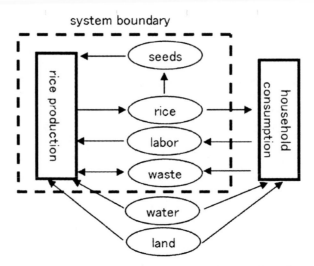

Fig. 2.4 The System with Rice Production and Household Consumption. The Area Circled by the Broken Line Refers to the System Boundary. The Flows Generated Inside the Boundary are Endogenous Flows, and the Remaining Flows are Exogenous.

Note that while the flows of water and land occur in Figure 2.4, there are no corresponding sectors that supply these inputs. An input, the supply of which does not occur in the system, is called exogenous or primary input. On the other hand, an input is called endogenous input when its supply occurs within the system. In other words, exogenous inputs refer to inputs with no corresponding sectors in the system under consideration. Henceforth, more frequent use will be made of the terms exogenous inputs than primary inputs because the same input can be both endogenous and exogenous, depending on the definition of the system under consideration, that is, whether it is inside the system (within the system boundary) or not.

In the economy under consideration, water and land are exogenous inputs, which are given from outside the system. This would correspond to the case where water is supplied by a river or rainfall, that is, by nature without any involvement of human activity. There is thus no sector in this economy that takes care of the water supply. As for land, the substantial work involved in its additional acquisition (for instance, land reclamation by drainage) would justify treating its level as given. Note also that of the inputs occurring in Figure 2.4, land is an item that cannot be used up in the production or consumption process: it is a durable, that can be used (for agricultural production) over an extensive length of time if proper care of it is taken (to keep its nutrient, physical, and chemical conditions, etc.). In the case of durables, it is useful to make a distinction between the flow of services that is generated by a durable and the stock of the durable itself. In the case of land, what is used in the production process is the land service.

The concept of service provided by durables, or capital service, can easily be understood if one considers the case of renting a car. The fee one pays for using the car is the price of the service provided by the car over a given period of time, but not the purchase price for owning the car itself. The car is not used up by its supply of service, and remains intact, except for a certain amount of wear, while gasoline and oil *are* used up, and transformed into outputs such as waste, or emissions. The inputs that are used up in production are called current inputs. Capital and labor services are counted as current inputs. The inputs occurring in a production process are current inputs.

Labor is an output of the household sector. To be exact, it is the labor service provided by household members. Hence, it may appear legitimate to treat labor as an endogenous input, the level of which depends on the amount of consumption. If one considers labor as the output of the household sector, and the consumption as the input that is necessary to reproduce labor, the household sector could formally be regarded as a production sector like any other production process. The household consumption process would then emerge as a process where foods and other consumption goods are transformed into human labor just as coal is transformed into electricity in electric power plants. In fact, in his 1951 book [13], Leontief proposed a model of this sort.

With the household sector integrated as a production process, one could obtain a system that is closed with regard to seeds, labor, and consumption. As will be shown later in Section 3.4.1, apart from the highly controversial nature of representing consumption behavior like any other production process, a system where the household sector is endogenized is not convenient for scenario analysis, which is of vital importance for a life cycle assessment (LCA) in particular, and IE in general. Except where otherwise stated, therefore, this book is concerned with the case where the household sector is exogenous.

In the current framework, the rice seeds thus occur as the only endogenous input, the level of which is to be determined within the system. In Figure 2.4, the square with dotted line represents the system boundary between the endogenous and exogenous sectors. The rice production emerges as the only endogenous sector, to the detailed analysis of which we now turn.

2.1.1.3 The Unit Process

Diagrammatic representation of the process in Figure 2.2 now needs to be quantified. Henceforth, the quantity of an endogenous input/output is denoted by x, and the quantity of an exogenous input/output by z. Suppose that a set of observed values were obtained for the process: $x' > 0$ kg of rice was produced (together with $w' \geq 0$ kg of waste) per year by use of x'_1 kg of rice seeds, z'_1 m^3 of water, z'_2 h of homogeneous labor work, and z'_3 m^2 of homogeneous land. Note that x' and z'_is refer to specific values (or realizations) of x and z_is, whereas the latter refer to variables, which can take any value.

Table 2.1 An Example of
Rice Production Process.

Inputs/outputs	Unit	Amount
Seeds	kg	3.8
Water	m^3	3,000
Labor	h	3,493
Land	m^2	1,000
Rice	kg	524
Waste	kg	1,572

Source: The values for seeds, rice, and
land are from [16]. All the remaining values are hypothetical ones.

This process can then be quantitatively represented by a 6×1 column array of
inputs and outputs as follows:

$$\text{Process}: \left(\frac{\text{inputs}}{\text{outputs}}\right) = \begin{pmatrix} x_1' & : \text{rice seeds (kg)} \\ z_1' & : \quad \text{water (m}^3) \\ z_2' & : \quad \text{labor (h)} \\ z_3' & : \quad \text{land (m}^2) \\ \hline x' & : \quad \text{rice (kg)} \\ w' & : \quad \text{waste (kg)} \end{pmatrix} \tag{2.1}$$

Because no rice can be obtained without using all four inputs, $x_1' > 0$ and $z_i' > 0$ for
all i.

For illustration, Table 2.1 gives a hypothetical numerical example of (2.1). The
values for the input of seeds and land, and the output of rice are based on the
Japanese data for rice production 2005 [16], according to which the sowing of 3.8 kg
per 100 m^2 yielded 524 kg of rice. The value for water is from [7]. The values for
the remaining inputs and outputs are hypothetical ones based on rough estimates,
and should not be taken seriously.

Denote by a_1 and b_i the average amount of rice seeds (kg) and the amount of
exogenous input i that were necessary to produce 1 kg of rice by the process P':

$$a_1 = \frac{x_1'}{x'}, \tag{2.2}$$

$$b_i = \frac{z_i'}{x'}, i = 1, 2, 3, \tag{2.3}$$

where b_1 refers to the amount of water (m^3) per kg of rice, b_2 to the amount of
labor (h) per kg of rice, and b_3 to the amount of land use (m^2) per kg of rice. The
coefficients a_1 and b_i are called the input coefficients. By definition, $a_1 > 0$, $b_i > 0$,
and the following holds

$$x_1' = a_1 x', \tag{2.4}$$

$$z_i' = b_i x', i = 1, 2, 3. \tag{2.5}$$

In general, (2.4) and (2.5) cannot be regarded as causal relationships between the amount of inputs and that of the product, because a_1 and b_i may depend on the level of production, and their values may change with the level of production. An increase in the production may require less than a proportional increase in some inputs, but may require more than a proportional increase in the use of other inputs. When a change (increase or decrease) in the amount of production is associated with the proportional change in all the inputs, the process is said to follow constant returns to scale: a doubling of the amount of all the inputs doubles the amount of rice, and the doubling of rice requires the doubling of all the inputs (see Section 4.2.2.3 below for further details on the concept of constant returns to scale).

Assume that the process (2.1) is subject to constant returns to scale. The input coefficients a_1 and b_i then become independent of the level of production, and hence (2.4) and (2.5) are no longer identities that hold for x', x'_1 and z'_i only, but can hold for any value of x, x_1, and z_i that might occur:

$$x_1 = a_1 x, \quad a_i > 0 \tag{2.6}$$

$$z_i = b_i x, \quad b_i > 0, \quad i = \{1,2,3\} \tag{2.7}$$

In economics, (2.6) and (2.7) refer to the factor demand or input demand functions (see Section 4.2.2.2 below for further details on microeconomic background). It is also assumed that there is a proportional relationship between the amount of rice production and waste:

$$w = g^{\text{out}} x \tag{2.8}$$

where $g^{\text{out}} > 0$ is a constant called the waste generation coefficient.

A unit process refers to the case where for a given process the level of production is at unity. For the process P', this corresponds to the case of $x = 1$. By use of a_1, b_i, and g^{out}, the unit process for P', say P, can be represented as follows:

$$P = \left(\frac{\text{inputs}}{\text{outputs}} \right) = \begin{pmatrix} a_1 & : & \text{rice seeds (kg)/rice (kg)} \\ b_1 & : & \text{water (m}^3\text{)/rice (kg)} \\ b_2 & : & \text{labor (h)/rice (kg)} \\ b_3 & : & \text{land (m}^2\text{/rice (kg)} \\ \hline 1 & : & \text{rice (kg)} \\ g^{\text{out}} & : & \text{waste (kg)/rice (kg).} \end{pmatrix} \tag{2.9}$$

Because of the assumption of constant returns to scale, P can be scaled up to any level of production. The level of output then refers to the level of activity by which the unit process is operated. For instance, (2.1) corresponds to the case where P is operated at the level of x'.

Table 2.2 gives an example of (2.9) that is obtained from the process data in Table 2.1. The production of 1,000 kg of rice would require 7 kg of rice seeds, 5,800 m^3 of water, 6,700 h of labor, and 1,900 m^2 of land.

Table 2.2 A Numerical
Example of Unit Process
for Rice Production.

Items	Notations	Units	Values
Inputs			
Seeds	a_1	kg/kg	0.007
Water	b_1	m³/kg	5.725
Labor	b_2	h/kg	6.667
Land	b_3	m²/kg	1.908
Outputs			
Rice		kg/kg	1.000
Waste	g^{out}	kg/kg	3.000

2.1.1.4 Productive Conditions

Up to now, the only condition that (2.9) had to satisfy was the positivity of its elements, that is, input coefficients. This condition, however, does not suffice for the
process to be "productive". A process is said to be productive if its operation yields
a surplus of the target output, the acquisition of which is the sole object of operation,
over the amount of inputs. In the present case where the economy consists of only
one production process, these conditions refer to the productiveness of the economy
as a whole, as well. Given that adequate amounts of exogenous inputs are available,
the condition of productiveness can be simply given by

$$x_1 < x \qquad (2.10)$$

or in terms of the input coefficient:

$$0 < a_1 < 1, \qquad (2.11)$$

which is called the productive condition.

If $a_1 \geq 1$, the amount of output (harvest) does not exceed the amount of
seeds used for input, and no rice will be left for the household to consume. This
corresponds to a situation where the economy is not productive. An unproductive
economy cannot survive (the people would starve to death), and hence would not
exist. Accordingly, for an economy to be productive, it has to meet (2.11). For the
unit process in Table 2.2, (2.11) is satisfied.

2.1.1.5 Intermediate Demand and Final Demand

Final Demand

Denoting the surplus of output over input in (2.10) by f, the following identity is
obtained:

$$x = x_1 + f \qquad (2.12)$$

This represents the allocation of x into two use (demand) categories that consist
of seeds for sowing and final consumption. The former is called the intermediate

demand, and the latter is called the final demand. Except for when otherwise stated, the final demand is assumed to be given from outside the model, that is, it is exogenous. Because the intermediate demand is explained by the model, it is endogenous. From (2.6), it follows that

$$x = \underbrace{a_1 x}_{\text{intermediate demand}} + \underbrace{f}_{\text{final demand}} . \qquad (2.13)$$

Productivity and the Level of Living

In economics, the very purpose of production is thought to consist in the satisfaction of final demand, that is, the wants of consumers. Independent of whether or not one is ready to accept this view, it is clear that f gives the material level of living standards of the economy (the issues of distribution among consumers and of the happiness of individuals, that is, the issue of whether an increase in the material level of consumption makes people happier, are not considered).

Rewriting (2.13) gives

$$f = (1 - a_1)x. \qquad (2.14)$$

For a given level of x, this gives the amount of f, that is, the material level of living that can be supported by x. The material level of living improves with a decrease in a_1, and deteriorates with an increase in a_1. A decrease in a_1 refers to an increase in the efficiency of production, or productivity.

2.1.2 The Leontief Quantity Model

We now turn to the derivation of the Leontief quantity mode, the most basic model in IOA, which can be used, among others, for estimating the amounts of exogenous inputs that are required to satisfy a given standard of living, and the effects on the level of production of a change in technology and or the consumer demand for products.

2.1.2.1 The Quantity Model and Leontief Inverse Coefficient

Provided that there is no limitation in the supply of exogenous inputs, we can consider the level of production that is necessary to satisfy a given level of final demand f under a given technology. Solving (2.14) for x yields:

$$x = (1 - a_1)^{-1} f. \qquad (2.15)$$

where the existence of the solution results from (2.11). This gives the level of output that is necessary to satisfy a given level of f, and is called the Leontief quantity model. The factor $(1 - a_1)^{-1}$ refers to the amount of rice that is required to fulfill a unit of final demand for it, and is called the Leontief inverse coefficient. Because the factor refers to the amount of product that is invoked by a unit of final demand, it is also called the output multiplier ([15], p.102).

For the unit process in Table 2.9, the multiplier becomes:

$$(1 - a_1)^{-1} = (1 - 0.007)^{-1} = 1.0073049, \tag{2.16}$$

which implies that 10^6 kg of final demand for rice invokes rice production of about 1,007,305 kg.

2.1.2.2 The Positivity Condition

Because of the productive condition (2.11)

$$(1 - a_1)^{-1} > 1 \tag{2.17}$$

and hence, the solution satisfies its positiveness, $x > 0$, for any $f > 0$. In other words, if the economy is productive, (2.14) always has a positive solution for any $f > 0$. On the other hand, if (2.14) has a positive solution for a particular $f > 0$, say, f^o, then the productive condition (2.11) follows. It is also true that if (2.14) has a positive solution for a particular $f^o > 0$, it has a positive solution for any $f > 0$. This can be stated as a theorem:

Theorem 2.1. *For (2.14), the following conditions (I), (II), and (HS) are identical.*

[I] For a particular value of final demand f, say $f' > 0$, (2.14) has a positive solution.
[II] For any positive value of f, (2.14) has a positive solution.
[HS] (2.11) holds.

From [II] it follows that an economy with positive final demand and production is always productive. In order to check for the productiveness of an economy with one production sector only, one does not need to check for (2.11). It suffices to see if its levels of consumption and production are positive.

2.1.2.3 Alternative Derivation of the Leontief Inverse Coefficient

Above, the Leontief inverse coefficient was obtained by solving (2.13). An alternative derivation based on the conversion of a series is shown, which will be useful to make clear its economic meaning.

In order to fulfill a unit of final demand for rice, at least one unit of it has to be produced. From (2.4), however, the production of a unit of rice requires the input of rice (for sowing) by a_1, which has to be produced in addition to the original one unit.

The additional production of rice by the amount of a_1, however, further requires the input of rice (for sowing) by a_1^2, that needs to be produced additional to $1 + a_1$. This process proceeds infinitely, and ends up in the sum of the following infinite series:

$$\underbrace{1}_{\text{direct requirement}} + \underbrace{a_1 + a_1^2 + \cdots}_{\text{indirect requirements}} = \sum_{t=0}^{\infty} a_1^t = (1 - a_1)^{-1} \qquad (2.18)$$

where the second equality is due to (2.11). The Leontief inverse coefficient thus refers to the amount of product that is necessary both directly and indirectly to fulfill a unit of final demand for it.

2.1.2.4 The Demand for Exogenous Inputs

With x given by (2.15), it is straightforward to obtain the amounts of the remaining inputs, exogenous inputs, z_i, that are necessary to realize f. Substituting from (2.7), they are given by

$$z_i = b_i (1 - a_1)^{-1} f, \quad i = 1, 2, 3. \qquad (2.19)$$

Taking labor input as an example, the term $b_2(1 - a_1)^{-1}$ refers to the amount of labor that is directly and indirectly required to deliver a unit of rice for the final demand. In fact, from (2.18) we have

$$\underbrace{b_2}_{\text{direct labor requirement}} + \underbrace{b_2(a_1 + a_1^2 + \cdots)}_{\text{indirect labor requirements}} = b_2(1 - a_1)^{-1} \qquad (2.20)$$

In other words, it refers to the amount of labor that is "embodied" in a unit of rice. Alternatively, (2.20) can be called the "employment multiplier" ([15], p.111) because it refers to the amount of employment that is invoked by a unit of final demand.

Similarly, the amount of water that is embodied in a unit of product 1 can be given by

$$\text{water m}^3/\text{rice kg} = b_1(1 - a_1)^{-1}, \qquad (2.21)$$

and the total amount (m^2) of a particular type of land that is occupied for the production of a unit of product by

$$\text{land m}^2/\text{rice kg} = b_3(1 - a_1)^{-1}. \qquad (2.22)$$

(2.21) corresponds to the concept of "virtual water" that refers to "the amount of water that is embedded in food or other products needed for its production" [25], and (2.22) to the concept of "ecological footprint" (EF) that refers to the amount of productive land required to support the consumption of a given population [22].

To the extent that the use of land and water for rice production represents environmental interventions, these equations give the relationship between environmental interventions, technology, and final demand. The input coefficients of exogenous inputs, b_1 and b_3 can then be called the intervention coefficients ([8], p.14).

For the unit process in Table 2.9, from (2.19), (2.21), and (2.22), the amount of water, labor, and land that are embodied in 1 kg of rice are respectively given by:

$$\text{Water (m}^3) = b_1(1-a_1)^{-1} = 5.725 \times 1.0073049 = 5.767, \qquad (2.23)$$

$$\text{Labor (h)} = b_2(1-a_1)^{-1} = 6.667 \times 1.0073049 = 6.715, \qquad (2.24)$$

$$\text{Land (m}^2) = b_3(1-a_1)^{-1} = 1.908 \times 1.0073049 = 1.922. \qquad (2.25)$$

It thus turns out that the 1 kg of rice embodies about 5.8 kg of water, 6.7 h of work, and 2 m^2 of (the service of) land.

2.1.2.5 Evaluating the Effects of a Change in Final Demand (Standard of Living) and in Technology

Within the current framework, the standard of living of households can be represented by the amount of rice that is available for final consumption, f. By use of (2.19), the effects on the demand for exogenous inputs of an exogenous change in f, Δf, say, due to an increase in the size of population, can easily be obtained:

$$\Delta z_i = b_i(1-a_1)^{-1}\Delta f, \quad i = 1,2,3. \qquad (2.26)$$

In a similar fashion, the effects of a given change in technology on the demand for exogenous inputs can also be obtained. Suppose for instance, that one is interested in comparing the effects on the demand for water of two production processes, say processes r and s. Denoting the input coefficients of these processes by superscripts r and s, the difference in the demand for water between them to satisfy a given f can be given by:

$$z_1^r - z_1^s = \left(b_1^r(1-a_1^r)^{-1} - b_1^s(1-a_1^s)^{-1}\right)f. \qquad (2.27)$$

Suppose that the unit process in Table 2.2 corresponds to process r, and that the process s requires less water, but has lower yield per seed with $b_1^s = 5.0$ and $a_1^s = .04$. It then follows from (2.23):

$$z_1^r - z_1^s = (5.767 - 5.0(1-.04)^{-1})f = 0.559f, \qquad (2.28)$$

which indicates that process s is more efficient than process r in water use even under consideration of lower yield.

2.1.2.6 Limited Supply of Exogenous Inputs

Implicit in (2.19) is the assumption that there is no immediate limitation in the supply of exogenous inputs that could hamper the realization of f. Human history, however, is full of instances where disruptions in the smooth supply of some exogenous inputs caused catastrophic impacts on the economy and human lives as well: a drought is a typical example. By use of the above framework, the possible effects of the limitation of particular exogenous inputs on the economy can also be considered.

Denoting by $*$ the given amount of supply of input i, the level of production is determined by:

$$x = \min\left(\frac{z_1^*}{b_1}, \frac{z_2^*}{b_2}, \frac{z_3^*}{b_3}\right).$$

(2.29)

which indicates that x is given by the minimum of the three arguments inside the parenthesis on the right hand side. The input item that gives the minimum is the scarcest input. The input of rice seeds does not occur in (2.29) because it is an endogenous input. If input 1 turns out to be the scarcest, the amount of rice production that is possible will be bounded from above by

$$x \le b_1^{-1} z_1^*.$$

(2.30)

and from (2.14) the amount of rice that is available for consumption, or the material standard of living in this economy, will be bounded from above by

$$f \le (1 - a_1) b_1^{-1} z_1^*,$$

(2.31)

The upper bound shows the maximum standard of living that is feasible under a given supply of the exogenous input and the state of technology.

Suppose that the available amounts of water, labor, and land are respectively given by $z_1^* = 10{,}000\,\text{m}^3$, $z_2^* = 15{,}000\,\text{h}$, and $z_3^* = 10{,}000\,\text{m}^2$ for the unit process in Table 2.9. It then follows from (2.29) that water is the scarcest input, and the maximal amount of rice production that is possible will be given by

$$\min\left(\frac{10000}{5.725}, \frac{15000}{6.667}, \frac{10000}{1.908}\right) = \min(1746, 2250, 5241) = 1746.$$

(2.32)

and the available amount for final consumption will be bounded from above by

$$f \le (1 - .007) \times 1746 = 1734.$$

(2.33)

2.1.3 Production, Income, and Consumption: The Input-Output Table

We now consider the flow of the product in this economy from the point of view of a system of national accounts that involves production, income, and expenditure.

2.1.3.1 Income and Its Source: The Case of No Land Ownership

In the current case, household consumption is the only item of expenditure. Because household consumption is paid out of household income, household income is the only item of income in this economy.

The source of household income is the compensation of services provided by the household. We first consider the simple case where there is no ownership of land: it can be used free of charge for rice production. Labor then occurs as the only item of exogenous inputs the use of which has to be compensated, and is hence the only source of income for the household. Because rice is the only product in this economy, labor has to be compensated by rice; rice is the form of income and the object of consumption at the same time.

The amount of product, x, thus needs to be distributed among seeds and labor. Of x, $a_1 x$ has to be left for (re)production. The rest $(1 - a_1)x$, can then be given to the households as compensation for the labor services they provided:

$$\underbrace{x}_{\text{the value of product}} = \underbrace{a_1 x}_{\text{the cost for seeds}} + \underbrace{(1 - a_1)x}_{\text{labor compensation}} \qquad (2.34)$$

Recall that in this economy rice is the standard of value, and hence its price, p, can be set at unity. The left hand side of (2.34) can therefore be regarded as the value of product, px. Analogously, the first term on the right hand side is the cost for seeds, px_1. The term $(1 - a_1)x$ can be interpreted as the value that was added by labor in the production process, and hence is also called value added, say v:

$$v = (1 - a_1)x \qquad (2.35)$$

From (2.14) and (2.35), the following identity follows between v and f:

$$\underbrace{v}_{\text{Value added=income}} = \underbrace{f}_{\text{final demand}} \qquad (2.36)$$

This corresponds to the identity of national accounts where the amount of income is equal to the amount of final demand.

Henceforth, the price of an endogenous input/output is denoted by p, and the price of an exogenous input is denoted by q. Denote by q_2 the wage per unit of hour, or the price of labor services, in terms of rice kg. Labor compensation per unit of rice is then given by $q_2 b_2$. Substitution of this into (2.35) gives

$$v = q_2 b_2 x = \pi x, \qquad (2.37)$$

where $\pi = q_2 b_2$ is called the rate of value added, because it gives the amount of value added per unit of product. Rewriting (2.37), an alternative representation of wage rate is obtained:

$$q_2 = v/(b_2 x) = (1 - a_1)/b_2. \qquad (2.38)$$

For the unit process in Table 2.9, (2.38) becomes

$$q_2 = \frac{(1 - .007)}{6.667\,\text{h/kg}} = 0.149\,\text{kg/h}, \tag{2.39}$$

which indicates that the wage is 0.149 kg of rice per hour.

2.1.3.2 Income and Its Source: The Case of Land Ownership

Consider next the case of land ownership, where the provision of land use for production has to be compensated for at the rate of q_3 units of rice per m^2 per year. Imagine, for instance, the case where rice production is done by tenant farmers, who pay the rents to land owners out of their harvest. As for the use of land for the habitat of households including both tenant farmers and land owners, however, its use is assumed free.

Adding this new income item (land rent), (2.35) now becomes:

$$v = q_2 b_2 x + q_3 b_3 x \tag{2.40}$$

and hence (2.34) becomes:

$$\underbrace{x}_{\text{the value of product}} = \underbrace{a_1 x}_{\text{the cost for seeds}} + \underbrace{q_2 b_2 x}_{\text{labor compensation}} + \underbrace{q_3 b_3 x}_{\text{land compensation}} \tag{2.41}$$

Note that q_2 is no longer given by (2.38). Recalling that $a_1 x$ refers to intermediate input, it can also be called intermediate cost. The remaining two terms on the right hand side can then be regarded as the cost for exogenous inputs. In short, (2.41) can be regarded as the equality between the value of output and the cost for production. From (2.12) and (2.40), it follows immediately that the accounting identity of national accounts (2.36) also holds in this case.

For (2.41), the rate of value added, π, is given by:

$$\pi = q_2 b_2 + q_3 b_3. \tag{2.42}$$

Because (2.41) gives the total cost of production, division of it by x gives the following expression for the cost per unit of output, that is, the unit cost:

$$1 = a_1 + \pi, \text{ or}$$
$$p = pa_1 + \pi, \tag{2.43}$$

where p refers to the price of rice, and the second equation follows from the fact that $p = 1$ due to rice serving as the unit of value.

2.1.3.3 An Input-Output Table

Above, the two fundamental identities of national accounts have been derived: one is (2.12), which refers to the allocation of product among different demand items, and the second is (2.41), which refers to the allocation of output among different cost items. By use of these identities, an IO table of the economy can be compiled.

Table 2.3 gives the result with the sector of origin (seller) listed on the left, and the same sector listed across the top as destinations (purchasers). Following (2.12), the second row of the table shows the allocation (sale) of the product to different demand categories, intermediate and final. On the other hand, following (2.41), the second column shows the cost elements, seeds, labor, and land. Note that both the sum of the three elements of the second column, and the sum of the two elements of the second row give rise to the amount of total output, which, at the same time, is also equal to the value of output, and the amount of total cost, as well. The total output is equal to the value of product because the unit of value is the price of rice, which is unity. The equality between the value of product and the cost follows from the fact that the surplus is distributed to people as labor-and-capital compensation.

The IO table in Table 2.3 depicts the flow of economic transactions in this economy, and that is what a standard IO table is supposed to do. There exist, however, other components of the system that are not taken into account in Table 2.3. This applies to water, land use for habitats, and waste, the prices of which are assumed to be zero. As will be discussed later in greater details (Chapter 5), the standard IO table referring to economic transactions only can be extended to incorporate the noneconomic environmental flow as well in several ways.

Table 2.4 shows an example, where Table 2.3 has been augmented by a panel referring to the flow of water, land use, and waste. In the panel, z_2, and f_{z_1} refer to the use of water in the rice production sector and in the household, f_{z_3} to the land use for dwellings. The way waste is recorded may require some explanation. As indicated in Figure 2.4, waste from both consumption and production can be used in production as compost. Denote the amount of waste generated in production by w^{out}, the amount used (recycled) in production by w^{in}, and the amount generated by

Table 2.3 The IO Table of an Economy with One Production Sector.

Rice	Intermediate demand rice producing sector	Final demand household	Total product
	x_1	f	x
Income (labor compensation)	$q_2 z_2$		
Income (land compensation)	$q_3 z_3$		
Total cost	x		

Units: kg of rice.

Table 2.4 An Example of Extended IO Table with Environmental Flows.

	Intermediate demand rice producing sector	Final demand household	Total
	Economic flows		
Rice (kg)	x_1	f	x
Labor compensation (kg of rice)	$q_2 z_2$		
Land compensation (kg of rice)	$q_3 z_3$		
	Environmental flows		
Water (m^3)	z_1	f_{z_1}	$z_1 + f_{z_1}$
Land (m^2)	z_3	f_{z_3}	$z_3 + f_{z_3}$
Waste (kg)	w	f_w	$w + f_w$

consumption by f_w. The amount of waste that remains unutilized is then given by:

$$\left(w^{\text{out}} - w^{\text{in}}\right) + f_w = w + f_w, \tag{2.44}$$

where w refers to the net amount of waste generated in production. In contrast to Table 2.3, where all the flows are measured by the same unit (the amount of rice), several physical units occur in Table 2.4. The latter is an example of a mixed units (hybrid) IO table.

It is important to note that the frequent allegation that "IOA has been developed with emphasis on the flow of money" ([8], p.118) is not fully correct, and that the use of mixed units is not a new phenomenon. In fact, the 1951 book by Leontief indicates that IOA was originally conceptualized in terms of physical units ([13], pp.35–37). Isard [10] gives an early example (from 1968) of the use of mixed-units IO, where attempts were made to extend the economic IO to include the marine ecological system (see Section 5.1.1 for details of the Isard model of economy and ecology).

2.1.4 Cost and Price: The Price Model

Let p be the price of rice. Solving (2.43) for p yields

$$p = \pi(1 - a_1)^{-1} \tag{2.45}$$

Substituting from (2.42), we have:

$$p = 1 \tag{2.46}$$

which reproduces the fact that rice serves as the base unit of value, that is, as the numeraire.

Consider now the case where labor is the only exogenous input with positive price. From (2.37), (2.45) then becomes:

$$p = q_2 b_2 (1 - a_1)^{-1} \tag{2.47}$$

From (2.20), we get the following interpretation of this equation:

$$\underbrace{p}_{\text{labor cost per unit of product}} = \underbrace{q_2 b_2}_{\text{direct labor cost}} + \underbrace{q_2 b_2 (a_1 + a_1^2 + \cdots)}_{\text{indirect labor costs}} \tag{2.48}$$

If one chooses to use the unit of labor (an hour of work) as the numeraire, that is, $q_2 = 1$, this reduces to

$$p(\text{hours}) = b_2 (1 - a_1)^{-1}. \tag{2.49}$$

The price of rice is then given by the amount of labor that is embodied in it (see (2.19)). This gives an IOA version of the "labor theory of value" according to which the values of commodities are related to the labor needed to produce them, a classic idea that can be traced back to the writing of William Petty in 1689 [4].

A classic economic idea of equal importance is the "land theory of value" or "the embodied land theory of value" due to the physiocrats, the founder and leader of which was Francois Quesnay (well known for his Le Tableau Economique), according to which the embodied land content of a commodity determined its value:

$$p(\text{land m}^2) = b_3 (1 - a_1)^{-1}. \tag{2.50}$$

In a similar fashion, the "water theory of value" can be given by

$$p(\text{water m}^3) = b_1 (1 - a_1)^{-1}. \tag{2.51}$$

Should some mineral resource, such as petroleum, emerge as the most scarce, and hence most valuable input, a "petroleum theory of value" would emerge, the computation of which can easily be done based on the present formula.

2.1.4.1 Effects of Changes in the Price of Exogenous Inputs

By use of (2.45), one can evaluate the effects of a change in the price of exogenous inputs. Writing $\Delta \pi$ for a change in π, say, due to an increase in q_2:

$$\Delta \pi = \Delta q_2 b_2 \tag{2.52}$$

the corresponding change in p is given by

$$\Delta p = \Delta \pi (1 - a_1)^{-1}. \tag{2.53}$$

Analogous to (2.27), the price model can also be used to evaluate the difference in the unit cost of alternative production processes. By use of (2.45), the difference in the unit cost between processes r and s, Δp^{rs}, under given prices of exogenous inputs can be represented by

$$\Delta p^{rs} = \sum_{i=1}^{3} b_i^r q_i (1 - a_1^r)^{-1} - \sum_{i=1}^{3} b_i^s q_i (1 - a_1^s)^{-1}. \tag{2.54}$$

If $\Delta p^{rs} > (<)0$, process s is more (less) efficient than process r under the given price of exogenous inputs. It is possible that a process that is found most efficient among a given alternative processes turns out to be less efficient when a different set of prices of exogenous inputs is used for cost evaluation.

2.2 The Two-Sector Model

This section considers the case of two endogenous sectors, where besides rice, a source of protein, say, fish, is also produced. It is shown how the basic concepts of the last section with one production sector can be extended to the case of two production sectors. No use is made of matrix algebra.

2.2.1 Production Processes

Consideration of the case of two producing sectors represents a substantial departure from the previous case of a one sector economy because of the introduction of an interdependence between the two endogenous sectors. The interdependence arises from the inter-sectoral flow of inputs and outputs between the two producing sectors. The presence of the inter-sectoral flow implies that this economy with two producing sectors cannot be represented as if it consisted of two separate economies, each producing and using just one of the products, and hence are technically disconnected from each other. An economy that consists of disconnected sectors only can be analyzed by use of the method in Section 2.1, however large the number of producing sectors may be.

The simple one product economy is now augmented with a second producing sector, an aqua farm, where fish are produced (raised, say in a pond), the process of which is also characterized by constant returns to scale. In the model of the previous section, rice was used to meet both the intermediate (seeds) and final demands (consumption). The same applies to fish as well: it is consumed by the households, but a portion of the fish that are raised have to be left for reproduction. The present case, however, is fundamentally different from the previous one because of the possibility of the flow of inputs not only within the same sector but across different sectors as well. Rice can be used in fish production as a feed. Fish, on the other hand, can also be used as fertilizer in rice production (dried sardine was once used as a fertilizer in Japan).

Accommodating for the inter-sectoral flows of inputs requires some extension of the notations used in the previous section. Henceforth, x_i and x_{ij} denote the amount of endogenous product i, and the amount of endogenous input i that enters the production process of product j. In a similar fashion, z_i and z_{ij} denote the amount of exogenous input i, and the amount of exogenous input i that enters the production process of product j. Accordingly, the input coefficients (2.2) and (2.3) are henceforth given by:

$$a_{ij} = x_{ij}/x_j > 0, \quad i = 1,2; j = 1,2, \tag{2.55}$$

$$b_{ij} = z_{ij}/x_j > 0, \quad i = 1,2,3; j = 1,2. \tag{2.56}$$

For instance, a_{21} refers to the amount of fish (kg) that is used per 1 kg of rice production, and a_{12} to the amount of rice (kg) that is used per 1 kg of fish production.

Extending (2.9) to accommodate the additional inputs, the unit process of rice- and fish production, P_1 and P_2, can respectively be given as follows:

$$P_1 \text{ (rice production)} = \begin{pmatrix} a_{11} : & \text{rice seeds (kg/kg)} \\ a_{21} : & \text{fish as fertilizer (kg/kg)} \\ b_{11} : & \text{water(m}^3/kg) \\ b_{21} : & \text{labor (h/kg)} \\ b_{31} : & \text{land(m}^2/\text{kg)} \\ \hline 1 : & \text{rice (kg)} \\ g_{11} : & \text{waste of type 1 (kg)} \end{pmatrix}, \tag{2.57}$$

$$P_2 \text{ (fish production)} = \begin{pmatrix} a_{12} : & \text{rice as feed (kg/kg)} \\ a_{22} : & \text{fish for reproduction (kg/kg)} \\ b_{12} : & \text{water(m}^3/kg) \\ b_{22} : & \text{labor (h/kg)} \\ b_{32} : & \text{land(m}^2/\text{kg)} \\ \hline 1 : & \text{fish (kg)} \\ g_{22} : & \text{waste of type 2 (kg)} \end{pmatrix}, \tag{2.58}$$

where the items above the line refer to the inputs while those below the line refer to the outputs. Waste is distinguished in two types by the sector of origin in order to accommodate possible difference in their properties. The generation of waste is represented by the waste generation coefficients g_{11} and g_{22}, where the former refers to the amount of waste of type 1 that is generated per unit of rice, and the latter to the amount of waste of type 2 that is generated per unit of fish production. For the sake of simplicity, it is assumed that both processes use the same type of labor and land.

The flow of inputs and outputs in this economy is now given by Figure 2.5, which extends Figure 2.4 for the case of one producing sector (waste and intra sectoral flows are omitted for the sake of simplification). With both rice and fish occurring as endogenous inputs, the levels of which are to be determined in the model, the system boundary (the area surrounded by the dotted line) is extended to cover both production processes.

Fig. 2.5 The Flow of Inputs and Outputs of an Economy with Two Producing Sectors, Rice and Fish Production. The Flow of Output of a Process Is Indicated by an Arrow Leaving the Box Referring to the Process, Whereas the Flow of Input into a Process Is Indicated by an Arrow Entering the Relevant Box. Rice Is Used in Fish Production as Feed, While Fish Is Used in Rice Production as Fertilizer.

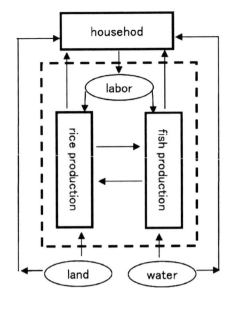

Table 2.5 A Numerical Example of Unit Processes for Rice and Fish Production.

Inputs/outputs	Units	Rice	Fish
Inputs			
Rice	kg/kg	0.007	0.08
Fish	kg/kg	0.320	0.09
Water	m³/kg	5.700	0.28
Labor	h/kg	0.110	0.94
Land	m²/kg	1.910	0.09
Outputs			
Rice	kg/kg	1	0
Fish	kg/kg	0	1
Waste 1	kg/kg	3.000	0
Waste 2	kg/kg	0	4.5

The input of fish into rice is based on the input of fertilizers into rice production taken from [16]. See also the reference of Table 2.9. The values for fish production are rough estimates partly based on Japanese IO table.

Table 2.5 gives a hypothetical numerical example of P_1 and P_2. For this example, $a_{21} = 0.320$ and $a_{12} = 0.08$, which indicates that the production of 1 kg of rice requires 0.32 kg of fish as fertilizer, and that the production of 1 kg of fish requires 0.08 kg of rice for feeding. It is also indicated that, compared to rice production, fish production requires less water and land, but requires a larger amount of labor.

2.2.2 The Leontief Quantity Model

2.2.2.1 Deriving the Leontief Quantity Model

From (2.55), the input demand for the two endogenous inputs can be given analogous to (2.6) as follows:

$$x_{ij} = a_{ij}x_j, \quad i,j = 1,2 \tag{2.59}$$

To be specific, x_{12} refers to the amount of rice used as feed in fish production, x_{22} to the amount of fish used for reproduction, and x_{21} to the amount of fish used as fertilizer for rice production. Following (2.11), it is assumed that

$$0 < a_{ii} < 1, i = 1,2. \tag{2.60}$$

This is a necessary condition for the productiveness, but is not a sufficient condition because it does not consider a_{12} and a_{21}.

Writing f_1 and f_2 for the final demand (household consumption) for rice and fish, respectively, the two sectors version of the equality between supply and demand (2.12) is given by

$$x_1 = x_{11} + x_{12} + f_1 \tag{2.61}$$
$$x_2 = x_{21} + x_{22} + f_2 \tag{2.62}$$

Substituting from (2.59), (2.61) and (2.62) become:

$$x_1 = a_{11}x_1 + a_{12}x_2 + f_1 \tag{2.63}$$
$$x_2 = a_{21}x_1 + a_{22}x_2 + f_2 \tag{2.64}$$

or

$$(1 - a_{11})x_1 - a_{12}x_2 = f_1 \tag{2.65}$$
$$-a_{21}x_1 + (1 - a_{22})x_2 = f_2 \tag{2.66}$$

For a given set of $a_{ij}, i,j = 1,2$ and $f_i, i = 1,2$, this system of equations can be solved for $x_i, i = 1,2$.

From (2.60), x_1 can be deleted from (2.65) and (2.66) by subtracting the former equation times $-a_{21}/(1 - a_{11})$ from the latter equation, which results in

$$((1 - a_{11})(1 - a_{22}) - a_{12}a_{21})x_2 = a_{21}f_1 + (1 - a_{11})f_2 \tag{2.67}$$

Analogously, subtraction of (2.66) times $-a_{12}/(1 - a_{22})$ from (2.65) results in

$$((1 - a_{11})(1 - a_{22}) - a_{12}a_{21})x_1 = (1 - a_{22})f_1 + a_{12}f_2. \tag{2.68}$$

From (2.60) and (2.55), the right hand sides of the above system of equations are positive for $f_i > 0, i = 1,2$. It then follows that in order for $x_i > 0, i = 1,2$, to hold, the expression inside the parenthesis of the left hand side has to be positive:

$$d = (1 - a_{11})(1 - a_{22}) - a_{12}a_{21} > 0. \qquad (2.69)$$

When (2.69) is met, solving (2.67) and (2.68) for x_1 and x_2 gives:

$$x_1 = \frac{1 - a_{22}}{d} f_1 + \frac{a_{12}}{d} f_2 = c_{11}f_1 + c_{12}f_2 \qquad (2.70)$$

$$x_2 = \frac{a_{21}}{d} f_1 + \frac{1 - a_{11}}{d} f_2 = c_{21}f_1 + c_{22}f_2 \qquad (2.71)$$

with $c_{ij}, i, j = 1, 2$, defined by the expression after the second equalities. Note that in the case of one production sector, where $d = 1 - a_{11}$ and $a_{12} = 0$, (2.70) reduces to (2.15) with $c_{11} = 1/(1 - a_{11})$. Accordingly, c_{ij}'s are the Leontief inverse coefficients, and the system of equations (2.70) and (2.71) give the Leontief quantity model for the present case of an economy with two production sectors.

For the unit processes in Table 2.5, (2.69) becomes:

$$d = (1 - 0.007) \times (1 - 0.09) - 0.08 \times 0.320 = 0.878, \qquad (2.72)$$

and (2.70) and (2.71) become:

$$x_1 = 1.037f_1 + 0.091f_2, \qquad (2.73)$$
$$x_2 = 0.365f_1 + 1.131f_2. \qquad (2.74)$$

2.2.2.2 Productiveness Conditions

The equations (2.60) and (2.69) refer to the productive conditions for the case of two endogenous sectors. What is the economic meaning of (2.69)? Dividing both sides of (2.69) by $(1 - a_{11})$, it becomes

$$(1 - a_{22}) - a_{21}(1 - a_{11})^{-1}a_{12} > 0 \qquad (2.75)$$

As was shown in the previous section, the term $(1 - a_{22})$ refers to the surplus of fish production over the amount needed as input for reproduction. The term a_{12} refers to the amount of rice that is required (as feed) to produce a unit of fish. Because $a_{12} > 0$, besides fish, rice is also required in fish production. Recall from (2.18) that $(1 - a_{11})^{-1}$ refers to the total amount of rice that is required to deliver a unit of rice for final use. The term $(1 - a_{11})^{-1}a_{12}$ then refers to the total (direct and indirect) amount of rice that is required to produce a unit of fish. Subtraction of a_{21} times this term from $(1 - a_{22})$ then gives the surplus of fish production over the amount of fish that is directly and indirectly required for fish production including the amount of fish needed for producing the rice that is used to feed the fish. It follows that, with (2.60) satisfied, (2.75) gives the productive condition of the fish production sector.

Following Dorfman et al. [5], who used coal and steel as examples in place of rice and fish, an alternative expression with the same content is that "if we add up the direct and indirect inputs of coal that go into a ton of output (the coal to make coal, the coal to make coal to make coal, the coal to make steel to make coal, the coal to make steel to make coal to make coal, the coal to make steel to make coal to make coal, the coal to make steel to make steel to make coal, etc., ad infinitum), that all this will be less than one ton" ([5], p.215).

Analogously, division of both sides of (2.75) by $(1-a_{22})$ yields the corresponding condition for the rice production sector

$$(1-a_{11}) - a_{12}(1-a_{22})^{-1}a_{21} > 0. \tag{2.76}$$

Following the same line of reasoning as above, it follows that, with (2.60) satisfied, (2.76) refers to the productive condition of the rice production sector: the output of rice that is obtained from the process exceeds the amount of rice that is needed for sowing and for feeding the amount of fish that is used as fertilizer in rice production. Because (2.75) and (2.76) are implied by (2.69), it follows that (2.69) and (2.60) give the productiveness condition of the economy with two producing sectors.

In the previous section, it was shown for the case of one production sector that the economy is productive when the level of production is positive for a positive final demand (Theorem 2.1), that is:

$$1-a_{11} > 0 \iff x > 0, \text{ where } f > 0. \tag{2.77}$$

From the above discussion, it follows that in the present case of two producing sectors the corresponding result is given by

$$1-a_{ii} > 0 \text{ and } d > 0 \iff x_i > 0, \text{ where } f_i > 0, \quad i = 1,2. \tag{2.78}$$

The following is the two sectors version of Theorem 2.1, which is a special case of the well known theorem by Nikaido [17]:

Theorem 2.2. *The following conditions (I), (II), and (HS) are equivalent:*

(I)   For a certain $f_i > 0, i = 1,2$, (2.63) and (2.64) have a nonnegative solution $x_i \geq 0, i = 1,2$.

(II)   For any $f_i \geq 0, i = 1,2$, (2.63) and (2.64) have a nonnegative solution.

(HS) $1-a_{ii} > 0, i = 1,2$ and $d > 0$.

Proof. Assuming that the condition $1-a_{ii} > 0, i = 1,2$ has already been established for the case of one producing sector, the proof is as follows:

(I) ⇒ (HS)   Because the right hand sides of (2.67) and (2.68) are positive, for $x_i > 0$ to hold, $d > 0$ has to hold.

(HS) ⇒ (II)   The right hand sides of (2.67) and (2.68) are nonnegative. Because the expression inside the parenthesis of the left hand sides is positive by assumption, $x_i \geq 0, i = 1,2$ has to hold.

(II) ⇒ (I)   This is obvious because the condition of (II) implies that of (I).

An important implication of the above statement is that in order to check for the productiveness of an economy, it not necessary to compute (2.69). From (I) it suffices to see if the level of production is positive when the final demands are positive.

Sufficient Conditions of Positive Solutions Due to Solow

There is a sufficient condition for the above theorem, which is known as the row (column) sum condition of Solow [20]:

Corollary to Theorem 2.2

(i)   If $\sum_{j=1,2} a_{ij} < 1, i = 1,2$, then (I), (II), and (HS) hold.

(ii)   If $\sum_{i=1,2} a_{ij} < 1, i = 1,2$, then (I), (II), and (HS) hold.

Proof.

(i)   Define $f_i = 1 - \sum_{j=1,2} a_{ij}$, then (2.65) and (2.66) have solutions $x_1 = x_2 = 1$. Hence, (I) is satisfied, and from Theorem 2.2, (II) and (HS) are also satisfied.

(ii) Replacing the position of a_{12} by a_{21} in (2.65) and (2.66), derive a new system of equations with regard to unknowns q_i with $r_i = 1 - \sum_i a_{ij} > 0$

$$(1 - a_{11})q_1 - a_{21}q_2 = r_1$$
$$-a_{12}q_1 + (1 - a_{22})q_2 = r_2 \tag{2.79}$$

The column sum of a_{ij} for (2.65) and (2.66) are identical with the row sum of a_{ij} for (2.79). By (i), (2.79) satisfies (I), (II), and (HS). Note that the same expression for d applies to (2.65) and (2.66), which implies that (HS) holds for them as well. From Theorem 2.2 it then follows that (I) and (II) also hold.

Is $a_{ij} < 1$ Necessary?

In the present example, all the products happen to be measured in the same mass unit, kg. This, however, is by no means a general rule to be observed. A simultaneous use of different units of measurement, such as kg, kWh, and J, is fully consistent with the above model. In such a case, it is possible that a_{ij} involves different physical units, and hence its value exceeds unity. It is important to note that the occurrence of $a_{ij} > 1, i \neq j$ does not imply violation of the productiveness condition. Consider for instance the following example

$$\begin{pmatrix} a_{11} & a_{12} \\ a_{21} & a_{22} \end{pmatrix} = \begin{pmatrix} 0.1 & 2.0 \\ 0.4 & 0.05 \end{pmatrix}, \tag{2.80}$$

While $a_{12} > 1$, the productiveness condition is satisfied, because $a_{ii} < 1$ and $d = 0.0055$.

2.2.2.3 The Meaning of Leontief Inverse Coefficients

The Leontief model is a demand driven model, where the production activity is invoked to meet the final demand. In fact, (2.15) shows for the case of one producing sector where the entire production of x is invoked by f. In the current case of two producing sectors, the corresponding expression is given by (2.70) and (2.71). Differing from the case of only one product in (2.15) with only one item of final demand, the production is now attributed to two final demand categories, the share of contribution of which is given by the Leontief inverse coefficients, c_{ij}. For instance, (2.70) indicates that production of rice x_1 was invoked to meet the final demand for both rice f_1 and fish f_2, with the contribution of the former given by $c_{11}f_1$ and the latter by $c_{12}f_2$. The term $c_{11}f_1$ refers to the total (direct and indirect) amount of rice that was necessary to meet the final demand for rice of f_1, while the term $c_{12}f_2$ refers to the total amount of rice that was produced to meet the final demand for fish of f_2.

The Leontief inverse coefficient c_{12} thus refers to the amount of rice that is required directly and indirectly to produce a unit of fish for the final demand, whereas c_{11} refers to the amount of rice that is required directly and indirectly to produce a unit of rice for the final demand. From the definition of c_{ii}:

$$c_{ii} = (1 - a_{ii}) / \big((1 - a_{11})(1 - a_{22}) - a_{12}a_{21} \big) \qquad (2.81)$$

and (2.69), it follows that

$$c_{ii} \geq 1, i = 1, 2, \qquad (2.82)$$

which corresponds to (2.17) for the case of one producing sector. This condition simply states that for delivering a unit of product i to the final demand at least one unit of it has to be produced.

By construction, $c_{ij} = 0$ when $a_{ij} = 0$. In the case of $a_{12} = 0$ but $a_{21} > 0$, that is, rice is not used to feed fish, but fish is used as fertilizer to grow rice, $c_{12} = 0$, and rice is produced for the satisfaction of the final demand for it alone, and is not affected by the final demand for fish. On the other hand, some portion of the production of fish is invoked to meet the final demand for rice. When there is no interdependence between the producing sectors, $a_{12} = a_{21} = 0$, and $c_{12} = c_{21} = 0$: the economy reduces to a case where two autonomous sectors coexist with no mutual interdependence.

2.2.2.4 The Demand for Exogenous Inputs

By use of the quantity model given by (2.70) and (2.71), it is straightforward to derive the two sectors counterpart of the demand equation for exogenous inputs (2.19):

$$
\begin{aligned}
z_i &= b_{i1}x_1 + b_{i2}x_2 \\
&= b_{i1}(c_{11}f_1 + c_{12}f_2) + b_{i2}(c_{21}f_1 + c_{22}f_2) \\
&= (b_{i1}c_{11} + b_{i2}c_{21})f_1 + (b_{i1}c_{12} + b_{i2}c_{22})f_2, \quad i = 1, 2, 3. \qquad (2.83)
\end{aligned}
$$

The first term of the last equation refers to the demand for exogenous input i that is attributable to the final demand for rice, while the second term does the same to that attributable to the final demand for fish.

In a similar fashion, the generation of waste can be associated with the final demand as follows:

$$w_i = g_{i1}^{\text{out}} x_1 + g_{i2}^{\text{out}} x_2$$

$$= g_{i1}^{\text{out}} (c_{11} f_1 + c_{12} f_2) + g_{i2}^{\text{out}} (c_{21} f_1 + c_{22} f_2)$$

$$= (g_{i1}^{\text{out}} c_{11} + g_{i2}^{\text{out}} c_{21}) f_1 + (g_{i1}^{\text{out}} c_{12} + g_{i2}^{\text{out}} c_{22}) f_2, \quad i = 1, 2, \tag{2.84}$$

where $g_{12} = g_{21} = 0$ in the present case.

2.2.2.5 Evaluating Effects of a Change in Lifestyle and Technology

Within the current framework, the lifestyle of households can be represented by the amount and composition of the final demand f_1 and f_2. By use of (2.83), the effects of a given change in lifestyle, say, Δf_1 and Δf_2, on the demand for exogenous inputs can easily be obtained:

$$\Delta z_i = (b_{i1} c_{11} + b_{i2} c_{21}) \Delta f_1 + (b_{i1} c_{12} + b_{i2} c_{22}) \Delta f_2, \quad i = 1, 2, 3. \tag{2.85}$$

The expression inside the first and second parentheses can respectively be called the input i multiplier of the final demand for products 1 and 2. Analogously, the expression inside the first and second parentheses of (2.84) can respectively be called the waste i multiplier of the final demand for products 1 and 2.

In a similar fashion, the effects of a given change in the technology in use can also be obtained. Suppose for instance, that one is interested in comparing the effects on the demand for exogenous inputs of two processes. The difference between processes r and s can be given by:

$$z_i^r - z_i^s = (b_{i1}^r c_{11}^r + b_{i2}^r c_{21}^r - b_{i1}^s c_{11}^s - b_{i2}^s c_{21}^s) f_1$$

$$+ (b_{i1}^r c_{12}^r + b_{i2}^r c_{22}^r - b_{i1}^s c_{12}^s - b_{i2}^s c_{22}^s) f_2 \tag{2.86}$$

where c_{ij}^r and c_{ij}^s are functions of a_{ij}^r and a_{ij}^s (recall (2.70) and (2.71)). It is important to remember that because of the integrated nature of the elements of a_{ij}'s and b_{ij}'s for a given process, as indicated by (2.9), any change in a subset of them would also be associated with a change in the other elements as well. In general, if two processes differ in terms of the input coefficients for endogenous inputs, they will be different for the exogenous inputs as well, and vice versa (see Section 2.3.5.1 for further details on this point).

2.2.3 The Price Model

We now turn to the derivation of the Leontief cost/price model.

2.2.3.1 Value of Product and Value Added

Write p_1 for the price of rice per kilogram, p_2 for the price of fish per kilogram, q_2 for the wage rate per hour, and q_3 for the price of the use of land per square meter. It is assumed that for a given input the same price applies to any user of it. In line with the previous section, no charge is incurred with regard to the use of water and the discharge of waste. Because the products then constitute the only items with nonzero prices, the value of output becomes equal to the value of product. The present counterpart of equality between the value of output and the value of inputs (2.41) is then given by:

$$\underbrace{p_i x_i}_{\text{value of output}} = \underbrace{p_1 x_{1i} + p_2 x_{2i}}_{\text{cost for endogenous inputs}} + \underbrace{q_2 z_{2i} + q_3 z_{3i}}_{\text{compensation for labor and land}}, \qquad (2.87)$$

$$= \left(p_1 a_{1i} x_i + p_2 a_{2i} x_i\right) + \left(q_2 b_{2i} x_i + q_3 b_{3i} x_i\right), i = 1, 2, \qquad (2.88)$$

and the counterpart of value added (2.40) by

$$v_i = q_2 b_{2i} x_i + q_3 b_{3i} x_i, i = 1, 2. \qquad (2.89)$$

The above expression for the value of output is based on the input or cost structure of a product. An alternative expression for the value of output can be obtained based on the output or supply structure of the product. Multiplication of p_1 to both sides of (2.61), and of p_2 to both sides of (2.62) gives:

$$p_i x_i = p_i x_{i1} + p_i x_{i2} + p_i f_i, i = 1, 2 \qquad (2.90)$$

Summing up (2.90) for $i = 1, 2$ and rewriting, gives:

$$p_1 x_1 + p_2 x_2 - \sum_{i,j=1,2} p_i x_{ij} = p_1 f_1 + p_2 f_2 \qquad (2.91)$$

Applying the analogous operation to (2.87) under consideration of (2.89) yields:

$$p_1 x_1 + p_2 x_2 - \sum_{i,j=1,2} p_i x_{ij} = v_1 + v_2 \qquad (2.92)$$

Because the left hand sides of (2.91) and (2.92) are identical, the present counterpart of the identity between expenditure and income (2.36) follows:

$$\underbrace{p_1 f_1 + p_2 f_2}_{\text{value of expenditure}} = \underbrace{v_1 + v_2}_{\text{income}}. \qquad (2.93)$$

2.2.3.2 The Leontief Price Model

Dividing both sides of (2.88) by the corresponding x_i gives the set of unit cost functions:

$$p_1 = p_1 a_{11} + p_2 a_{21} + \pi_1 \tag{2.94}$$

$$p_2 = p_1 a_{12} + p_2 a_{22} + \pi_2, \tag{2.95}$$

where π_i refers to the rate of value added for sector i:

$$\pi_i = q_2 b_{2i} + q_3 b_{3i}, \quad i = 1, 2, \tag{2.96}$$

the one sector counterpart of which is given by (2.43). The equations for q_2 and q_3 are missing because they refer to exogenous inputs whose prices are not determined by the model, but are given from outside the model. In preparation for solving (2.94) and (2.95) for the endogenous prices p_1 and p_2, rewrite them:

$$p_1(1 - a_{11}) - p_2 a_{21} = \pi_1 \tag{2.97}$$

$$-p_1 a_{12} + p_2(1 - a_{22}) = \pi_2. \tag{2.98}$$

Comparing (2.97) and (2.98) with their quantity counterparts (2.65) and (2.66), one notices that the occurrence of the coefficients a_{21} and a_{12} is transposed between them: while a_{12} (a_{21}) occurs in the equation for x_1 (x_2) in the quantity model, it occurs in the equation for p_2 (p_1) in the price model.

Analogous to the case of the quantity model discussed above, (2.97) and (2.98) can be solved for p_1 and p_2. Subtracting minus times $a_{12}/(1 - a_{11})$ of (2.97) from (2.98) deletes p_1 and gives:

$$p_2\big((1 - a_{11})(1 - a_{22}) - a_{21}a_{12}\big) = \pi_1 a_{12} + \pi_2(1 - a_{11}) \tag{2.99}$$

Analogously, subtracting minus times $a_{21}/(1 - a_{22})$ (2.98) from (2.97) deletes p_2:

$$p_1\big((1 - a_{11})(1 - a_{22}) - a_{21}a_{12}\big) = \pi_1(1 - a_{22}) + \pi_2 a_{21}. \tag{2.100}$$

Note that the expression inside the parenthesis of the left hand side corresponds to d given by (2.69). Assuming that the positive condition is satisfied, the solution is given by:

$$p_1 = \pi_1 \frac{1 - a_{22}}{d} + \pi_2 \frac{a_{21}}{d} = \pi_1 c_{11} + \pi_2 c_{21}, \tag{2.101}$$

$$p_2 = \pi_1 \frac{a_{12}}{d} + \pi_2 \frac{1 - a_{11}}{d} = \pi_1 c_{12} + \pi_2 c_{22}. \tag{2.102}$$

This is the Leontief price model for the case of two producing sectors. Comparison with the quantity model given by (2.70) and (2.71) reveals that the price model is characterized by the transposed occurrence of c_{12} and c_{21}.

In essence, the price model states that the price of an endogenous product is determined by the price of exogenous inputs and technology. Just as the quantity model attributes each output to the final demand of origin, the price model attributes the cost to the price of exogenous inputs, which for p_1 can be given by:

$$p_1 = \underbrace{\pi_1 c_{11}}_{\text{cost of exogenous inputs used for rice}} + \underbrace{\pi_2 c_{21}}_{\text{cost of exogenous inputs used for fish}} \tag{2.103}$$

Consider now the case where labor is the only exogenous input with a positive price, and an hour of work is used as the numeraire, that is, $q_2 = 1$ and $q_3 = 0$. The price model then reduces to

$$p_1 = b_{21}c_{11} + b_{22}c_{21} \tag{2.104}$$

$$p_2 = b_{21}c_{12} + b_{22}c_{22}. \tag{2.105}$$

In this labor theory of value economy (see Section 2.1.4), rice and fish are priced by the hours of work embodied in them. For instance, the price of 1 kg of rice, p_1, is equal to the sum of the hours of work required for producing 1 kg of rice for final use, $b_{21}c_{11}$, and the hours of work required for producing the amount of fish that is required to produce 1 kg of rice for final use, $b_{22}c_{21}$. Application to the unit processes in Table 2.57 gives

$$p_1 = 0.110 \times 1.037 + 0.94 \times 0.365 = 0.457, \tag{2.106}$$

$$p_2 = 0.110 \times 0.091 + 0.94 \times 1.131 = 1.073. \tag{2.107}$$

which indicates that fish is more than twice as expensive as rice when evaluated by the amount of embodied labor.

In a similar fashion, the price model for the land theory of value economy (see Section 2.1.4) can also be obtained.

2.2.3.3 Effects of a Change in the Price of Exogenous Inputs

Analogous to (2.85), the effects of an exogenous change in the price of exogenous inputs can also be obtained. Writing the change in the price of exogenous inputs by $\Delta \pi_i$, it follows from (2.101) and (2.102):

$$\Delta p_1 = \Delta \pi_1 c_{11} + \Delta \pi_2 c_{21} \tag{2.108}$$

$$\Delta p_2 = \Delta \pi_1 c_{12} + \Delta \pi_2 c_{22}. \tag{2.109}$$

2.2.4 The IO Table

Following Section 2.1.3.3, an IO table (IOT) of the economy can be obtained by use of the identities referring to the allocation of products among different demand items (2.61) and (2.62), and the cost structure (2.87).

2.2.4.1 The Monetary Table

In the case of a single product, the adding up of the cost elements is straightforward because the product can serve as the unit of value, and accordingly all the economic transactions can be measured by the amount of that product. In the case of two (or more) products, however, this is no longer the case because of the presence of alternative units of value. Accordingly, in (2.87) each of the cost items is multiplied by the corresponding price. In a monetary IO table, the sum of column elements (total cost) must equal the sum of row elements (total expenditure) by the identity of national accounts. Consequently, the elements of (2.61) and (2.62) also have to be multiplied by the corresponding prices:

$$p_i x_i = p_i x_{i1} + p_i x_{i2} + p_i f_i, i = 1, 2. \tag{2.110}$$

Recording the elements occurring in (2.87) column-wise, and the elements occurring in (2.110) row wise, the two sector counterpart of the IO table is obtained (Table 2.6). For a given endogenous sector, the sum of its row elements gives the total value of its product, whereas the sum of its column elements gives the total value of inputs or cost. The total value of final demand is equal to the total value of income due to the identity (2.93).

2.2.4.2 The Physical Table

By definition, Table 2.6 depicts the flow of economic transactions only, and neglects all the other flows that have no economic cost. In the previous section, an attempt was made to extend the IO table in value terms to accommodate these flows, and an example of a physical IO table was shown (Table 2.4). Table 2.7 gives a two sector counterpart of such an extension, where 'waste 3' refers to waste generated from household consumption.

All the elements in Table 2.7 are in physical units. Due to the coexistence of different units, the column-wise adding up of the elements does not make much sense except that the adding up of material inputs (rice, fish, and water) as a mass can serve as a measure of total material input (the mass balance between inputs

Table 2.6 The IO Table of an Economy with Two Production Sectors in Value Units.

	Rice	Fish	Final demand	Value of product
Rice	$p_1 x_{11}$	$p_1 x_{12}$	$p_1 f_1$	$p_1 x_1$
Fish	$p_2 x_{21}$	$p_2 x_{22}$	$p_2 f_2$	$p_2 x_2$
Income: labor	$q_2 z_{21}$	$q_2 z_{22}$		
Income: land	$q_3 x_{31}$	$q_3 z_{32}$		
Total cost	$p_1 x_1$	$p_2 x_2$		

Table 2.7 An Extended IO Table with Physical Flows for the Case of Two Production Sectors.

	Rice	Fish	Final demand	Total
Rice (kg)	x_{11}	x_{12}	f_1	x_1
Fish (kg)	x_{21}	x_{22}	f_2	x_2
Water (m^3)	z_{11}	z_{12}	f_{z_1}	z_1
Labor (h)	z_{21}	z_{22}		z_2
Land use (m^2)	z_{31}	z_{32}	f_{z_3}	z_3
Waste 1 (kg)	w_1^{out}			w_1
Waste 2 (kg)		w_2^{out}		w_2
Waste 3 (kg)			w_3^{out}	w_3

and outputs in the strict sense of MFA does not hold here because of the neglect of other flows, say, those associated with photosynthesis, that are of importance in the growth of plants and animals). On the other hand, the row-wise addition simply refers to (2.61) and (2.62), and is well defined.

2.2.4.3 Value or Physical Tables?

Multiplication of the elements occurring in the physical Table 2.7 by the corresponding prices gives the value Table 2.6. On the other hand, dividing the flows in value units in the value table by the corresponding prices, one obtains the part of the physical table with nonzero prices. So far as the flows with nonzero (positive) prices are concerned, there is thus a one to one correspondence between an IO table in value units and its physical counterpart.

Up until now, our starting point has been the unit process represented by a set of input- and output coefficients. Operating these processes to meet the requirements for a given final demand then generates the inter-sectoral flow of inputs and outputs, an accounting system of which is given by the IO table. The IO table has thus been derived by use of information on unit processes and the final demand. In reality, however, collecting data on unit processes for the economy as a whole is still a very challenging task, if not impossible. The IO tables currently available are, except for minor exceptions, not in physical units, but in value units, because they are compiled by the use of data on economic transactions in monetary units. Accordingly, it is usually the case that the input coefficients are derived from a given IO table measured in value units.

Write $\bar{a}_{ij}$ for the input coefficients obtained from the IO table in value units (Table 2.6):

$$\bar{a}_{ij} = \frac{p_i x_{ij}}{p_j x_j} = \frac{p_i}{p_j} a_{ij}, i,j = 1,2, \tag{2.111}$$

which can be called monetary input coefficients to distinguish them from their physical counterparts a_{ij}. It is of interest to see if the use of $\bar{a}_{ij}$ in place of a_{ij} in calculations such as (2.85) and (2.108) results in any discrepancies to the results based on the latter, that is, the original physical input coefficients. Consider if any alternation to the quantity model (2.70) and (2.71) is required when the monetary coefficients are used. Write $\bar{d}$ for the counterpart of d in (2.69). It then follows from (2.111)

$$\bar{d} = (1 - \frac{p_1}{p_1}a_{11})(1 - \frac{p_2}{p_2}a_{22}) - \frac{p_1}{p_2}a_{12}\frac{p_2}{p_1}a_{21} = d, \qquad (2.112)$$

that is, d remains unaltered by the use of value coefficients. Furthermore, the value table counterpart of (2.70) and (2.71) is given by:

$$p_1x_1 = \frac{1 - (p_2/p_2)a_{22}}{d}p_1f_1 + \frac{p_1a_{12}/p_2}{d}p_2f_2$$

$$= c_{11}p_1f_1 + c_{12}p_1f_2 \qquad (2.113)$$

$$p_2x_2 = \frac{p_2a_{21}/p_1}{d}p_1f_1 + \frac{1 - p_1a_{11}/p_1}{d}p_2f_2$$

$$= c_{21}p_2f_1 + c_{22}p_2f_2, \qquad (2.114)$$

which are identical to the original quantity model up to the multiplication by the product price.

Because the use of $\bar{a}_{ij}$ in place of a_{ij} does not change c_{ij}, the results obtained by (2.85) and (2.108) remain the same irrespective of whether one uses value coefficients or physical coefficients. Furthermore, the physical amount of exogenous inputs required to produce given final deliveries (2.83) will be the same irrespective of whether a physical model is used or a monetary model extended by physical exogenous inputs is used. This is a reproduction based on elementary calculation of the result by Weitz and Duchin [23].

By definition, the monetary input coefficients obtained from (2.111) are subject to an additional condition on its column sum:

$$\sum_{i=1,2} \bar{a}_{ij} < 1, \qquad (2.115)$$

which indicates that the column sum condition of Solow (**Corollary to Theorem 2.2**) is satisfied. The productiveness condition is thus automatically satisfied for the input coefficients obtained from monetary IO tables with nonnegative inputs.

For a practitioner of IOA, it is of great convenience and relief that one can use readily available monetary IO tables in place of physical tables. This convenience, however, is premised on the following two conditions:

1. The relative prices remain *constant for all the inputs* over the period or space under consideration, say $t \in T$:

$$p_i(t)/p_j(t) = \theta_{ij}, \quad t \in T, \qquad (2.116)$$

where θ_{ij} is a constant that does not depend on t, and T refers to the set of time or space, that is, $T = \{t_1, t_2, \cdots\}$.

2. The same price of an input applies to all its users:

$$p_{ij} = p_i, \quad \forall i, j, \tag{2.117}$$

where p_{ij} refers to the price of input i that applies when it is used in sector j, and $\forall i, j$ refers to all i and j.

The IO Table in Value Units When the Relative Prices Do Not Remain Constant

First, consider the implication of (2.116), assuming that the second condition (2.117) is satisfied. It then follows from (2.111) that there is a one-to-one correspondence between the physical coefficients and value coefficients. By an appropriate choice of units of measurement, it is then possible to regard the value input coefficients as physical coefficients. The monetary value, say in US dollars, of x_{ij}, which is expressed in kg, is $p_i x_{ij}$. If one chooses as a new unit of measurement the amount of input i that can be purchased out of \$ 1 (\$1-worth), then $p_i x_{ij}$ becomes the physical quantity in terms of this new unit. The same applies to $p_j x_j$ as well. Accordingly, $\bar{a}_{ij}$ can be regarded as physical coefficients.

In order to consider the case where (2.116) is not satisfied, denote the level of relative prices of inputs that prevailed at the time (in the space) when (where) $\bar{a}_{ij}$'s were obtained from a value IO table by $(p_i(t_1)/p_j(t_1))$:

$$\bar{a}_{ij}(t_1) = \frac{p_i(t_1)}{p_j(t_1)} a_{ij}, \tag{2.118}$$

where a_{ij} itself is assumed constant over T. Suppose that then the relative prices have changed (or the model is applied to a different space unit with different relative prices) to a different level given by $(p_i(t_2)/p_j(t_2))$. The equality between $\bar{d}$ and d still holds for $\bar{a}_{ij}(t_2)$, because only $p_i(t_1)$s appear in the definition of $\bar{d}$. With the $\bar{a}_{ij}$'s based on the level of relative prices at t_1 continuing to be used, but the final demand evaluated by the actual level of relative prices at t_2, the right hand sides of (2.70) and (2.71) become:

$$\frac{1-a_{22}}{d} p_1(t_2) f_1 + \frac{a_{12}}{d} \frac{p_1(t_1)}{p_2(t_1)} p_2(t_2) f_2$$
$$= c_{11} p_1(t_2) f_1 + c_{12} \frac{p_1(t_1)}{p_2(t_1)} p_2(t_2) f_2 \tag{2.119}$$

$$\frac{a_{21}}{d} \frac{p_2(t_1)}{p_1(t_1)} p_1(t_2) f_1 + \frac{1-a_{11}}{d} p_2(t_2) f_2$$
$$= c_{21} \frac{p_2(t_1)}{p_1(t_1)} p_1(t_2) f_1 + c_{22} p_2(t_2) f_2. \tag{2.120}$$

which do not reduce to (2.113) and (2.114) unless $p_1(t_1)/p_2(t_1) = p_1(t_2)/p_2(t_2)$. If $p_1(t_1)/p_2(t_1) > p_1(t_2)/p_2(t_2)$, for instance, the use of value based input coefficients would overestimate the effect of f_2 on x_1, and underestimate the effect of f_2 on x_1.

The above problem arises because of the use of different price levels in the calculation of monetary values. This problem can easily solved by bringing the prices to the same level, that is, by the use of price indices. Multiplying $p_1(t_2)f_1$ by $p_1(t_1)/p_1(t_2)$ and $p_2(t_2)f_2$ by $p_2(t_1)/p_2(t_2)$, the value of final demand evaluated at the price at t_2 is transformed into the value evaluated at t_1. With this transformation, (2.119) and (2.120) reduce to the right hand side of (2.113) and (2.114).

The IO Table in Value Units When the Price of Input Is User Specific

Consider next the case where the second condition (2.117) is not satisfied, while (2.116) is satisfied. The corresponding IO table is given by Table 2.8 where $\bar{p}_i$ is given by

$$\bar{p}_i = \Big(\sum_{j=1,2} x_{ij}p_{ij} + f_i p_{if} \Big) / x_i, \tag{2.121}$$

where p_{if} refers to the price of i that applies to the final demand sector. The identity between income and expenditure (2.93) holds for this IO table because it does not assume the absence of cell specific input prices. It is, however, not possible to convert Table 2.8 into a physical table as in Table 2.7. The input coefficients in value terms (2.111) then become:

$$\bar{a}_{ij} = \frac{p_{ij}}{\bar{p}_j}a_{ij}. \tag{2.122}$$

Substituting (2.122) into (2.69), it turns out that the equality between d and $\bar{d}$ as in (2.112) no longer holds:

$$\bar{d} = \Big(1 - \frac{p_{11}}{\bar{p}_1}a_{11}\Big)\Big(1 - \frac{p_{22}}{\bar{p}_2}a_{22}\Big) - \frac{p_{12}}{\bar{p}_2}a_{12}\frac{p_{21}}{\bar{p}_1}a_{21} \neq d, \tag{2.123}$$

and the counterparts of (2.70) and (2.71) become:

$$\bar{p}_1 x_1 = \frac{1 - a_{22}}{\bar{d}}\bar{p}_1 f_1 + \frac{a_{12}}{\bar{d}}p_{12}f_1 \tag{2.124}$$

$$\bar{p}_2 x_2 = \frac{a_{21}}{\bar{d}}p_{21}f_1 + \frac{1 - a_{11}}{\bar{d}}\bar{p}_2 f_2, \tag{2.125}$$

which do not reduce to (2.113) and (2.114). In this case, the use of monetary coefficients produces results which diverge from those obtained from physical coefficients.

Table 2.8 IO Table with User Specific Price of Inputs.

	Rice	Fish	Final demand	Value of product
Rice	$p_{11}x_{11}$	$p_{12}x_{12}$	$p_{1f}f_1$	$\bar{p}_1 x_1$
Fish	$p_{21}x_{21}$	$p_{22}x_{22}$	$p_{2f}f_2$	$\bar{p}_2 x_2$
Income labor	$q_{21}z_{21}$	$q_{22}z_{22}$		
Income land	$q_{31}z_{31}$	$q_{32}z_{32}$		
Total cost	$\bar{p}_1 x_1$	$\bar{p}_2 x_2$		

The presence of cell specific input prices, a violation of (2.117), thus has severe consequences for the use of IOTs in value units. The following can be mentioned as possible reasons for the possible occurrence of cell specific input price:

1. Product mix: Aggregation of heterogeneous inputs and/or of heterogeneous processes: the product mix delivered by sector i to sector j may vary significantly from the product mix delivered to sector k [23] ([21], p.48).
2. Indirect taxation: For a given product, different rules of tax exemptions may apply to different users ([21], p.48).
3. Trade and transport services: These services will vary as between different classes of users: business users do not buy in retail markets ([21], p.48).

The issues related to product-mix can be coped with by increasing the resolution of the IOT, that is, by using more detailed sectoral classifications. The issue of indirect taxes is dealt with by recording the entries (in a value term IOT) on a factor cost rather than on a market price basis (see [21], p.51, for details). The issue of trade and transport services merits a separate discussion.

2.2.4.4 Trade and Transport Services: Producers' and Purchasers' Prices

With the exception of an economy based on barter exchange and limited to a small geographical area, any flow of inputs and outputs requires trade and transport services. Consumers purchase in retail markets, while producers do so in wholesale markets. Any input and output have to be transported from suppliers' sectors to users. In other words, even for a simple economy with two products, the actual flow will not take place without involving the sectors supplying trade and transport services.

Because up to now the issue of trade and transport has not been taken into account, the product price p_i does not include the costs associated with them, such as trade margins and transport costs. The price of a product that does not include trade margins and transport costs is called the producers' price. The price of a product that includes these costs is called the purchasers' price. The purchasers' price of a product is the actual price that is charged to its users. Write t_{ij} for the sum of trade margin (mark-up) and transport cost associated with the use of product i by sector j. Suppose that the issue of product mix and indirect taxation have been properly dealt with, and hence for the producers' price, p_i, (2.117) holds. The actual expenditure of sector j for product i is then given by

$$p_{ij}x_{ij} = (p_i + t_{ij})x_{ij}, \qquad (2.126)$$

where p_{ij} is the purchaser's price of product i.

There are two issues to be settled with regard to the treatment of trade and transport in IOT. First, if one literally tries to depict in an IOT the fact that any user of a product purchases it from the trade sector, the IOT would look like Table 2.9: all the products are first delivered to the trade and transport sector, and then distributed to

Table 2.9 Monetary IO Table with a Direct Representation of Trade and Transport.

	Rice	Fish	T&T	Final demand	Value of product
Rice	0	0	$\sum_j p_{1j}x_{1j}$	0	$\sum_j p_{1j}x_{1j}$
Fish	0	0	$\sum_j p_{2j}x_{2j}$	0	$\sum_j p_{2j}x_{2j}$
T&T	$\sum_i p_{i1}x_{i1}$	$\sum_i p_{i2}x_{i2}$	0	$\sum_i p_{if}x_{if}$	$\sum_{ij} p_{ij}x_{ij}$

T&T refers to trade and transport. The flow of exogenous inputs and total cost are not shown.

Table 2.10 Monetary IO Table with an Indirect Representation of Trade and Transport: Purchasers' Prices.

	Rice	Fish	T&T	Final demand	Trade and transport costs	Value of product
Rice	$p_{11}x_{11}$	$p_{12}x_{12}$	0	$p_{1f}x_{1f}$	$-\sum_j t_{1j}x_{1j}$	p_1x_1
Fish	$p_{21}x_{21}$	$p_{22}x_{22}$	0	$p_{2f}x_{2f}$	$-\sum_j t_{2j}x_{1j}$	p_2x_2
T&T	0	0	0	0	0	

T&T refers to trade and transport. The flow of exogenous inputs and total cost are not shown.

Table 2.11 Monetary IO Table with an Indirect Representation of Trade and Transport: Producers' Prices.

	Rice	Fish	T&T	Final demand	Value of product
Rice	p_1x_{11}	p_1x_{12}	0	p_1x_{1f}	p_1x_1
Fish	p_2x_{21}	p_2x_{22}	0	p_2x_{2f}	p_2x_2
T&T	$\sum_i t_{i1}x_{i1}$	$\sum_i t_{i2}x_{i2}$	0	$\sum_i t_{if}x_{if}$	$\sum_{ij} t_{ij}x_{ij}$

T&T refers to trade and transport. The flow of exogenous inputs and total cost are not shown.

individual users. Consequently, all the entries that refer to the flow of inputs/outputs among sectors are zero except for those involving the trade and transport sector. Because the provision of information about this inter-sectoral flow is the principal aim of an IOT, this method of directly representing the flow of inputs is not acceptable. Note that this is not specific to a monetary IOT, but holds for a physical IOT as well.

One possible way to solve this shortcoming of the direct method is to record the inter-sectoral flow of rice and fish as if they were purchased directly from producing sectors as in Table 2.10. Because the actual prices paid by the users of inputs are the purchasers' prices, the monetary flow is evaluated by p_{ij}. The result is a monetary IOT with user specific input prices.

An alternative indirect method of representation is given by Table 2.11. It is similar to Table 2.10 in its representation of the inter-sectoral flow, but differs from it in its evaluation. The flow of rice and fish is evaluated by producers' prices. The fare and margins associated with the input of rice and fish are no longer included in their inputs, but occur as the input from the trade and transport sector. Owing to this feature, an IOT based on this method is called an IOT at producers' prices. Compared to the above two methods, this method has two advantages. First, it restores the inter-sectoral flow of inputs and outputs, the principal aim of an IOT. Secondly, it preserves price homogeneity, and hence product homogeneity as well, in the rows of the table ([21], p.51). The IOT at producers' prices is the most widely used form of monetary IOTs.

Note that Tables 2.10 and 2.11 deal with a highly simplified case where there was no nonzero entry to the columns referring to the T&T sector (this follows Stone [21]). For a real example of Tables 2.10 and 2.11, see Table 2.12 in 2.3.4.3.

2.3 The n Sectors Model

Practical application of IOA deals with IOTs with the number of producing sectors in the range of dozens to hundreds. This makes the use of matrix algebra indispensable for IOA. For the ease of readers who are unfamiliar with matrix algebra, we recapitulate the results for the case of the two endogenous sectors in Section 2.2 by use of matrices. An important virtue of using matrices in IOA is that once a matrix representation has been obtained for the case of $n = 2$ endogenous sectors, it is applicable to the case of any $n \geq 1$.

2.3.1 Matrix Notations

The IOT in Table 2.6 consists of a rectangular array of numbers with five rows and four columns with $p_1 x_{11}$ occurring as the first row element of the first column, and $q_3 z_{32}$ occurring as the fourth element of the second column. A rectangular array of numbers (scalars) arranged in m rows and n columns is called a matrix of order $m \times n$. Table 2.6 gives a 5×4 matrix of the flow of inputs and outputs. In matrix algebra, an $m \times 1$ matrix is a called column vector or simply vector, and a $1 \times n$ matrix is called a row vector [19]. The column array of the fish sector is a 5×1 column vector, while the row array of the same sector is a 1×4 row vector. Henceforth, the term column and row will be omitted in referring to a vector when it is accompanied by its order.

Denote by P the matrix obtained by arranging the unit processes P_1 (2.57) and P_2 (2.58) in the following way, and define its submatrices A, B, D, and G^{out}:

$$P = \begin{pmatrix} a_{11} & a_{12} \\ a_{21} & a_{22} \\ \hline b_{11} & b_{12} \\ b_{21} & b_{22} \\ b_{31} & b_{32} \\ \hline 1 & 0 \\ 0 & 1 \\ \hline g_{11}^{\text{out}} & 0 \\ 0 & g_{22}^{\text{out}} \end{pmatrix} = \begin{pmatrix} A \\ B \\ D \\ G^{\text{out}} \end{pmatrix} \tag{2.127}$$

where A is the 2×2 matrix of endogenous input coefficients, B is the 3×2 matrix of exogenous input coefficients, D is the 2×2 matrix of product output coefficients, which refer to the amount of products that are obtained by the operation of the process per unit, and G^{out} is the 3×2 matrix of waste output coefficients. Henceforth, omitting the term "endogenous", A is simply called the matrix of input coefficients. For the unit processes in Table 2.5:

$$A = \begin{pmatrix} 0.007 & 0.08 \\ 0.320 & 0.09 \end{pmatrix}, B = \begin{pmatrix} 5.70 & 0.28 \\ 0.11 & 0.94 \\ 1.91 & 0.09 \end{pmatrix}, G^{\text{out}} = \begin{pmatrix} 3.0 & 0 \\ 0 & 4.5 \end{pmatrix}. \quad (2.128)$$

The expression $A = (a_{ij})$ will be used to state that A is a matrix with a_{ij} as its (i, j)th element. It will also be convenient to refer to the (i, j)th element as $(A)_{ij}$, that is, $a_{ij} = (A)_{ij}$. Similarly, the jth column of A will be denoted by $(A)_{\cdot j}$, and the ith row of A by $(A)_{i \cdot}$.

A matrix with the same number of rows as columns is called a square matrix. The matrices A, and D (and G^{out} in (2.128)) are square matrices. A square matrix is called a diagonal matrix when its elements are zero except for the diagonal elements. The matrices D and G^{out} in (2.128) are diagonal matrices. The element $d_{ij} = (D)_{ij}$ refers to the amount of product i that is generated per unit of production of product j. Because P refers to the matrix of unit processes, $d_{ii} = 1, i = 1, 2$, by definition. A diagonal matrix with all its diagonal elements equal to unity is called an identity matrix. The matrix D is an identity matrix. If there is a by-product, however, where the production process of product i produces product $j \neq i$ as well, $d_{ij} > 0$, and D is no longer diagonal (the issue of by-products will be dealt with in Section 3.2).

2.3.1.1 Multiplication and Addition of Matrices

Let there be two matrices $Q = (q_{ij})$ and $R = (r_{ij})$ of orders $m \times n$ and $o \times p$.

Addition of Matrices

The sum of two matrices Q and R is defined if they have the same number of rows and the same number of columns; in this case

$$Q + R = (q_{ij} + r_{ij}). \quad (2.129)$$

Scalar Multiplication of a Matrix

The multiplication of Q by a scalar α gives an $m \times n$ matrix:

$$\alpha Q = (\alpha q_{ij}). \quad (2.130)$$

Multiplication of Matrices

The premultiplication of R by Q is defined only if the number of columns of Q is equal to the number of rows of R. Thus, if $n = o$, then $T = QR$ will be the $m \times p$ matrix $T = (t_{ij})$ with

$$t_{ij} = \sum_{k=1}^{n} q_{ik} r_{kj}. \tag{2.131}$$

Note that the right hand side involves the ith row elements of Q, $(Q)_{i\cdot}$ and the jth column elements of R, $(R)_{\cdot j}$. Accordingly, (2.131) can also be represented by

$$t_{ij} = (Q)_{i\cdot}(R)_{\cdot j}. \tag{2.132}$$

In the special case where R is a diagonal matrix, this becomes

$$t_{ij} = q_{ij} r_{jj} \tag{2.133}$$

Transpose of a Matrix

An $n \times m$ matrix $Z = (z_{ij})$ with $(Z)_{ij} = (Q)_{ji}$ is called the transpose of Q, and is denoted by $Q^\top$. By definition, $(Q^\top)^\top = Q$. The following holds for the transpose of the product of matrices:

$$(QR)^\top = (R^\top Q^\top) \tag{2.134}$$

$$(\alpha Q)^\top = \alpha Q^\top, \tag{2.135}$$

where α is a scalar.

2.3.1.2 The Equality between Supply and Demand

Write x for the 2×1 vector of products

$$x = \begin{pmatrix} x_1 \\ x_2 \end{pmatrix}. \tag{2.136}$$

Following (2.131), the post-multiplication of A by x gives the 2×1 vector

$$Ax = \begin{pmatrix} a_{11}x_1 + a_{12}x_2 \\ a_{21}x_1 + a_{22}x_2 \end{pmatrix}, \tag{2.137}$$

which refers to the intermediate demand for endogenous products.

Denote by f the 2×1 vector of final demand for endogenous products:

$$f = \begin{pmatrix} f_1 \\ f_2 \end{pmatrix}, \tag{2.138}$$

where $f_i > 0, i = 1, 2$. From (2.137) and (2.138), the matrix representation of the equality between supply and demand (2.63) and (2.64) is derived:

$$x = Ax + f. \tag{2.139}$$

Subtraction of Ax from both sides of (2.139) gives

$$(I - A)x = f, \tag{2.140}$$

which is the matrix representation of (2.65) and (2.66), where I is the identity matrix of order 2:

$$I = \begin{pmatrix} 1 & 0 \\ 0 & 1 \end{pmatrix}. \tag{2.141}$$

2.3.2 Inversion of a Matrix and the Quantity Model

If (2.140) can be solved for x, the matrix representation of the quantity model would be obtained. A nonnegative solution $x \geq 0$ for $f > 0$ is obtained if there exists the inverse matrix of $(I - A)$, $(I - A)^{-1}$, such that:

$$(I - A)(I - A)^{-1} = I \tag{2.142}$$

2.3.2.1 Inversion of a Square Matrix

If there exists a square matrix R, which satisfies for a square matrix Q:

$$QR = I \tag{2.143}$$

R is called the inverse matrix of Q, and is denoted by Q^{-1}. Two simple examples are:

1. Q is a diagonal matrix of order m:

$$Q^{-1} = \begin{pmatrix} q_{11}^{-1} & 0 & \cdots & 0 \\ 0 & q_{22}^{-1} & \cdots & 0 \\ \cdots & \cdots & \cdots & \cdots \\ 0 & 0 & \cdots & q_{mm}^{-1} \end{pmatrix}. \tag{2.144}$$

2. The case of $m = 2$:

$$Q^{-1} = \frac{1}{|Q|} \begin{pmatrix} q_{22} & -q_{12} \\ -q_{21} & q_{11} \end{pmatrix}, \tag{2.145}$$

where

$$|Q| = q_{11}q_{22} - q_{12}q_{21} \neq 0. \tag{2.146}$$

$|Q|$ is called the determinant of Q. When $|Q| = 0$, Q^{-1} does not exist. In fact, $|Q| \neq 0$ is the necessary and sufficient condition for its existence. A square matrix with a nonzero determinant is called nonsingular: a nonsingular matrix is invertible.

The following are (nonexclusive) properties of the inverse matrix (see [19] for a proof):

1. The inverse of a matrix Q is unique.
2. $QQ^{-1} = I$.
3. The inverse of a product is the product of the inverse taken in reverse order

$$(AB)^{-1} = B^{-1}A^{-1} \tag{2.147}$$

provided B^{-1} and A^{-1} exist.
4. A repetitive application of the above rule gives:

$$(ABC)^{-1} = \left(A(BC)\right)^{-1} = (BC)^{-1}A^{-1} = C^{-1}B^{-1}A^{-1}. \tag{2.148}$$

5. The inverse of a transpose is the transpose of the inverse:

$$(Q^{\top})^{-1} = (Q^{-1})^{\top}. \tag{2.149}$$

2.3.2.2 The Quantity Model

Replacing Q by $(I - A)$ in (2.145) results in the following expression:

$$(I - A)^{-1} = \begin{pmatrix} (1 - a_{11}) & -a_{12} \\ -a_{21} & (1 - a_{22}) \end{pmatrix}^{-1} = \frac{1}{|I - A|} \begin{pmatrix} (1 - a_{22}) & a_{12} \\ a_{21} & (1 - a_{11}) \end{pmatrix} \tag{2.150}$$

where

$$|I - A| = (1 - a_{11})(1 - a_{22}) - a_{12}a_{21} > 0. \tag{2.151}$$

The inequality follows from the productiveness condition (2.69). If the input coefficients matrix A is productive, $(I - A)^{-1}$ exists. Note that in (2.150) the expression after the second equality coincides with the inverse coefficients c_{ij} that occur in (2.70) and (2.71). Using matrix notations, the Leontief quantity model given by (2.70) and (2.71) is thus compactly represented by

$$x = (I - A)^{-1}f. \tag{2.152}$$

In line with (2.18), where $(1 - a_1)^{-1}$ is obtained as the infinite sum of the elements a_1^i, $i = 0, 1, \cdots$, it can also be established for a nonsingular $I - A$:

$$(I - A)^{-1} = I + A + A^2 + \cdots = \sum_{i=0}^{\infty} A^i, \qquad (2.153)$$

(see [17] for a proof). Because all the elements of A are nonnegative, (2.153) implies that all the elements of the Leontief inverse matrix will be nonnegative, as well.

In order to see the economic meaning of (2.153), consider the case of delivering a unit of product 1 to the final demand, which gives the following counterpart of (2.18):

$$\begin{pmatrix} 1 \\ 0 \end{pmatrix} + \begin{pmatrix} a_{11} \\ a_{21} \end{pmatrix} + A \begin{pmatrix} a_{11} \\ a_{21} \end{pmatrix} + A^2 \begin{pmatrix} a_{11} \\ a_{21} \end{pmatrix} + \cdots . \qquad (2.154)$$

or by use of the rule of matrix multiplication

$$\begin{pmatrix} 1 \\ 0 \end{pmatrix} + A \begin{pmatrix} 1 \\ 0 \end{pmatrix} + A^2 \begin{pmatrix} 1 \\ 0 \end{pmatrix} + A^3 \begin{pmatrix} 1 \\ 0 \end{pmatrix} + \cdots . \qquad (2.155)$$

The second term refers to the additional production of products 1 and 2 that are required for the production of a unit of product 1, the third term refers to the second additional production that is invoked by the first additional production, the third term to the third additional production invoked by the second additional production, and so on. For the production of product 2, the corresponding expression becomes:

$$\begin{pmatrix} 0 \\ 1 \end{pmatrix} + A \begin{pmatrix} 0 \\ 1 \end{pmatrix} + A^2 \begin{pmatrix} 0 \\ 1 \end{pmatrix} + A^3 \begin{pmatrix} 0 \\ 1 \end{pmatrix} + \cdots \qquad (2.156)$$

For the case of simultaneous satisfaction of a unit of final demand for both products, it then follows:

$$\begin{pmatrix} 1 & 0 \\ 0 & 1 \end{pmatrix} + A \begin{pmatrix} 1 & 0 \\ 0 & 1 \end{pmatrix} + A^2 \begin{pmatrix} 1 & 0 \\ 0 & 1 \end{pmatrix} + \cdots ,$$
$$= I + A + A^2 + \cdots = (I - A)^{-1}. \qquad (2.157)$$

2.3.2.3 Obtaining the Input Coefficients Matrix A from the IOT

Write $\hat{x}$ for a 2×2 diagonal matrix with x as its diagonal elements:

$$\hat{x} = \begin{pmatrix} x_1 & 0 \\ 0 & x_2 \end{pmatrix} . \qquad (2.158)$$

Henceforth, "ˆ" above a vector refers to the diagonal matrix with the elements of the vector occurring as the diagonal elements. Denoting by $X = (x_{ij})$ the 2×2 matrix of endogenous flows:

$$X = \begin{pmatrix} x_{11} & x_{12} \\ x_{21} & x_{22} \end{pmatrix} , \qquad (2.159)$$

it then follows

$$X = A\hat{x}. \tag{2.160}$$

Accordingly, provided $\hat{x}$ is nonsingular, that is, all the elements of x are positive (negative production makes no sense), the matrix A can be obtained from x and X as

$$A = X\hat{x}^{-1}. \tag{2.161}$$

2.3.3 Exogenous Inputs and Waste Generation

2.3.3.1 The Demand for Exogenous Inputs and Waste Generation

In line with (2.83), using (2.152) the demand for exogenous inputs, the vector of which is denoted by z, is given by

$$z = B(I - A)^{-1}f, \tag{2.162}$$

and the generation of waste is given, in line with (2.84), by

$$w = G^{\text{out}}(I - A)^{-1}f. \tag{2.163}$$

In line with (2.86), the effects of a change in technology on the demand for exogenous inputs are given by

$$z' - z'' = \left(B'(I - A')^{-1} - B''(I - A'')^{-1}\right)f \tag{2.164}$$

A special case of (2.162) with the elements of B referring to land only corresponds to the IO calculation of an ecological footprint that was first initiated by Bicknell et al. [1], and further elaborated, among others, by [12, 14, 24]. In the pioneering application to New Zealand, Bicknell et al., [1] found that New Zealand's ecological footprint did not exhaust the amount of ecologically productive land currently available within the country, whereas many overseas countries, by contrast, were subject to extreme ecological deficits.

Another special case of (2.162) is the estimation of virtual water flows which is obtained with the elements of B referring to water only. Sánchez-Chóliz and Duarte [18] used this type of model to analyze the environmental impacts of the Spanish economy by way of water and atmospheric pollution, and found, among others, that the Spanish economy avoids a great deal of pollution by importing inputs, which pollute where they are produced. Guan and Hubacek [6] applied this type of model to account for virtual water flows between North and South China, and found that water-scarce North China predominantly produces and exports water intensive products but imports nonwater-intensive commodities.

2.3.3.2 The Maximum Level of Final Demand for Given Exogenous Inputs

In Section 2.1.2.6, the issue of a limited supply of exogenous inputs was discussed. In the case of multiple sectors, however, this issue becomes substantially more complex because of the presence of sectoral interdependence. While one can still use (2.30) to obtain the upper level of production that could be achieved by the maximum use of the exogenous inputs that are available to each sector, no consideration is made of inter-sectoral interdependence. Therefore, a simple matrix extension of (2.31) is not applicable. In particular, there is no guarantee that the maximum level of production that is determined by the available exogenous inputs is able to support nonnegative final demand. Dealing with this issue calls for the solution of an optimization problem subject to inequality constraints. This is a rather advanced topic, and will be dealt with in Section 3.5.1.

2.3.4 Cost and Price

2.3.4.1 The Price Model

The starting point of the price model is the equality between the value of outputs and the value of inputs (2.87), which can be written:

$$\left(p_1 x_1 \ p_2 x_2\right) = \left(p_1 \ p_2\right) \begin{pmatrix} x_{11} \ x_{12} \\ x_{21} \ x_{22} \end{pmatrix} + \left(v_1 \ v_2\right) \tag{2.165}$$

or

$$p\widehat{x} = pX + v = pX + qB\widehat{x} \tag{2.166}$$

where q refers to the row vector of the price of exogenous inputs.

Post-multiplication of $\widehat{x}^{-1}$ to both sides yields:

$$p = pX\widehat{x}^{-1} + qB = pA + qB \tag{2.167}$$

where use is made of (2.161) and (2.134). Solving for p yields

$$p = qB(I - A)^{-1} \tag{2.168}$$

This represents the Leontief price model. In line with (2.108), the effects of a change in q on the price of endogenous inputs are given by

$$\Delta p = \Delta q B(I - A)^{-1} \tag{2.169}$$

2.3.4.2 Physical and Value Coefficients

Writing $\widehat{p}$ for the diagonal matrix obtained from p:

$$\widehat{p} = \begin{pmatrix} p_1 & 0 \\ 0 & p_2 \end{pmatrix},$$ (2.170)

the equality between supply and demand in value terms (2.90) can be given by

$$\widehat{p}x = \widehat{p}Ax + \widehat{p}f.$$ (2.171)

From (2.111), the value counterpart of A, $\bar{A} = (\bar{a}_{ij})$, is given by:

$$\bar{A} = \widehat{p}X\widehat{x}^{-1}\widehat{p}^{-1} = \widehat{p}A\widehat{p}^{-1}$$ (2.172)

Substituting from (2.172), (2.171) becomes:

$$\widehat{p}x = \widehat{p}A\widehat{p}^{-1}\widehat{p}x + \widehat{p}f = \bar{A}\widehat{p}x + \widehat{p}f,$$ (2.173)

which gives the counterpart of (2.139) in value units.

Solving (2.173) for the value of product, $\widehat{p}x$, gives

$$\widehat{p}x = (I - \bar{A})^{-1}\widehat{p}f$$ (2.174)

It was shown above (Section 2.2.4.3) for the case of $n = 2$ that when the relative prices are constant the solution of the value model yields the same results as the physical model. The matrix counterpart of this result is to be derived. Substituting from the definition of $\bar{A}$ given by (2.172), it follows that

$$(I - \bar{A})^{-1} = \left(I - \widehat{p}A\widehat{p}^{-1}\right)^{-1} = \left(\widehat{p}\widehat{p}^{-1} - \widehat{p}A\widehat{p}^{-1}\right)^{-1} = \left(\widehat{p}(I - A)\widehat{p}^{-1}\right)^{-1}$$

$$= (\widehat{p}^{-1})^{-1}(I - A)^{-1}\widehat{p}^{-1} = \widehat{p}(I - A)^{-1}\widehat{p}^{-1},$$ (2.175)

where the fourth equality is due to property (2.148) of the inverse matrix. Substitution of (2.175) into (2.174) then gives

$$\widehat{p}x = \widehat{p}(I - A)^{-1}\widehat{p}^{-1}\widehat{p}f = \widehat{p}(I - A)^{-1}f.$$ (2.176)

Premultiplying both sides by $\widehat{p}^{-1}$, (2.176) reduces to the original quantity model (2.152). This is a reproduction of the result obtained by Weisz and Duchin who state "The 2 models [a monetary input-output model and a physical input-output model] are the same except for the change of unit operation, and the vector of unit prices provides the information needed for the change of unit" ([23], p.536).

As has been shown in Section 2.2.4.3 (p.40), when p is user specific, the above result is no longer applicable.

Table 2.12 Japanese IO Table for 2000 at Producers' and Purchasers' Prices.

	AGR	MTL	MCN	OMnf	Utl	T&T	Oth	FD	T&T*	Prd
					Producers' prices					
AGR	1,558	0	0	8,427	0	9	1,488	2,887	0	14,370
MTL	18	13,696	9,610	2,479	34	296	9,452	1,164	0	36,750
MCN	79	122	45,439	455	30	262	7,953	74,256	0	128,596
OMnf	2,366	2,948	8,175	47,397	3,636	3,852	43,131	32,689	0	144,194
Utl	92	1,243	1,577	3,561	1,623	1,394	8,611	8,902	0	27,004
T&T	920	2,095	7,678	11,263	720	1,217	18,681	66,233	0	108,807
Oth	1,261	3,630	16,160	16,296	5,673	26,593	96,201	333,351	0	499,165
V	8,075	13,016	39,956	54,315	15,289	76,804	312,027			
Prd	14,370	36,750	128,596	144,194	27,004	110,428	497,544			

	AGR	MTL	MCN	OMnf	Utl	T&T	Oth	FD	T&T*	Prd
					Purchasers' prices					
AGR	1,626	0	0	10,440	0	18	2,321	6,013	−6,049	14,370
MTL	23	15,200	11,136	2,829	42	341	11,669	2,070	−6,561	36,750
MCN	81	147	49,864	580	37	343	9,805	97,485	−29,745	128,596
OMnf	3,208	3,502	9,866	56,143	4,335	4,895	56,820	71,507	−66,081	144,194
Utl	92	1,243	1,577	3,561	1,623	1,394	8,611	8,902	0	27,004
T&T	0	0	0	0	0	0	0	0	108,807	108,807
Oth	1,265	3,641	16,196	16,326	5,678	26,634	96,291	333,504	−370	499,165
V	8,075	13,016	39,956	54,315	15,289	76,804	312,027			
Prd	14,370	36,750	128,596	144,194	27,004	110,428	497,544			

Units: 1,000 million Japanese yen. AGR: Agriculture, forestry and fishery, MTL: Metal and metal products, MCN: Machinery, OMnf: Other manufacturing and mining, Utl: Utilities, T&T: Trade and transportation, Oth: Others, V: Value added, Prd: Production, FD: Final demand, T&T*: Trade and transportation margins.

2.3.4.3 'Producers' and 'Purchasers' IOTs: An Example

In Section 2.2.4.4 above, two types of monetary IOTs were introduced which differ in the way trade and transport costs (margins) associated with the flow of goods and services are accounted for, that is, the IOT at 'producers prices' (Table 2.11) and the IOT at 'purchasers prices' (Table 2.10). Table 2.12 gives a real example of these tables with the upper panel referring to the IOT at producers' prices, and the lower panel referring to the IOT at purchasers' prices. Note that in real IOTs such as these the columns referring to the trade & transport sector are not zeros, while they were so in simplified IOTs as in Tables 2.10 and 2.11. These column elements refer to the inputs into the trade & transport sector, which are inherent to the activity of the sector, and hence are not transferred to other sectors according to the way margins are accounted for.

2.3.5 Structural Decomposition Analysis

In the Leontief quantity model, the level of production is determined by the Leontief inverse matrix and the vector of final demand. A given change in the level of production, say, over time, can then be decomposed into the change in technology (represented by the Leontief inverse matrix) and the final demand. This decomposition is called structural decomposition analysis to which we now turn.

2.3.5.1 SDA: Quantity Model

When there is a change in the final demand, say Δf, while A remains unchanged, the resulting change in the amount of production is, following (2.85), given by:

$$\Delta x = (I - A)^{-1} \Delta f, \qquad (2.177)$$

where

$$\Delta f = \begin{pmatrix} \Delta f_1 \\ \Delta f_2 \end{pmatrix} \qquad (2.178)$$

In (2.177), the change in x is attributed solely to a change in f, because A is assumed to remain constant. Consider now the case where A also changes.

To be concrete, suppose that we deal with historical data on x for two time periods, t and s with $s > t$, that is, $x(t)$ and $x(s)$. Because A as well as f may be different for t and s, the change in x that occurred between t and s can be represented as

$$\Delta x = x(s) - x(t) = (I - A(s))^{-1} f(s) - (I - A(t))^{-1} f(t), \qquad (2.179)$$

which can be rewritten as

$$\begin{aligned} \Delta x &= \Big[C(s) - C(t) \Big] f(s) + C(t) \Big[f(s) - f(t) \Big] \\ &= \Delta C f(s) + C(t) \Delta f \end{aligned} \qquad (2.180)$$

where $C(t) = (I - A(t))^{-1}$. This breaks down the change in x into the change in its factors, technology ΔC and final demand Δf. This 'decomposition' into its factors of the observed change in endogenous variables based on IOA is called structural decomposition analysis (SDA) [3]. SDA is widely used because of its ability to allow for the quantification of the underlying sources of change in a variety of variables (see [9] and [11] for a recent application to environmental issues).

Applying (2.180) to (2.162) gives the structural decomposition of the change in the demand for exogenous inputs:

$$\Delta z = z(s) - z(t) = \underbrace{\Delta (BC) f(s)}_{\text{change in technology}} + \underbrace{B(t) C(t) \Delta f}_{\text{change in final demand}} \qquad (2.181)$$

where

$$\Delta (BC) = B(s) C(s) - B(t) C(t) \qquad (2.182)$$

Analogously, applying the same procedure to (2.163), the change in the amount of waste generation over time can be decomposed as follows:

$$\Delta w = w(s) - w(t) = \underbrace{\Delta(G^{\mathrm{out}}C)f(s)}_{\text{change in technology}} + \underbrace{G^{\mathrm{out}}(t)C(t)\Delta f}_{\text{change in final demand}} \tag{2.183}$$

2.3.5.2 SDA: Price Model

The application of SDA is not limited to the quantity model, but is also applicable to the price model. In the Leontief price model (2.168), the price of endogenous sectors p depends on the matrices A and B, and the price of exogenous inputs q. Because the technology matrices A and B, and q are assumed independent from each other, it makes sense to decompose the observed change in p into the change in these two factors.

Applying SDA to (2.168) in a manner analogous to (2.180) gives the following decomposition of Δp

$$\begin{aligned}
\Delta p &= p(s) - p(t) \\
&= q(s)B(s)C(s) - q(t)B(t)C(t) \\
&= q(s)\Big(B(s)C(s) - B(t)C(t)\Big) + \Big(q(s) - q(t)\Big)B(t)C(t) \\
&= \underbrace{q(s)\Delta\Big(BC\Big)}_{\text{change in technology}} + \underbrace{\Delta q\,B(t)C(t)}_{\text{change in the price of exogenous inputs}}
\end{aligned} \tag{2.184}$$

2.3.5.3 Some Remarks on the Use of SDA

Two remarks are due on the application of SDA. The first point refers to the nonuniqueness of the decomposition, which results from the fact that the weights attached to a specific factor to measure its contribution is not unique. Taking the decomposition given by (2.180) for illustration, this problem refers to the fact that there is an alternative way of decomposition:

$$\begin{aligned}
\Delta x &= \Big[C(s) - C(t)\Big]f(t) + C(s)\Big[f(s) - f(t)\Big] \\
&= \Delta C f(t) + C(s)\Delta f
\end{aligned} \tag{2.185}$$

where ΔC is weighted by $f(t)$ instead of $f(s)$, and Δf is weighted by $C(s)$ instead of $C(t)$.

The use of different weights will yield different contributions of a given factor, and there is no reason for preferring one decomposition in favor of the other. This is the nonuniqueness problem (see [3] for further details). This problem can be coped with, albeit ad hoc, by using the mean value of weights:

$$\Delta x = \Delta C \frac{1}{2}\Big(f(t) + f(s)\Big) + \frac{1}{2}\Big(C(s) + C(t)\Big)\Delta f \tag{2.186}$$

Analogously, the following decomposition can be used for the price model (2.184) to avoid the nonuniqueness problem

$$\Delta p = \frac{1}{2}\Big(q(s)+q(t)\Big)\Delta\Big(BC\Big)+\Delta q\frac{1}{2}\Big(B(s)C(s)+B(t)C(t)\Big) \qquad (2.187)$$

The second remark is a warning against a mechanical application of SDA to the case of multiple factors without paying due attention to their independence. Consider for example the following 'extension' of (2.181)

$$\Delta z = \Delta BC(s)f(s)+B(t)\Delta C f(s)+B(t)C(t)\Delta f \qquad (2.188)$$

where the contribution of $\Delta(BC)$ is now further decomposed into its 'factors' ΔB and ΔC. It is important to recall that each element of the inputs and outputs of a given unit process constitutes an integral part of the process, and hence cannot be altered independent from the other elements. Accordingly, the elements of A (and hence C) and B cannot be altered independently from each other.

Suppose for instance that a process P_j that is available in time s uses less water (exogenous input k) than one that was available in time t:

$$b(t)_{kj} > b(s)_{kj} \qquad (2.189)$$

It is then highly unlikely that all the elements of $a(t)_{ij}$ and $a(s)_{ij}$ will be the same: the saving of water may have been brought about by a new design or the use of different materials, which manifest themselves as a change in certain elements of a_{ij}. Stated differently, a unit process where all the elements are the same except for b_{kj} will not exist. If this is the case, the first two terms on the right hand side of (2.188) represent a decomposition into factors which do not exist in the real world.

Due care should be taken in interpreting the result of SDA extended to the case of multiple factors, if there is a possibility of interdependence among the factors.

2.3.6 IOA for the Case of $n \geq 2$

So far, we have been concerned with deriving matrix representations of the results obtained for the case of two endogenous sectors in Section 2.2. Compared with the case of only one endogenous sector in Section 2.1, which is straightforward and requires only elementary calculation, the case of two sectors with a system of equations may appear rather clumsy. When the same approach is applied to the case of three endogenous sectors or larger, most readers will find themselves overwhelmed by the clumsiness of the expressions. An important virtue of using matrices consists in getting rid of this clumsiness without losing any of its logical content, a good example of which is the system of equations (2.65) and (2.66) being represented by a single equation (2.140).

Table 2.13 Basic Equations of IOA in Terms of Scalars and Matrices.

Scalar		Matrix	
$x = a_1 x + f$	(2.13)	$\mathbf{x} = \mathbf{Ax} + \mathbf{f}$	(2.139)
$(1 - a_1)x = f$	(2.14)	$(\mathbf{I} - \mathbf{A})\mathbf{x} = \mathbf{f}$	(2.140)
$x = (1 - a_1)^{-1} f$	(2.15)	$\mathbf{x} = (\mathbf{I} - \mathbf{A})^{-1} \mathbf{f}$	(2.152)
$z = b(1 - a_1)^{-1} f$	(2.19)	$\mathbf{z} = \mathbf{B}(\mathbf{I} - \mathbf{A})^{-1} \mathbf{f}$	(2.162)
$p = p a_1 + \pi$	(2.43)	$\mathbf{p} = \mathbf{pA} + \pi$	(2.168)
$p = (1 - a_1)^{-1} \pi$	(2.45)	$\mathbf{p} = \pi(\mathbf{I} - \mathbf{A})^{-1}$	(2.168)
$\Delta x = (1 - a_1)^{-1} \Delta f$		$\Delta \mathbf{x} = (\mathbf{I} - \mathbf{A})^{-1} \Delta \mathbf{f}$	
$\Delta p = (1 - a_1)^{-1} \Delta \pi$		$\Delta \mathbf{p} = \Delta \pi (\mathbf{I} - \mathbf{A})^{-1}$	

The bold face letters on the right panel refer to matrices or vectors, while the normal letters on the left panel refer to scalars.

Table 2.13 compares major expressions obtained for the one sector model in Section 2.1 with those based on matrices for the case of two sectors obtained above. In the table, vectors and matrices are indicated by bold face letters to distinguish them from scalars (this convention is not used in the rest of this book). Looking at Table 2.13, one realizes the presence of a strong similarity between the expressions in terms of scalars on the left hand side and the expression in terms of matrices on the right hand side. In fact, except that "1" and lower case letters on the left hand side are replaced by "$\mathbf{I}$" and upper case letters on the right hand side, they are practically identical. Once one gets used to expressions in terms of matrices, it is no longer necessary to be bothered with the clumsy expressions involving systems of equations as in Section 2.2. Another important virtue of the use of matrices is that once the system of equations of a given order, say 2, is represented in matrices, the representation is applicable to a system of any order. The matrix equations listed on the right hand side of Table 2.13 are applicable to the IOA of any number of endogenous sectors.

2.3.6.1 Positivity Condition

The general version of Theorem 2.2 that is applicable to the case of any $n > 0$ is the following due to Fukukane Nikaido [17]:

Theorem 2.3. *Let A be an $n \times n$ matrix, with all its elements nonnegative. The following three conditions are equivalent.*

(WS) Weak solvability: For a particular $n \times 1$ vector $f > 0$, $(I - A)x = f$ has a solution $x \geq 0$.

(SS) Strong solvability: For any $f \geq 0$, $(I - A)x = f$ has a solution $x \geq 0$.

(HS) Hawkins Simon: All principal minors of $(I - A)$ are positive:

$$1 - a_{11} > 0, \quad \begin{vmatrix} 1 - a_{11} & -a_{12} \\ -a_{21} & 1 - a_{22} \end{vmatrix} > 0, \cdots, \quad \begin{vmatrix} 1 - a_{11} & -a_{12} & \cdots & -a_{1n} \\ -a_{21} & 1 - a_{22} & \cdots & -a_{2n} \\ & & \cdots\cdots\cdots & \\ -a_{n1} & -a_{n2} & \cdots & 1 - a_{nn} \end{vmatrix} > 0$$

Proof. See [17].

For the matrix of input coefficients that is obtained from an IOT with no negative elements via (2.161), (WS) is always satisfied when all the elements of f are positive (if not, it can easily be made positive by consolidating sectors with zero final demand with those with positive final demand). From Theorem 2.3 it then follows that the solution of the quantity model (2.152) for this matrix of input coefficients will always be nonnegative for any nonnegative vector of final demand. It is not necessary to check for (HS) in order to know whether its solution is nonnegative. It is, however, important to notice that this powerful result does not hold when some elements of the matrix of input coefficients are negative, which is not uncommon when there are by-products, an issue which will be addressed below, in Section 3.2.

2.4 Exercise with Excel

Excel provides an easy and readily accessible way for applying IOA to real data. This section illustrates the use of Excel for some basic applications of IOA. For illustration, Japanese IO tables for 2000 (IO tables with 13 and 32 endogenous sectors) are used, which can be obtained from the Internet:

- The 13-sector tables: `http://www.stat.go.jp/english/data/io/2000/zuhyou/sec013.xls`
- The 32-sector tables: `http://www.stat.go.jp/english/data/io/2000/zuhyou/sec032.xls`

2.4.1 Basic Analysis

Using the 13-sector IO table (Sheet1 in sec013.xls) we first consider the calculation of the matrix of input coefficients, A, and the Leontief inverse matrix, $(I - A)^{-1}$.

2.4.1.1 Calculating the Matrix of Input Coefficients A

Method 1: Using matrix multiplications

This method involves creating the diagonal matrix of x (that occurs transposed as a vector of 13×1 in the file).

1. Typing $=\mathtt{IF\,(ROW\,(\,)\,-33=COLUMN\,(\,)\,-3\,,C\$27\,,0\,)}$ into cell **C33** (henceforth, the term "cell" will be omitted), copying, and pasting it into range **C33:O45** (henceforth, the term "range" will be omitted) gives the diagonal matrix $\hat{x}$.

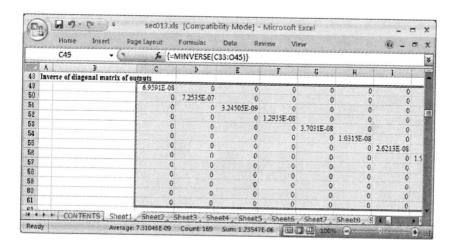

2. The inversion of $\hat{x}$ involves the use of an array formula, which works with an array, or series, of data values rather than a single data value. Suppose that the inverse matrix is to be placed in C49:O61. Select the range, and type the following formula in C49

```
=MINVERSE(C33:O45)
```

Entering this formula as an array formula (press the CTRL, SHIFT and ENTER keys at the same time rather than just ENTER) gives the result:

3. The final step involves the multiplication of the intermediate flow matrix X that occurs in C6:O18 from the right by $\hat{x}^{-1}$. Entering

```
=MMULT(C6:O18,C49:O61)
```

in C65:O77 as an array formula gives the result

Steps 2 and 3 can be combined into a single step by nesting them:
`=MMULT(C6:O18,MINVERSE(C33:O45))`

Method 2: Further simplification

The above calculation can further be simplified to a single step. Suppose that A is to be placed in `C81:O93`. Entering
`=(C6:O18)/(C27:O27)`
as an array formula into the range yields the result in one step:

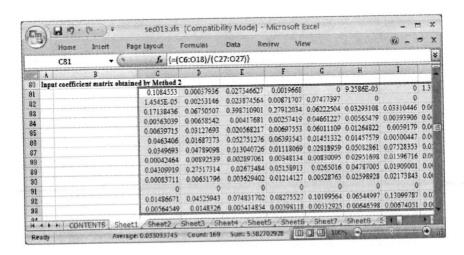

Note that the above formula involves the division of a 13×13 matrix by a 1×13 vector, which is not defined in matrix algebra. Still, the above formula enables the calculation of $X\hat{x}^{-1}$, because in Excel the slash/stands for an element-by-element division of the two arrays. If the two arrays happen to be of different sizes (the numbers of the rows and columns relevant for the multiplication are not equal), the array with a smaller size is enlarged by duplicating its rows and columns to make the multiplication possible.

2.4.1.2 Calculating the Leontief Inverse Matrix

Unlike software specialized in matrix calculations, such as MATLAB, Excel does not have a function to calculate an identity matrix. Accordingly, it has to be generated. Two methods are shown below.

Method 1: Calculation using a stored identity matrix

1. Suppose that an identity matrix of order 13 is to be placed in C97:O109. First, note that entering =IF (ROW()-97=COLUMN()-3,1,0) into C97 returns 1, which corresponds to the second argument, because the logical expression ROW()-97=COLUMN()-3 is true. Copying C97, and pasting it to C97:O109 yields the identity matrix:

2. Entering =MINVERSE((C97:O109)-(C81:O93)) into C113:O125 as an array formula then gives the Leontief inverse matrix:

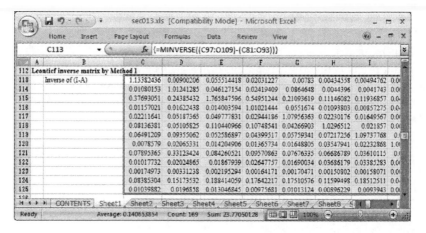

In `MINVERSE`, the term `C97:O109` refers to the identity matrix, while the term `C81:O93` refers to the input coefficient matrix.

A modified form of the formula to enter the identity matrix in Step 1 is
`=IF(ROW()-ROW($C$97)=COLUMN()-COLUMN($C$97),1,0)`
This prevents the formula from becoming invalid when rows and/or columns are deleted or inserted.

Note that the values of the Leontief inverse matrix obtained here differ from those occurring in Sheet4 of the original data file, because of the difference in the way imports are accounted for (see Section 3.1).

Method 2: Calculating the Leontief inverse matrix without storing the identity matrix

The above method is simple, but suffers from the fact that the identity matrix has to be prepared and stored in advance. Making use of the following formula as an array formula, one can skip this step:
`=MINVERSE((ROW()-129=COLUMN()-3)-(C81:O93))`

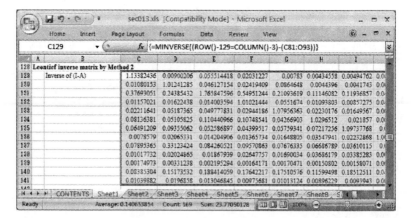

2.4.1.3 The Leontief Quantity Model

The amount of production that is induced by the final demand (2.152) is obtained by entering the following formula into C145:C157 as an array formula:
=MMULT(C113:O125,AE6:AE18),
where the first argument refers to the inverse matrix obtained above, and the second argument to the total final demand.

Suppose that the number of employees (Table 2.14) occur in C162:O162 (as a 1×13 row vector). One obtains the labor input coefficients by entering =C162/C27 into C164, copying and pasting it to D164:O164. Entering
=MMULT(C164:O164,C145:C157)
gives the amount of employment that is induced by the final demand.

Table 2.14 Number of Employed Persons.

Sectors	Number of employed persons
01 Agriculture, forestry and fishery	5,569,678
02 Mining	47,442
03 Manufacturing	11,034,273
04 Construction	6,572,311
05 Electric power, gas and water supply	631,611
06 Commerce	13,987,846
07 Finance and insurance	1,874,102
08 Real estate	698,521
09 Transport	3,186,040
10 Communication and broadcasting	770,243
11 Public administration	2,010,732
12 Services	21,858,281
13 Activities not elsewhere classified	48,368

2.4.1.4 The Price Model

Suppose that the value-added ratios occur in C170:O170. The solution of the price model (2.168) is given by =MMULT(C170:O170,C129:O141)
The calculated prices are all equal to unity, because the amounts of goods and services are measured in 1 yen's worth (see Section 2.2.4.3).

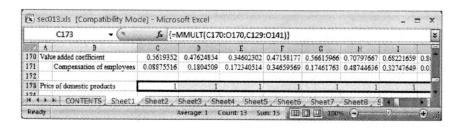

Next, consider the case where the wage rates are increased by 10%. Changing the formula in C170 as =C26/C27+0.1*C171, copying the cell, and paste it into D170:O170 gives the result:

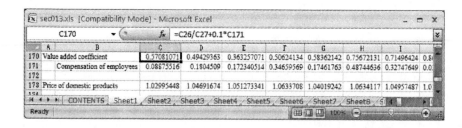

2.4.2 Consolidating of Sectors

In the application of IOA one often encounters the need for consolidating (aggregating) several sectors of a given IOT. A method of consolidation is illustrated for the case of consolidating the 32-sector table into a 13-sector one (reproducing the 13-sector table). Tables 2.15 and 2.16 give the classification of sectors for the two tables, and the correspondence between them.

Table 2.15 The Classification of Row Sectors.

	The 32-sector table			The 13-sector table
C	Sectors	S	C	Sectors
1	Agriculture, forestry and fishery	1	1	Agriculture, forestry and fishery
2	Mining	2	2	Mining
3	Foods	3	3	Manufacturing
4	Textile products	3	3	Manufacturing
5	Pulp, paper and wooden products	3	3	Manufacturing
6	Chemical products	3	3	Manufacturing
7	Petroleum and coal products	3	3	Manufacturing
8	Ceramic, stone and clay products	3	3	Manufacturing
9	Iron and steel	3	3	Manufacturing
10	nonferrous metals	3	3	Manufacturing
11	Metal products	3	3	Manufacturing
12	General machinery	3	3	Manufacturing
13	Electrical machinery	3	3	Manufacturing
14	Transportation equipment	3	3	Manufacturing
15	Precision instruments	3	3	Manufacturing
16	Miscellaneous manufacturing products	3	3	Manufacturing
17	Construction	4	4	Construction
18	Electricity, gas and heat supply	5	5	Electric power, gas and water supply
19	Water supply and waste management services	5	5	Electric power, gas and water supply
20	Commerce	6	6	Commerce
21	Financial and insurance	7	7	Finance and insurance
22	Real estate	8	8	Real estate
23	Transport	9	9	Transport
24	Communication and broadcasting	10	10	Communication and broadcasting
25	Public administration	11	11	Public administration
26	Education and research	12	12	Services
27	Medical service, health and social security and nursing care	12	12	Services
28	Other public services	12	12	Services
29	Business services	12	12	Services
30	Personal services	12	12	Services
31	Office supplies	3	3	Manufacturing
32	Activities not elsewhere classified	13	13	Activities not elsewhere classified
33	Total of intermediate sectors	14	33	Total of intermediate sectors
35	Consumption expenditure outside households (row)	15	35	Consumption expenditure outside households (row)
36	Compensation of employees	16	36	Compensation of employees

(continued)

Table 2.15 (continued)

The 32-sector table		The 13-sector table		
C	Sectors	S	C	Sectors
37	Operating surplus	17	37	Operating surplus
38	Depreciation of fixed capital	18	38	Depreciation of fixed capital
39	Indirect taxes (except custom duties and commodity taxes on imported goods)	19	39	Indirect taxes (except custom duties and commodity taxes on imported goods)
40	(less) Current subsidies	20	40	(less) Current subsidies
52	Total of gross value-added sectors	21	52	Total of gross value added sectors
55	Domestic production (gross inputs)	22	55	Domestic production (gross inputs)
56	Net domestic product at factor cost	23	56	Net domestic product at factor cost
57	Gross domestic product	24	57	Gross domestic product

Note: C and S stand for the code and sequential number, respectively.

Table 2.16 The Classification of Column Sectors.

The 32-sector table		The 13 sector table		
C	Sectors	S	C	Sectors
1	Agriculture, forestry and fishery	1	1	Agriculture, forestry and fishery
⋮	⋮			⋮
32	Activities not elsewhere classified	13	13	Activities not elsewhere classified
33	Total of intermediate sectors	14	33	Total of intermediate sectors
35	Consumption expenditure outside households (column)	15	35	Consumption expenditure outside households (column)
36	Consumption expenditure (private)	16	36	Consumption expenditure (private)
37	Consumption expenditure of government	17	37	Consumption expenditure of government
38	Gross domestic fixed capital formation (public)	18	38	Gross domestic fixed capital formation
39	Gross domestic fixed capital formation (private)	18	38	Gross domestic fixed capital formation
40	Increase in stocks	19	40	Increase in stocks
41	Total domestic final demand	20	41	Total domestic final demand
42	Total domestic demand	21	42	Total domestic demand
43	Exports	0		[eliminated]
44	Balancing sector	0		[eliminated]
45	Exports total	22	45	Exports total
46	Total final demand	23	46	Total Final demand
47	Total demand	24	47	Total demand
48	(less) Imports	25	48	(less) Imports
49	(less) Custom duties	26	49	(less) Custom duties
50	(less) Commodity taxes on imported goods	27	50	(less) Commodity taxes on imported goods
51	(less) Total imports	28	51	(less) Total imports
52	Total of final demand sectors	29	52	Total of final demand sectors
55	Domestic production (outputs)	30	55	Domestic production (outputs)
57	Gross domestic expenditure	31	57	Gross domestic expenditure

Note: C and S stand for the code and sequential number, respectively. The column sector classification of industrial sectors is omitted because it is the same as the classification of row sectors.

1. Open Book sec032.xls, insert a new sheet before Sheet1, and name it consolidation.
2. Make consolidation the active sheet.
3. Create a {0,1}-matrix that represents the correspondence of the row sectors between the original table and the consolidated table. First, enter the information on the consolidation of row sectors in Table 2.15 into A3:E45. Second, enter the sequential numbers 1–24, which refer to the number of rows in the consolidated table, into F2:AC2. Third, enter =IF($C3=F$2,1,0) into F3. Finally, copy F3, and paste it into F3:AC45. Then, assign the name RCons to F3:AC45.

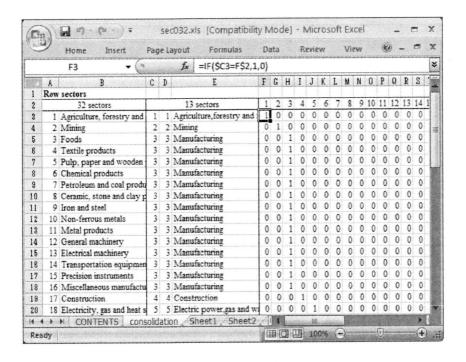

4. In a similar fashion, create a {0,1}-matrix that represents the correspondence of the column sectors between the original table and the consolidated table. First, enter the information on the consolidation of column sectors in Table 2.16 into AE3:AI55. Second, enter the sequential numbers 1–31, which refer to the number of columns in the consolidated table, into AJ2:BN2. Third, enter =IF($AG3=AJ$2,1,0) into AJ3. Finally, copy AJ3, and paste it into AJ3:BN55. Then, assign the name CCons to AJ3:BN55.

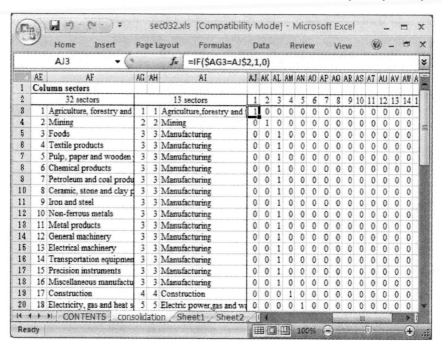

5. Assign the name IOT32 to Sheet1!C6:BC48, which gives a 43-by-53 matrix of the 32-sector table including its fourth quadrant. Enter zeros into AJ39:BC48 in order to remove empty cells from Sheet1!C6:BC48.

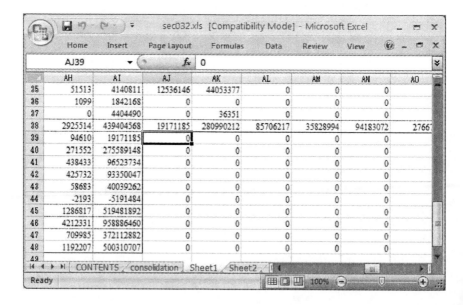

6. Insert a new sheet named IOT13 before consolidation, and enter the labels referring to the 13-sector table (by copy and paste).

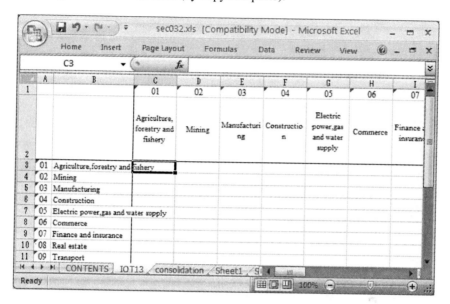

7. Entering into C3:AG26 the following formula
=MMULT(TRANSPOSE(RCons),MMULT(IOT32,CCons))
gives the 13-sector table:

References

1. Bicknell, K., Ball, R., Cullen, R., & Bigsby, H. (1998). New methodology for the ecological footprint with an application to the New Zealand economy. *Ecological Economics, 27,* 149–160.
2. Brunner, P., & Rechberger, H. (2003). *Practical handbook of material flow analysis.* Boca Raton, FL: Lewis Publishers.
3. Dietzenbacher, E. (1998). Structural decomposition techniques: Sense and sensitivity. *Economic Systems Research, 10(4),* 307–323.
4. Dooley, P. (2005). *The Labour theory of value.* London: Routledge.
5. Dorfman, R., Samuelson, P., & Solow, R. (1958). *Linear programming and economic analysis.* New York: McGraw-Hill.
6. Guan, D., & Hubacek, K. (2007). Assessment of regional trade and virtual water flows in China. *Ecological Economics, 61,* 159–170.
7. The web page of Gunma prefecture, Japan. Accessed March 30, 2008. http://www.pref.gunma. jp/e/01/cb/cbnouson/keikaku/hukutokuhon.htm
8. Heijungs, R., & Suh, S. (2002). *The computational structure of Life Cycle Assessment.* Dordrecht: Kluwer.
9. Hoekstra, R., & van den Bergh, J. (2002). Structural decomposition analysis of physical flows in the economy. *Environmental and Resource Economics, 23,* 357–378.
10. Isard, W. et al. (1968). On the linkage of socio-economic and ecological systems. *Papers in Regional Science, 21(1),* 79–99.
11. Kagawa, S., & Inamura, H. (2001). A structural decomposition of energy consumption based on a hybrid rectangular input-output framework: Japan's case. *Economic Systems Research, 13(4),* 339–363.
12. Lenzen, M., & Murray, S. (2001). A modified ecological footprint method and its application to Australia. *Ecological Economics, 37,* 229–255.
13. Leontief, W. (1951). *The structure of American economy 1919–1939.* New York: Oxford University Press.
14. McDonald, G., & Patterson, G. (2004). Ecological footprints and interdependencies of New Zealand regions. *Ecological Economics, 50,* 49–67.
15. Miller, R., & Blair, P. (1985). *Input-output analysis.* Englewood Cliffs, NJ: Prentice-Hall.
16. The Ministry of Agriculture, Forestry and Fisheries, Statistics Department (2008). *The cost of production investigation.* http://www.tdb.maff.go.jp/toukei. Cited January 2008.
17. Nikaido, F. (1969). *Convex structures and economic theory.* New York: Academic Press (English translation of the original publication in Japanese in 1960).
18. Sánchez-Chóliz, J., & Duarte, R. (2005). Water pollution in the Spanish economy: Analysis of sensitivity to production and environmental constraints. *Ecological Economics, 53,* 325–338.
19. Schott, J. (1997). *Matrix analysis for statistics.* New York: Wiley.
20. Solow, R. (1952). On the structure of linear models. *Econometrica, 20(1),* 29–46.
21. Stone, R. (1961). *Input-output and national accounts.* Paris: The Organization for European Economic Development.
22. Wackernagel, M., & Rees, W. (1996). *Our ecological footprint.* Gabriora Islands: New Society Publishers.
23. Weisz, H., & Duchin, F. (2006). Physical and monetary input-output analysis: What makes the difference? *Ecological Economics, 57,* 534–541.
24. Wiedmann, T., Minx, J., Barrett, J., & Wackernagel, M. (2006). Allocating ecological footprints to final consumption categories with input-output analysis. *Ecological Economics, 56,* 28–48.
25. World Water Council Virtual Water. http://www.worldwatercouncil.org/index.php?id=866

Chapter 3
Extensions of IOA

Abstract The basic IO model introduced in Chapter 2 needs some extensions before it can effectively be put to a practical application. This applies particularly to regional extensions and the treatment of by-product. Regional extensions are necessary to account for the plain fact that economies are connected to each other thorough the trans boundary flows of goods and services, just as the sectors of an economy are connected to each other through the flow of goods and services. Waste and emissions, which are objects of a great concern in industrial ecology, are typical examples of by-product. Alternative ways to accounting for it in IOA are discussed, including one based on the make and use matrices. In Chapter 2 a brief mention was made of the possibility of representing the household consumption just like any other production process. Section 3.4 deals with the closing of the IO model by making endogenous the final demand sectors including not only household consumption but capital formation as well. Other topics to be dealt with include the introduction of inequalities to allow for the presence of resource constraints, the identification of a fundamental structure of production based on some rearrangements of sector orderings, and a variant of IOA model called Ghosh model.

3.1 Regional Extensions

Production processes and consumers are distributed throughout the world, which consists of spatial units, such as nations, states, cities, and villages. Interdependence of production processes implies that the operation of a production process or consumption activity in a certain part of the world can invoke the flow of inputs and outputs over many different spatial units. In other words, the production processes in various geographical parts of the world are mutually interconnected, the extent of which is certainly growing due to significant increases in the efficiency of communications and transport. This section deals with the spatial extension of the previous models to accommodate the issue of interdependence among different spatial units.

In the following, the term "region" will be used as a general term to refer to a space unit, including a nation like India, a group of nations like the EU, a state like Connecticut (US), a city like Tokyo, and so on.

3.1.1 A Two-Region Open Model

First, consider a simple model which consists of two regions, the region under consideration called region a (or the domestic region), and the rest of the world (ROW). The term "a two-region open model" refers to the fact that the model deals with a single endogenous region: the activity of the ROW is exogenously given, and is not explained in the model. The inputs of products produced in the ROW are reckoned as imports of region a from the ROW, while the demand for domestic products by the ROW are reckoned as exports from region a to the ROW. A regional model is regionally closed when all the regions that occur in the model are endogenous in the sense that the interregional flows of products are explained just like the inter sectoral flows of products. When this is not the case, the model is regionally open. The current model is regionally open. Spatial extension thus calls for the introduction of the flow of imports and exports into the previous models.

3.1.1.1 Competitive and Noncompetitive Imports

Imports can be described as competitive or complementary (noncompetitive) according as the products in question are or are not produced domestically in region a ([28], p.55). Consider the case where nations are chosen as the regional unit. For Japan, iron ore and sheep wool are noncompetitive imports, because she has no corresponding domestic processes. On the other hand, for Australia, both iron ore and sheep wool are domestic products. Tropical agricultural products such as cacao, coffee, and pepper will be noncompetitive imports for countries in high latitudes, while salmon, cod, and wool will be noncompetitive imports for low-latitude countries. For most OECD countries, ISO (International Organization for Standardization) standardized manufacturing products can be mentioned as examples of competitive imports: they are domestically produced, but are also imported. Depending on the location of the region under consideration, products such as electricity and water can also be competitive imports. The distinction between competitive and noncompetitive imports is a relative one which depends on the definition of the domestic region.

Suppose that there are k noncompetitive imports, the demand for which is denoted by a $k \times 1$ vector z^m. Write f^x for the $n \times 1$ vector of domestic final demand for endogenous products/competitive imports, f^{zm} for the $k \times 1$ vector of domestic final demand for noncompetitive imports, and x^e for the $n \times 1$ vector of exports (foreign demand for domestic products). The vector of final demand f is then given by

$$f = \begin{pmatrix} f^x \\ f^{zm} \end{pmatrix} + \begin{pmatrix} x^e \\ 0 \end{pmatrix} \tag{3.1}$$

Depending on their source of origins, the elements of f^x are further divided into those produced domestically, f^{xd}, and competitive imports, f^{xm}:

$$f^x = f^{xd} + f^{xm}. \tag{3.2}$$

Noncompetitive Imports

First, we consider the case where all imports are noncompetitive. Accordingly, A includes domestic inputs only, and $f^{xm} = 0$. For a given region, a noncompetitive import can be reckoned as an exogenous input because of the absence of corresponding domestic production process. In the unit process P in (2.127), a noncompetitive import can be introduced as an additional row of the matrix B, the jth column element of which refers to the amount of the input of the import per unit of production of product j. Write $B_m = (b_{ij})$ for the $k \times n$ matrix of input coefficients referring to noncompetitive imports. Following (2.162), the demand for noncompetitive imports that is invoked by a given final demand can then be given by:

$$z^m = B_m (I - A)^{-1} (f^{xd} + x^e) + f^{zm} \tag{3.3}$$

Note that f^{zm} does not invoke any domestic production because of the absence of corresponding domestic sectors. The first term on the right hand side refers to the import that is indirectly required for the production of domestic products, while the second term refers to that directly required by the final demand.

In a similar fashion, the effects of a change in the price of noncompetitive imports on the price of domestic products can be analyzed by (2.169) with the price of noncompetitive imports, q_m, occurring as the elements of q:

$$\Delta p = \Delta q_m B_m (I - A)^{-1} \tag{3.4}$$

Competitive Imports

We next consider the case where all imports are competitive. By definition, competitive imports are indistinguishable from their domestic counterparts, and occur as row elements of A mixed with their domestic counterparts: $(A)_{ij}$ consists of input i coming from both domestic and imported sources. The same applies to the final demand f as well, each element of which consists of both domestic and imported sources as in (3.2). Writing x^d for the $n \times 1$ vector of domestic products and x^m for the corresponding vector of competitive imports, the quantity balance equation (2.139) becomes:

$$x^d = Ax^d + f^x + x^e - x^m. \tag{3.5}$$

The import x^m is not exogenously given, but depends on domestic production x^d and final demand f^x. Define the share of competitive imports in the total domestic demand for i, μ_i, by:

$$\mu_i = \frac{x_i^m}{\sum_j a_{ij}x_j^d + f_i^x}, i = 1, \cdots, n,$$

or

$$\widehat{\mu} = \widehat{x^m}\widehat{(Ax^d + f^x)}^{-1} \tag{3.6}$$

The export x^e is excluded from the denominator because it is unlikely that a product that is imported is exported as such. By definition

$$0 \le \mu_i \le 1. \tag{3.7}$$

When there is no competitive import of product i, $\mu_i = 0$, and the entire demand for the product is met by the domestic source. On the other hand, $\mu_i = 1$ follows when there is no domestic production of i except to meet the demand for export; $\mu_i = 1$ does not imply $x_i = 0$.

Assuming that μ is a given vector of constants, (3.6) can be rewritten as:

$$x^m = \widehat{\mu}(Ax^d + f^x). \tag{3.8}$$

Substituting from (3.8), (3.5) becomes

$$x^d = Ax^d + f^x - \widehat{\mu}(Ax^d + f^x) + x^e = (I - \widehat{\mu})Ax^d + (I - \widehat{\mu})f^x + x^e, \tag{3.9}$$

which can be solved for x^d as

$$x^d = \left(I - (I - \widehat{\mu})A\right)^{-1}\left((I - \widehat{\mu})f^x + x^e\right). \tag{3.10}$$

This is the Leontief quantity model extended to accommodate the presence of competitive imports. In (3.10), the matrix $(I - \widehat{\mu})A$ refers to the matrix A adjusted for the share of imports with the inputs of competitive imports set equal to zero. Analogously, $(I - \widehat{\mu})f^x$ gives the domestic components of the domestic final demand, say f^{xd}. No adjustment is made for the export x^e because it is assumed to consist of domestic products only.

From (3.8) and (3.10), the demand for imports is then given by:

$$x^m = \widehat{\mu}A\left(I - (I - \widehat{\mu})A\right)^{-1}\left((I - \widehat{\mu})f^x + x^e\right) + \widehat{\mu}f^x, \tag{3.11}$$

where the first and the second terms respectively refer to the intermediate and final demand for imports.

The Mixed Case

The case where both competitive and noncompetitive imports coexist will be the one that is most relevant to reality. In this case, (3.10) remains the quantity model for x, and hence the demand for competitive imports also remains to be given by (3.11). Combining (3.3) and (3.10), the demand for noncompetitive imports is given by

$$z^m = B_m \left(I - (I - \widehat{\mu})A \right)^{-1} \left((I - \widehat{\mu})f^x + x^e \right) + f^{zm} \tag{3.12}$$

From (2.162), the effects on the demand for noncompetitive imports of a change in exports, Δx^e, can be given by:

$$\Delta z^m = B_m \left(I - (I - \widehat{\mu})A \right)^{-1} \Delta x^e. \tag{3.13}$$

Furthermore, in analogy to (3.4), the effects on the price of domestic products of a change in q_m will be given by

$$\Delta p = \Delta q_m B_m \left(I - (I - \widehat{\mu})A \right)^{-1} \tag{3.14}$$

An IO Table of Competitive Imports Type

As an example, Table 3.1 shows an aggregated version of the Japanese IOT (input-output table) for 2000, with seven producing sectors, and four final demand sectors. The IOT is of competitive imports type, as is indicated by the presence of the column referring to the negative of imports (for each sector, the row sum gives the amount of its production). Table 3.2 shows the μ given by (3.6), which indicates that 30% of the supply of the products of the primary sector was imports, while for the construction, and utilities sectors there was no import.

Table 3.1 Japanese IO Table for Year 2000: An IO Table of Competitive Imports Type.

	Endogenous sectors							Final demand				
	PRM	FWP	CHM	MMC	CST	UTL	SRV	CNS	INV	STK	EXP	IMP
PRM	1,563	8,215	6,543	1,026	826	2,019	1,351	3,959	189	763	83	−10,788
WP	1,369	15,573	1,214	2,736	4,057	191	17,535	38,115	1,032	−15	1,123	−10,024
CHM	1,018	3,513	16,628	9,046	6,973	1,070	14,571	9,489	0	−33	5,427	−6,343
MMC	169	1,584	1,185	71,387	10,550	420	10,583	17,789	38,690	−587	40,036	−17,909
CST	90	218	399	670	199	1,259	6,144	0	68,331	0	0	0
UTL	135	1,353	2,093	2,892	539	1,623	9,466	8,873	0	0	31	−2
SRV	2,673	12,769	11,448	31,037	17,709	5,134	118,640	307,642	21,770	149	10,786	−9,096
V	8,731	29,680	21,849	55,102	36,458	15,289	352,373					
x	15,748	72,905	61,360	173,896	77,311	27,004	530,662					

Units: Billion Japanese yen. CNS: Consumption expenditure, INV: Gross domestic fixed capital formation, STK: Increase in stocks, EXP: Exports total, IMP: (the negative of) Total imports, V: Value added, x: domestic production. See Table 3.2 for other notations.

Table 3.2 The Share of Competitive Imports in Total Domestic Demands: μ.

PRM	Primary industries	0.296
FWP	Food, wood, textiles, and paper	0.054
CHM	Chemicals, glass, cement, pottery	0.048
MMC	Metals, machinery and other manufacturing	0.044
CST	Construction	0.000
UTL	Utilities	0.000
SRV	Service and others	0.007

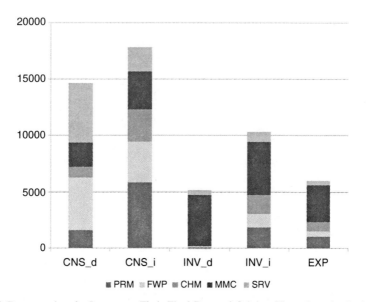

Fig. 3.1 Decomposing the Imports to Their Final Demand Origins. The _d's and _i's Attached to Each Final Demand Category (see Table 3.1 for the Notations) Respectively Refer to the Direct Effects and Indirect Effects. The Changes in the Stock Were Omitted Because of Their Small Sizes.

By use of (3.11), it is possible to decompose the imports to their final demand origins (Figure 3.1). In the figure, the direct effects refer to $\hat{\mu} f^x$ in (3.11), and the indirect effects refer to the rest (note that in actual computation f^x was not a vector, but a 7×3 matrix). It is found that for both consumption and capital formation the indirect effects exceed the direct effects, and that the indirect effects were distributed over a larger number of product items than the direct effects. Furthermore, about 11% of Japanese imports were indirectly induced by the Japanese exports.

3.1.2 A Two-Region Closed Model

Consider next the case where the world consists of two regions, region a and region b, while there is no ROW. In other words, the model is regionally closed. An important implication of this is that the trade between the two regions is mutually interrelated with the export of region a (b) to region b (a) equal to the import of region b (a) from a (b).

3.1.2.1 The Case of Noncompetitive Imports Only: The Isard Model

First, consider the case, where the products traded between the two regions consist of noncompetitive imports only. Within the framework of a single region model, the products of region a are exogenous inputs for region b, while the products of region b are exogenous to region a. Within the framework of a two-region model, however, where regions a and b are integrated via the trade flow, all the imports become endogenous inputs. Therefore, it is no longer appropriate to use the matrix B, that is reserved for exogenous inputs, to denote the input of noncompetitive imports.

Denote by A^{aa}, x^a, f^{aa}, and x^{ea} the matrix of domestic input coefficients, the vector of domestic production, the vector of the final demand for domestic products, and the vector of the export in region a, respectively. Applying the same rule of notations to region b as well, the production in the two regions can be represented by:

$$x^a = A^{aa}x^a + f^{aa} + x^{ea} \tag{3.15}$$

$$x^b = A^{bb}x^b + f^{bb} + x^{eb} \tag{3.16}$$

Similarly, denote by $(A^{ab})_{ij}$ the input coefficient of product i made in region a into sector j in region b, and by f^{ab} the vector of final demand in region b for the products made in a. Write x^{ma} for the vector of import of region a, which is equal to the export from region b to a, x^{eb}. Applying similar notations also to region b, the export and import will be given by

$$x^{ea} = x^{mb} = A^{ab}x^b + f^{ab}, \tag{3.17}$$

$$x^{eb} = x^{ma} = A^{ba}x^a + f^{ba}, \tag{3.18}$$

where the first term of the right-hand side after the second equality refers to the export for intermediate use and the second term to that for final use. Substitution of these expressions into (3.15) and (3.16) removes exports, and yields:

$$\begin{pmatrix} x^a \\ x^b \end{pmatrix} = \begin{pmatrix} A^{aa} & A^{ab} \\ A^{ba} & A^{bb} \end{pmatrix} \begin{pmatrix} x^a \\ x^b \end{pmatrix} + \begin{pmatrix} f^{aa} + f^{ab} \\ f^{ba} + f^{bb} \end{pmatrix}, \tag{3.19}$$

which can be solved for production in the two regions simultaneously

$$\begin{pmatrix} x^a \\ x^b \end{pmatrix} = \begin{pmatrix} I - A^{aa} & -A^{ab} \\ -A^{ba} & I - A^{bb} \end{pmatrix}^{-1} \begin{pmatrix} f^{aa} + f^{ab} \\ f^{ba} + f^{bb} \end{pmatrix}$$

$$= \begin{pmatrix} C^{aa} & C^{ab} \\ C^{ba} & C^{bb} \end{pmatrix} \begin{pmatrix} f^{aa} + f^{ab} \\ f^{ba} + f^{bb} \end{pmatrix}. \tag{3.20}$$

Because this type of model was first proposed by Isard [12], this model is called the Isard model.

The amount of the demand for imports can then be obtained from (3.17) and (3.18). The effects on x^a of an increase in region a's final demand for the products of region b, Δf^{ba}, are given by:

$$\Delta x^a = C^{ab} \Delta f^{ba}. \tag{3.21}$$

3.1.2.2 The Case of Competitive Imports Only: The Chenery–Moses Model

In the case of competitive imports, it is necessary to consider balance equation (3.5) for each region:

$$x^a = A^a x^a + f^a + x^{ea} - x^{ma} \tag{3.22}$$

$$x^b = A^b x^b + f^b + x^{eb} - x^{mb}, \tag{3.23}$$

where A^i and f^i, $i = a, b$, now include imports. By analogy with (3.6), define the share of imports by origins in total domestic demand of each region as

$$\widehat{\mu^a} = \widehat{x^{ma}} \left(\widehat{A^a x^a + f^a} \right)^{-1} \tag{3.24}$$

$$\widehat{\mu^b} = \widehat{x^{mb}} \left(\widehat{A^b x^b + f^b} \right)^{-1} \tag{3.25}$$

Assuming that the μ's are exogenously given, this can be rewritten as

$$x^{ma} = \widehat{\mu^a}(A^a x^a + f^a) = x^{eb}, \tag{3.26}$$

$$x^{mb} = \widehat{\mu^b}(A^b x^b + f^b) = x^{ea}, \tag{3.27}$$

where, for each of these equations, the second equality results from the fact that the model is regionally closed. Substitution into (3.22) and (3.23) gives

$$
\begin{aligned}
x^a &= A^a x^a + f^a + \widehat{\mu^b}(A^b x^b + f^b) - \widehat{\mu^a}(A^a x^a + f^a) \\
&= (I - \widehat{\mu^a})A^a x^a + (I - \widehat{\mu^a})f^a + \widehat{\mu^b}(A^b x^b + f^b)
\end{aligned}
\tag{3.28}
$$

$$
\begin{aligned}
x^b &= A^b x^b + f^b + \widehat{\mu^a}(A^a x^a + f^a) - \widehat{\mu^b}(A^b x^b + f^b) \\
&= (I - \widehat{\mu^b})A^b x^b + (I - \widehat{\mu^b})f^b + \widehat{\mu^a}(A^a x^a + f^a),
\end{aligned}
\tag{3.29}
$$

which, upon rearrangement, becomes:

$$
\begin{pmatrix} x^a \\ x^b \end{pmatrix} = \begin{pmatrix} (I - \widehat{\mu^a})A^a & \widehat{\mu^b}A^b \\ \widehat{\mu^a}A^a & (I - \widehat{\mu^b})A^b \end{pmatrix} \begin{pmatrix} x^a \\ x^b \end{pmatrix} + \begin{pmatrix} (I - \widehat{\mu^a})f^a + \widehat{\mu^b}f^b \\ (I - \widehat{\mu^b})f^b + \widehat{\mu^a}f^a \end{pmatrix}. \tag{3.30}
$$

Solving this gives the following quantity model:

$$\begin{pmatrix} x^a \\ x^b \end{pmatrix} = \begin{pmatrix} I - (I - \widehat{\mu^a})A^a & -\widehat{\mu^b}A^b \\ -\widehat{\mu^a}A^a & I - (I - \widehat{\mu^b})A^b \end{pmatrix}^{-1} \begin{pmatrix} (I - \widehat{\mu^a})f^a + \widehat{\mu^b}f^b \\ (I - \widehat{\mu^b})f^b + \widehat{\mu^a}f^a \end{pmatrix}. \qquad (3.31)$$

Because this type of model (albeit in a regionally open form) was first proposed by Chenery [3] and Moses [23] independently, it is called the Chenery–Moses model. Compared with the Isard model (3.20), this model is characterized by a substantial saving of data requirements. While information on A^{aa}, A^{ab}, A^{ba}, A^{bb}, f^{aa}, f^{ba}, f^{bb}, and f^{ab} is required in the Isard model, it is not required in the Chenery–Moses model, but is estimated by

$$\begin{aligned} A^{aa} &= (I - \widehat{\mu^a})A^a, & A^{bb} &= (I - \widehat{\mu^b})A^b, & A^{ab} &= \widehat{\mu^b}A^b, & A^{ba} &= \widehat{\mu^a}A^a, \\ f^{aa} &= (I - \widehat{\mu^a})f^a, & f^{bb} &= (I - \widehat{\mu^b})f^b, & f^{ab} &= \widehat{\mu^b}f^b, & f^{ba} &= \widehat{\mu^a}f^a. \end{aligned} \qquad (3.32)$$

With an increase in μ^a and μ^b, the fraction of domestic demand that is met by domestic production decreases. In the extreme case of all the elements of μ^a and μ^b being unity, the whole production is directed for export, and the entire domestic final demand is satisfied by import.

3.1.2.3 The Mixed Case

The fact that the Chenery–Moses type model (3.31) was derived for competitive imports does not exclude its application to the case of noncompetitive imports. If $\mu_i^a = 1$ and $\mu_i^b = 0$, product i is noncompetitive imports for region a as there is no corresponding domestic production to satisfy the domestic final demand for it, that is, $x_i^a = 0$ but $f_i^a > 0$. Analogously, product j is noncompetitive imports for region b, if $\mu_j^b = 1$ and $\mu_j^a = 0$. By extending the dimension of x, A, and f to accommodate the presence of noncompetitive imports, (3.31) can just as well be applied to a mixed case involving both competitive and noncompetitive imports.

3.1.3 A Three-Region Model: An Open Model

Finally, consider the case of extending the above two-region model, which is spatially closed, to a spatially open one by incorporating the ROW. This is a three-region model consisting of two endogenous regions, regions a and b, and one exogenous region, the ROW. Given the model with two endogenous regions, its extension to the case with more than three endogenous regions will be straightforward. With an increase in the extent of spatial extension, the area on earth covered by endogenous regions increases, while the area falling under the category of ROW decreases. In the ultimate state of regional extension, the fraction of ROW would disappear, and a fully closed world model would be obtained (see [5] for world models based on IOA). A spatially open model with two endogenous regions is a general one that can deal with all these cases with a straightforward extension.

3.1.3.1 The Isard-Type Model

Write A^{Ra} for the coefficients matrix of the input of the products of the ROW into the producing sectors in region a, f^{Ra} for the final demand in region a for the products of the ROW, and x^{aR} for the export of the products of region a to the ROW. The analogous rule of notations applies to A^{Rb}, f^{Rb}, and so on. Extending (3.15) and (3.16), the Isard-type model for the case of three regions is then given by

$$x^a = A^{aa}x^a + A^{ab}x^b + f^{aa} + f^{ab} + x^{aR}$$
$$x^b = A^{ba}x^a + A^{bb}x^b + f^{ba} + f^{bb} + x^{bR} \qquad (3.33)$$
$$x^{Ra} + x^{Rb} = A^{Ra}x^a + A^{Rb}x^b + f^{Ra} + f^{Rb}.$$

Note that the matrices, $A^{iR}, i = R, a, b$, do not occur because this is an open model where the economic activity in the ROW is exogenous.

Solving this for x^a and x^b in a manner analogous to (3.20) gives the Isard-type open model for three regions:

$$\begin{pmatrix} x^a \\ x^b \end{pmatrix} = \begin{pmatrix} I - A^{aa} & -A^{ab} \\ -A^{ba} & I - A^{bb} \end{pmatrix}^{-1} \begin{pmatrix} f^{aa} + f^{ab} + x^{aR} \\ f^{ba} + f^{bb} + x^{bR} \end{pmatrix}, \qquad (3.34)$$

with the import from the ROW given by

$$x^{Ra} + x^{Rb} = (A^{Ra}\ A^{Rb}) \begin{pmatrix} I - A^{aa} & -A^{ab} \\ -A^{ba} & I - A^{bb} \end{pmatrix}^{-1} \begin{pmatrix} f^{aa} + f^{ab} + x^{aR} \\ f^{ba} + f^{bb} + x^{bR} \end{pmatrix} + f^{Ra} + f^{Rb} \quad (3.35)$$

3.1.3.2 The Chenery–Moses-Type Model

Extending (3.22) and (3.23), the quantity balance now becomes:

$$x^a = A^a x^a + f^a + x^{ab} + x^{aR} - x^{ba} - x^{Ra}, \qquad (3.36)$$
$$x^b = A^b x^b + f^b + x^{ba} + x^{bR} - x^{ab} - x^{Rb} \qquad (3.37)$$

Extending (3.24) and (3.25), define the (vector of) fraction of imports by region of origin:

$$\widehat{\mu^{ba}} = \widehat{x^{ba}}(\widehat{A^a x^a + f^a})^{-1},$$
$$\widehat{\mu^{Ra}} = \widehat{x^{Ra}}(\widehat{A^a x^a + f^a})^{-1},$$
$$\widehat{\mu^{Rb}} = \widehat{x^{Rb}}(\widehat{A^b x^b + f^b})^{-1}, \qquad (3.38)$$
$$\widehat{\mu^{ab}} = \widehat{x^{ab}}(\widehat{A^b x^b + f^b})^{-1},$$

where $(\mu^{ab})_i$ refers to the fraction in the total demand for i in region b of import from region a. Assuming that μ's are exogenously given constants, this can be rewritten as

$$
\begin{aligned}
x^{ba} &= = \widehat{\mu^{ba}}(A^a x^a + f^a), \\
x^{ab} &= = \widehat{\mu^{ab}}(A^b x^b + f^b), \\
x^{Ra} &= \widehat{\mu^{Ra}}(A^a x^a + f^a), \\
x^{Rb} &= \widehat{\mu^{Rb}}(A^b x^b + f^b).
\end{aligned}
\tag{3.39}
$$

Substitution into (3.36) and (3.37) yields

$$
\begin{aligned}
x^a &= (I - \widehat{\mu^{ba}} - \widehat{\mu^{Ra}})A^a x^a + (I - \widehat{\mu^{ba}} - \widehat{\mu^{Ra}})f^a \\
&\quad + \widehat{\mu^{ab}}(A^b x^b + f^b) + x^{aR}
\end{aligned}
\tag{3.40}
$$

$$
\begin{aligned}
x^b &= (I - \widehat{\mu^{ab}} - \widehat{\mu^{Rb}})A^b x^b + (I - \widehat{\mu^{ab}} - \widehat{\mu^{Rb}})f^b \\
&\quad + \widehat{\mu^{ba}}(A^a x^a + f^a) + x^{bR},
\end{aligned}
\tag{3.41}
$$

the solution of which gives the following quantity model of Chenery–Moses type:

$$
\begin{pmatrix} x^a \\ x^b \end{pmatrix} = \begin{pmatrix} I - (I - \widehat{\mu^{ba}} - \widehat{\mu^{Ra}})A^a & \widehat{\mu^{ab}}A^b \\ -\widehat{\mu^{ba}}A^a & I - (I - \widehat{\mu^{ab}} - \widehat{\mu^{Rb}})A^b \end{pmatrix}^{-1}
$$
$$
\begin{pmatrix} (I - \widehat{\mu^{ba}} - \widehat{\mu^{Ra}})f^a + \widehat{\mu^{ab}}f^b + x^{aR} \\ (I - \widehat{\mu^{ab}} - \widehat{\mu^{Rb}})f^b + \widehat{\mu^{ba}}f^a + x^{bR} \end{pmatrix}
\tag{3.42}
$$

Substitution into the expressions for x^{Ra} and x^{Rb} in (3.39) then gives the Chenery–Moses counterpart of (3.35):

$$
x^{Ra} + x^{Rb} = \begin{pmatrix} \widehat{\mu^{Ra}} & \widehat{\mu^{Rb}} \end{pmatrix} \left[\begin{bmatrix} A^a \\ A^b \end{bmatrix} \Omega^{-1} \zeta + \begin{pmatrix} f^a \\ f^b \end{pmatrix} \right]
\tag{3.43}
$$

where Ω and ζ respectively refer to the expressions inside the first and second parenthesis on the right of (3.42).

3.1.4 International IO Tables

A regional IO model is called an international IO model when the unit of region refers to a country, and the underlying IO data are called international IOT. The history of international IOTs can be traced back to the work of Ronald Wonnacott on the US–Canada bilateral table (of the Chenery–Moses type) around 1960 [33]. When it comes to more organizational compilation of international IOTs on a regular basis, one should note the remarkable activities that have taken place in Japan. Starting with the publication in 1976 of a Korea–Japan IOT, the first international

IOT of an Isard type ever published, the Institute of Developing Economics (IDE), a Japanese public research institute, has been a pioneer in the compilation of international IOTs, which include bilateral IOTs for US–Japan, Thailand–Japan, China–Japan, Singapore–Japan, Malaysia–Japan, Indonesia–Japan, Philippines–Japan, and Taiwan–Japan, as well as multilateral IOTs involving all these countries [30]. Another active body in Japan is the statistics division of Ministry of Economy, Trade and Industry (METI), which has compiled bilateral IOTs for US–Japan, UK–Japan, Germany–Japan, and France–Japan, as well as multilateral IOTs involving the US, the UK, Germany, France, ASEAN countries, Korea, China, Taiwan, and Japan [19].

The US–Japan international IO table provides a good empirical example of the two-region open model, to the illustration of which we now turn.

3.1.4.1 An Example: The Japan–US International Table

The Isard-Type Model

Table 3.3 gives the U.S.–Japan IO table for year 2000 in an aggregated version of seven producing sectors [19]. Using the notations in (3.33), the structure of Table 3.3 can be represented as follows:

	Intermediate demand		Final demand		
	Japan	US	Japan	US	ROW
Japan	$X^{JJ}(=A^{JJ}\hat{x}^J)$	$X^{JU}(=A^{JU}\hat{x}^U)$	f^{JJ}	f^{JU}	x^{JR}
U.S.	$X^{UJ}(=A^{UJ}\hat{x}^J)$	$X^{UU}(=A^{UU}\hat{x}^U)$	f^{UJ}	f^{UU}	x^{UR}
ROW	$X^{RJ}(=A^{RJ}\hat{x}^J)$	$X^{RU}(=A^{RU}\hat{x}^U)$			

$$(3.44)$$

For instance $(X^{UU})_{31} = 272$ refers to the amount of input of U.S. chemical products into U.S. primary sectors, $(X^{JU})_{47} = 35$ to the amount of input of Japanese metal & machinery products into the U.S. service sectors, and $(X^{RU})_{13} = 658$ to the amount of input of the ROW primary products into the U.S. chemical sectors. Because the inputs are distinguished by the source of origins, the table is of the Isard (noncompetitive import) type.

Table 3.4 gives the Leontief inverse coefficients matrix of the Isard type (3.34). Several interesting features of the interdependence of the U.S.–Japan economy emerge. For instance, the satisfaction of a unit of Japanese final demand for food/paper invokes in total 0.12 units of primary (agricultural) products in Japan and 0.0107 units in the U.S., whereas the U.S. final demand for food/paper invokes 0.166 units in the U.S. and only 0.0002 in Japan: the final demand for food in Japan relies significantly on the U.S. agricultural sector, but not vice versa. On the other hand, the U.S. final demand for metal and machinery invokes 0.037 units of production in the Japanese metal and machinery sector, whilst the corresponding effect of the Japanese final demand for the metal and machinery sector in the U.S. remains at 0.0219 units.

Table 3.3 The 2000 Japan–U.S. Input-Output Table: Isard Type.

	J1	J2	J3	J4	J5	J6	J7	U1	U2	U3	U4	U5	U6	U7	f^{JJ}	f^{JU}	x^{JR}	Total
	X^{JJ}							X^{JU}							Final demand			
J1 Primary	77	580	60	7	78	5	138	0	0	0	0	0	0	0	520	0	8	1,475
J2 Food/paper	108	966	124	215	322	20	1,651	0	7	0	1	1	0	2	2,799	6	87	6,309
J3 Chemicals	115	317	857	753	683	117	1,140	1	5	35	10	2	0	19	703	35	405	5,197
J4 Metal, machinery	13	127	76	3,375	948	5	953	6	4	4	407	13	1	35	4,260	646	2,539	13,412
J5 Construction	8	20	39	68	17	117	562	0	0	0	0	0	0	0	6,340	0	0	7,171
J6 Utility	15	123	204	295	51	56	841	0	0	0	0	0	0	1	822	1	1	2,410
J7 Service	269	1,224	974	2,589	1,665	499	10,895	2	5	13	44	3	1	81	29,048	148	680	48,141

	J1	J2	J3	J4	J5	J6	J7	U1	U2	U3	U4	U5	U6	U7	f^{UJ}	f^{UU}	x^{UR}	Total
	X^{UJ}							X^{UU}							Final demand			
U1 Primary	4	39	4	1	0	3	5	490	1,344	990	61	72	493	167	9	509	218	4,409
U2 Food paper	0	35	2	1	4	0	19	174	1,629	197	238	487	17	3,107	46	5,400	595	11,951
U3 Chemicals	1	4	30	8	4	1	16	272	573	1,246	770	874	90	2,001	12	2,067	972	8,941
U4 Metal, machinery	0	1	0	166	4	0	13	125	292	192	4,038	961	55	2,107	188	7,336	3,858	19,339
U5 Construction	0	0	0	0	0	0	0	13	27	25	47	9	83	914	0	7,990	1	9,110
U6 Utility	0	0	0	0	0	0	0	101	209	235	245	55	34	1,582	0	1,820	9	4,290
U7 Service	1	18	8	27	3	1	62	828	2,351	1,757	3,986	2,414	1,092	24,469	119	70,548	3,030	110,713

	J1	J2	J3	J4	J5	J6	J7	U1	U2	U3	U4	U5	U6	U7	f^{RJ}	f^{RU}	Total
	X^{RJ}							X^{RU}							Final demand		
R1 Primary	14	81	467	88	1	165	34	117	95	658	7	2	248	67	62	80	2,186
R2 Food, paper	2	137	5	4	48	0	58	7	368	20	29	60	2	183	315	1,050	2,289
R3 Chemicals	7	15	183	22	18	10	80	41	76	365	101	78	10	320	117	726	2,167
R4 Metal, machinery	1	3	4	493	20	1	26	31	74	70	1,625	169	13	755	593	2,978	6,858
R5 Construction	0	0	0	0	0	0	0	0	0	0	0	0	0	0	0	0	0
R6 Utility	0	0	0	0	0	0	0	1	1	1	2	0	0	10	0	11	26
R7 Service	1	15	13	41	23	7	366	12	30	84	135	15	4	796	241	675	2,458

Unit: 100 billion U.S. dollar. Source: [19] Primary: Agriculture, forestry, fishing, mining; Food, paper: Food, textiles, pulp, paper and wooden products, publishing and printing; Chemicals: Chemical products, petroleum and coal products, plastics, rubber and leather products, ceramics, stone and clay products; Metal, machinery: Steel and steel products, non-steel metals and products other metal products, general machinery, electric machinery, transportation equipment, precision instruments, other manufactured products.

Table 3.4 The 2000 Japan–U.S. Matrix of Leontief Inverse Coefficients.

		J1	J2	J3	J4	J5	J6	J7	U1	U2	U3	U4	U5	U6	U7
J1	Primary	1.0683	0.1201	0.0213	0.0081	0.0227	0.0079	0.0107	0.0001	0.0002	0.0002	0.0003	0.0001	0.0000	0.0000
J2	Food paper	0.1102	1.2114	0.0522	0.0466	0.0807	0.0300	0.0588	0.0003	0.0011	0.0006	0.0016	0.0005	0.0002	0.0002
J3	Chemicals	0.1201	0.1017	1.2200	0.1099	0.1490	0.0798	0.0492	0.0014	0.0016	0.0062	0.0044	0.0018	0.0007	0.0007
J4	Metal. machinery	0.0293	0.0484	0.0377	1.3535	0.1952	0.0239	0.0415	0.0039	0.0028	0.0030	0.0368	0.0067	0.0020	0.0018
J5	Construction	0.0126	0.0119	0.0164	0.0146	1.0110	0.0550	0.0180	0.0001	0.0001	0.0002	0.0005	0.0001	0.0000	0.0000
J6	Utility	0.0247	0.0375	0.0573	0.0431	0.0271	1.0346	0.0284	0.0002	0.0002	0.0005	0.0013	0.0003	0.0001	0.0001
J7	Service	0.3259	0.3825	0.3432	0.3939	0.4210	0.3282	1.3451	0.0029	0.0030	0.0055	0.0156	0.0041	0.0016	0.0022
U1	Primary	0.0046	0.0107	0.0029	0.0015	0.0014	0.0020	0.0010	1.1514	0.1660	0.1609	0.0217	0.0407	0.1424	0.0155
U2	Food paper	0.0015	0.0091	0.0018	0.0015	0.0018	0.0007	0.0013	0.0680	1.1818	0.0515	0.0340	0.0848	0.0275	0.0461
U3	Chemicals	0.0027	0.0035	0.0090	0.0034	0.0025	0.0014	0.0013	0.0979	0.0901	1.1886	0.0721	0.1370	0.0495	0.0351
U4	Metal. machinery	0.0011	0.0022	0.0016	0.0219	0.0043	0.0011	0.0014	0.0562	0.0558	0.0507	1.2788	0.1534	0.0369	0.0369
U5	Construction	0.0001	0.0002	0.0001	0.0002	0.0001	0.0000	0.0000	0.0076	0.0074	0.0081	0.0071	1.0063	0.0238	0.0117
U6	Utility	0.0003	0.0007	0.0005	0.0006	0.0003	0.0002	0.0002	0.0361	0.0334	0.0423	0.0249	0.0207	1.0196	0.0218
U7	Service	0.0047	0.0121	0.0072	0.0122	0.0049	0.0027	0.0040	0.3487	0.3894	0.3817	0.3810	0.4554	0.4048	1.3289

See Table 3.3 for the notations.

The Chenery–Moses-Type Model

Given the Isard-type IO table such as Table 3.3, deriving its Chenery–Moses counterpart is straightforward (Table 3.5). In Table 3.5 the matrix of intermediate flow X and the US and Japanese final demand contain imports, and are obtained from the variables in Table 3.3 as follows:

$$X^j = \sum_{i=J,U,R} X^{ij}, \; j = J, U \tag{3.45}$$

$$f^j = \sum_{i=J,U,R} f^{ij}, \; j = J, U \tag{3.46}$$

Table 3.6 gives the estimates of the fraction of imports by region of origin (3.38).

3.1.4.2 Issues of International IO Tables in Monetary Units: Exchange Rates and Purchasing Power Parities

Common to real examples of international IOTs is the fact that they are based on local IOTs measured in local currencies. Write p^a and p^b for the price vectors in countries a and b expressed in respective local currencies. From (2.172), the matrices of input coefficients obtained from these IOTs, say $\bar{A}^a$ and $\bar{A}^b$, are related to their physical counterparts as follows:

$$\bar{A}^a = \widehat{p^a} A^a \widehat{p^a}^{-1}, \tag{3.47}$$

$$\bar{A}^b = \widehat{p^b} A^b \widehat{p^b}^{-1}. \tag{3.48}$$

Write ε_{ab} for the rate of exchange of two currencies. If the same relative prices prevail for all the products in the two countries, ε_{ab} becomes equal to the ratio of their purchasing power:

$$\frac{p_i^b}{p_i^a} = \varepsilon_{ab}, \forall i, \tag{3.49}$$

or in a matrix form

$$p^b = \varepsilon_{ab} p^a. \tag{3.50}$$

This corresponds to the case where the purchasing power parity (PPP) theory holds exactly: a theory developed by Cassel, according to which exchange rates are in equilibrium when the domestic purchasing power of currencies is the same (see [25] for a survey of the literature on PPP). In this case, any product, whether it is 1 kWh of electricity, a liter of gasoline, a pound of bread, or a liter of mineral water, costs the same in the two countries, once one takes the exchange rate into account. Substitution of (3.50) into (3.48) gives

$$\bar{A}^b = \varepsilon_{ab} \widehat{p^a} A^b \widehat{p^a}^{-1} \varepsilon_{ab}^{-1} = \widehat{p^a} A^b \widehat{p^a}^{-1}, \tag{3.51}$$

Table 3.5 The 2000 Japan–U.S. Input-Output Table: Chenery–Moses Type.

	Inter-industry flows							Final demand			Import		total[a]
	X^J							f^J	f^{IU}	x^{JR}	x^{UJ}	x^{RJ}	
J1 Primary	94	700	530	96	79	173	177	591	2	8	65	912	1,475
J2 Food/paper	110	1,138	131	220	374	21	1,728	3,160	17	87	107	570	6,309
J3 Chemicals	123	336	1,069	783	705	128	1,236	832	107	405	76	451	5,197
J4 Metal, machinery	14	130	81	4,034	972	7	992	5,041	1,116	2,539	373	1,142	13,412
J5 Construction	8	20	39	68	17	117	562	6,340	0	0	0	0	7,171
J6 Utility	15	123	204	295	51	56	841	822	2	1	0	0	2,410
J7 Service	271	1,257	994	2,657	1,691	507	11,323	29,407	299	680	237	707	48,141
	X^U							f^U	f^{UJ}	x^{UR}	x^{IU}	x^{RU}	
U1 Primary	607	1,439	1,648	68	74	741	235	590	65	218	2	1,274	4,409
U2 Food/paper	181	2,004	217	268	548	19	3,292	6,456	107	595	17	1,719	11,951
U3 Chemicals	314	654	1,645	881	954	101	2,341	2,827	76	972	107	1,717	8,941
U4 Metal, machinery	162	370	266	6,070	1,143	70	2,897	10,960	373	3,858	1,116	5,717	19,339
U5 Construction	13	27	25	47	9	83	914	7,990	0	1	0	0	9,110
U6 Utilities	102	211	236	247	55	34	1,593	1,832	0	9	2	26	4,290
U7 Service	842	2,386	1,855	4,165	2,433	1,097	25,347	71,372	237	3,030	299	1,751	110,713

[a]The total is obtained by counting the imports as negative entries. See Table 3.3 for the notations.

Table 3.6 Estimates of μ^{ab} for the US–Japan IO Table.

	μ^{UJ}	μ^{JU}	μ^{RJ}	μ^{RU}
Primary	0.019	0.010	0.267	0.191
Food/paper	0.014	0.007	0.075	0.117
Chemicals	0.013	0.007	0.079	0.149
Metal, machinery	0.029	0.013	0.089	0.199
Construction	0.000	0.000	0.000	0.000
Utility	0.000	0.000	0.000	0.006
Service	0.005	0.002	0.014	0.016

See Table 3.3 for the notations.

which shows that the monetary coefficients matrices $\bar{A}^a$ and $\bar{A}^b$ can just as well be regarded as physical matrices A^a and A^b. In particular, when the physical matrices happen to be the same in the two countries, the same is true with monetary matrices as well, and vice versa. This is a spatial extension of the result obtained by Weisz and Duchin [32] (see Section 2.3.4.2) with regard to the equivalence of a monetary IO model and a physical IO model up to the change of unit operation.

In reality, however, ample evidence suggests that (3.50) does not hold (see, for instance, OECD Purchasing Power Parities Data [27]), and has to be replaced by:

$$p^b = \widehat{\zeta} p^a, \tag{3.52}$$

where ζ is a vector with $\zeta = (\zeta_1, \cdots, \zeta_n)$. Substitution of (3.52) into (3.51) gives:

$$\bar{A}^b = \widehat{\zeta} \widehat{p^a} A^b \widehat{p^a}^{-1} \widehat{\zeta}^{-1}, \tag{3.53}$$

which does not simplify to the expression after the second equality of (3.51) unless all the elements of ζ have the same value. Even if the physical matrices happen to be the same in the two countries, the monetary matrices would be different owing to differences in relative prices.

Fortunately, however, (3.53) has a much simpler structure than the case of user-specific input prices discussed in Section 2.2.4.3. Note that ζ refers to the bilateral price ratios in local currencies, which are, in principle, observable, and some estimates of which are reported on a regular basis by international organizations such as the OECD [27]. If data on ζ are available, applying them to (3.53) gives the desired matrix:

$$\widehat{\zeta}^{-1} \bar{A}^b \widehat{\zeta} = \widehat{p^a} A^b \widehat{p^a}^{-1}. \tag{3.54}$$

The actual implementation of this straightforward calculation, however, is hampered by the fact that for some goods or services the data on ζ are rather difficult to obtain. This applies in particular to intermediate goods like metals, chemicals, and semis, which constitute the majority of the inter-industry flow. This is so because published data on PPP are mostly focused on final products (see [27]). If available, the (complementary) use of physical data is desirable because they are free of these problems.

3.2 By-Product

Up to now, our analysis has been focused on a production process with a single product. Products refer to the items of outputs with a positive price. The issue of joint production, where more than one output is obtained in production, has not been considered except for the joint generation of waste. This section is devoted to the extension of the previous results to accommodate the presence of by-product, which belongs neither to primary products nor to waste.

3.2.1 Defining By-Product and Waste

It is important to distinguish at the outset joint products, which are technology-grounded, from multiple products, which are organization-grounded. Joint production refers to the case where the quantities of two or more outputs are technically interdependent, and are not separable to processes each of which produce a single product. In other words, joint production belongs to the technological characteristics of a production process. Some common examples are feathers from poultry processing, sawdust and bark from the processing of logs into lumber, gypsum from flue gas desulfurization, fly ash from the combustion of coal, sludge from wastewater treatments, electric power from waste incineration plant, as well as gold and silver from copper- and lead electrolysis.

On the other hand, multiple production refers to the coexistence in an organizational unit (factory, shop, corporation, etc.) of individual processes, and hence is not technologically but organizationally grounded. Some common examples are motorcycles made by auto manufacturers, cake made by a bakery, hotels run by airlines, and extension schools run by universities. Because the quantities of multiple outputs are not technically interdependent, they can be decomposed into individual processes with no multiple production. As such, issues of multiple production can be coped with by appropriate definition of sectoral classification (see [28]). Because of this feature, the issue of multiple production will not be further discussed.

Above, by-product was characterized by saying that "it belongs neither to primary products nor to waste". Unfortunately, this definition is a loose one, because the definition of waste that is tenable for distinguishing waste from by-product is still missing. Waste definition is not straightforward, and can be subject to considerable debate [6]. Trying to make a general distinction between waste and by-product runs the risk of falling into metaphysics. For the sake of analytical convenience, the following definition of by-product "in a narrow sense", by-product (I), is introduced:

Definition 3.1 (by-product (I)). An output from a given production sector is by-product (I), when it is not the primary product of the process, but is related technically to its production, and happens to be the primary product of another production sector.

Except for the use of the numeric identification (I), this definition follows that of Stone ([28], p.39). Note that this excludes joint outputs which are not produced in any other sector as primary products. Accordingly, of the examples mentioned above, feathers, sawdust and bark, sewage sludge, and fly ash are not by-product (I) because of the absence of processes where these outputs are obtained as primarily products. On the other hand, electricity from waste incineration plants, and gold and silver from copper- and lead smelting processes are counted as by-product (I) because of the presence of a corresponding primary production process: silver and gold can be obtained from concentrate. Henceforth, items of by-product that do not belong to 'by-product (I)' are termed 'by-product (II)'.

Definition 3.2 (by-product (II)). An output from a given production sector is by-product (II), when it is not the primary product of the process, but is related technically to its production, and there is no producing sector where it is primarily produced.

The outputs from a given production process can then be exclusively classified into the following categories:

$$
\text{Outputs} \begin{cases} \text{products} \begin{cases} \text{primary product} \\ \text{by-product} \begin{cases} \text{by-product (I)} \\ \text{by-product (II)} \end{cases} \end{cases} \\ \text{waste} \\ \text{emissions} \end{cases} \tag{3.55}
$$

Following the Basel convention, waste refer to those items of joint outputs "which are disposed of or are intended to be disposed of or are required to be disposed of by the provisions of national law" [6]. It should be noticed that this definition of waste does not exclude the possibility of its reuse and recycling. Due to this possibility, the distinction between by-product (II) and waste is blurred, and may fluctuate depending on market conditions: items that are waste under a sluggish economic condition can become products under booming economic conditions. The output items that belong neither to products nor to waste, are emissions, examples of which are CO_2, NO_x and SO_x in the exhaust gases of automobiles driven by an internal combustion engine.

3.2.2 The Leontief Quantity Model with By-Product

Recall that in the definition (2.127) of unit process P the matrix of output coefficients, D, is the identity matrix because of the absence of by-product. In the presence of by-product (I) given by definition 3.1, however, this no longer holds. Denoting by

$d_{ij} \geq 0$ the amount of product i that is obtained as a by-product in the unit production process of product j, a general form of the output matrix D is given by:

$$D = \begin{pmatrix} 1 & d_{12} & \cdots & d_{1n} \\ d_{21} & 1 & \cdots & d_{2n} \\ \cdots & \cdots & \cdots & \cdots \\ d_{n1} & d_{n2} & \cdots & 1 \end{pmatrix} \tag{3.56}$$

Note that, by definition, for each by-product (I) there is a column in D that refers to the sector where the by-product occurs as the primary product. This feature makes D a square matrix.

Subtracting matrix A from matrix D gives the matrix of net output coefficients:

$$D - A = \begin{pmatrix} 1 - a_{11} & d_{12} - a_{12} & \cdots & d_{1n} - a_{1n} \\ d_{21} - a_{21} & 1 - a_{22} & \cdots & d_{2n} - a_{2n} \\ \cdots & \cdots & \cdots & \cdots \\ d_{n1} - a_{n1} & d_{n2} - a_{n2} & \cdots & 1 - a_{nn} \end{pmatrix}, \tag{3.57}$$

which is a general version of $(I - A)$ that allows for the presence of by-product. In this matrix, an output is indicated by a positive number, while an input is indicated by a negative number. In the presence of by-product, some off-diagonal elements of the matrix can take positive values, whereas in their absence diagonal elements only are positive. In LCA, this matrix is called the "technology matrix" ([8], p.14). It is noteworthy that, while the matrix $(I - A)$ was previously obtained in the course of solving the quantity-balancing equation (2.139), $(D - A)$ has been obtained directly from process information without resorting to any balancing equation.

With the matrix $(I - A)$ replaced by this generalized one, the Leontief quantity model that allows for the presence of by-product is given by:

$$x = (D - A)^{-1} f. \tag{3.58}$$

An alternative representation which resembles the original quantity model will be

$$x = (I - (A - D + I))^{-1} f = (I - \breve{A})^{-1} f, \tag{3.59}$$

where

$$\breve{A} = A - D + I. \tag{3.60}$$

The matrix $\breve{A}$ refers to the matrix of input-coefficients extended for the presence of by-product, where a by-product occurs as a negative input into the sector in which it is actually produced and as a negative output of the sector in which it occurs as the primary product.

This way of introducing by-product into the input-coefficients is called the negative input method of Stone, taking the name of its inventor ([28], p.19). In this method, an increase in the final demand for the primary product of a sector with a by-product would increase the supply of the by-product, and would reduce its supply from the sector that produces it as the primary product.

3.2.3 A Numerical Example

For illustration, consider an example where A is given by

$$A = \begin{pmatrix} 0.0 & 0.0 & 0.15 \\ 0.4 & 0.0 & 0.3 \\ 0.1 & 0.1 & 0.0 \end{pmatrix}, \tag{3.61}$$

with

$$I - A = \begin{pmatrix} 1 & 0.0 & -0.15 \\ -0.4 & 1 & -0.3 \\ -0.1 & -0.1 & 1 \end{pmatrix}, \tag{3.62}$$

and the Leontief inverse matrix:

$$(I - A)^{-1} = \begin{pmatrix} 1.02 & 0.01 & 0.15 \\ 0.45 & 1.03 & 0.37 \\ 0.14 & 0.10 & 1.05 \end{pmatrix}. \tag{3.63}$$

Suppose now that 0.3 units of product 1 is obtained as by-product per unit of production of product 2:

$$D = \begin{pmatrix} 1 & 0.3 & 0 \\ 0 & 1 & 0 \\ 0 & 0 & 1 \end{pmatrix} \tag{3.64}$$

From (3.58) and (3.59) the net output, or technology, matrix then becomes

$$D - A = I - \breve{A} = \begin{pmatrix} 1 & 0.3 & -0.15 \\ -0.4 & 1 & -0.3 \\ -0.1 & -0.1 & 1 \end{pmatrix}, \tag{3.65}$$

with the generalized version of the Leontief inverse matrix given by

$$(I - \breve{A})^{-1} = \begin{pmatrix} 0.89 & -0.26 & 0.05 \\ 0.39 & 0.91 & 0.33 \\ 0.12 & 0.06 & 1.03 \end{pmatrix} \tag{3.66}$$

While the off-diagonal elements of $(I - A)$ are nonpositive, this is no longer the case for its counterpart with by-product $(I - \breve{A})$. This leads to a remarkable difference in the Leontief inverse matrix. While all the elements of $(I - A)^{-1}$ are nonnegative, $(I - \breve{A})^{-1}$ contains a negative element. The final delivery of a unit of product 2 generates product 1 by 0.26 units as by-product, and reduces its production in Sector 1 by the same amount. If Sector 1 is a power generation sector using coal as fuel and Sector 2 is a waste incineration sector, this example corresponds to the case of power from waste replacing a portion of power from coal, which may contribute to a reduction in the emission of CO_2 into the atmosphere.

3.2.4 Implications for Positivity Conditions

While the above procedure represents a straightforward way to introduce by-product into the IO model, this generalization can cause problems. The point is that the Leontief inverse matrix $(I - \check{A})^{-1}$ now contains some negative elements, and hence Theorem 2.3 is no longer applicable. Accordingly, nonnegativity of the solution (3.58) is no longer guaranteed. For instance, in the above example, when the final demand for product 2 is five times larger than that for product 1, the output of product 1 can become negative:

$$
\begin{pmatrix} x_1 \\ x_2 \\ x_3 \end{pmatrix} = \begin{pmatrix} 0.89 & -0.26 & 0.05 \\ 0.39 & 0.91 & 0.33 \\ 0.12 & 0.06 & 1.03 \end{pmatrix} \begin{pmatrix} 1 \\ 5 \\ 4 \end{pmatrix} = \begin{pmatrix} -0.20 \\ 6.30 \\ 4.61 \end{pmatrix}. \tag{3.67}
$$

Since the amount of output cannot take a negative value, this represents a weakness of the model.

A negative amount of output, while not possible in reality, does occur in the model (3.58) when there is an excess supply of by-product over the demand for it. This happens because in the model the supply of by-product is driven not by the demand for it, but is determined by the demand for the primary product alone. In reality, however, an excess supply of by-product will be subjected to waste management, that is, there is an adjustment mechanism through waste management. This adjustment process of waste management is not taken into account in the present model. The weakness of the model does not consist in the occurrence of a nonzero off-diagonal element in D, that is, the Stone method, but in the neglect of this adjustment mechanism. This point will be dealt with in greater detail in Section 5.3.2.4.

3.3 The Model Based on Use and Make Matrices

An alternative way to take into account the presence of by-product is the method based on a Use and Make matrix or commodity-by-industry input-output accounts [22, 29]. Henceforth in this section, the term sector is used synonymously with industry, and the term product is used synonymously with commodity.

3.3.1 U and V Matrices, and Related Identities

Suppose that there are m products and n sectors with $m \geq n$. Denote by u_{ij} the amount of product i that occurs as an input in sector j, and by v_{ij} the amount of product j that is produced by sector i. Write χ for the $m \times 1$ vector of the output of products, and f for $m \times 1$ vector of the final demand for products. By definition, the following holds:

$$\chi = (\chi_i) = \left(\sum_{k=1}^{n} v_{ki}\right) = V^\top \iota_n \tag{3.68}$$

$$\chi = \left(\sum_{k=1}^{n} u_{ik} + f_i\right) = U\iota_n + f. \tag{3.69}$$

where ι_n refers to an $n \times 1$ vector of unity. Write $B = (b_{ij})$ for an $n \times m$ matrix of the fraction of total production of product j in the economy produced by sector i:

$$B = (b_{ij}) = (v_{ij}/\chi_j) = V\hat{\chi}^{-1}, \tag{3.70}$$

where b_{ij} is referred to as the commodity output proportion ([22], p.165). Assuming that B thus defined is a constant matrix, V is given by

$$V = B\hat{\chi} \tag{3.71}$$

Let p be the $m \times 1$ vector of commodity prices, and y^* be the $n \times 1$ vector of value of outputs:

$$y^* = \left(\sum_{j=1}^{m} v_{ij}p_j\right) = V p. \tag{3.72}$$

Note that y^* is defined only for monetary data because the sum of different products in mass, that is, $\sum_j v_{ij}$ (for example, the sum of motorcycles and automobiles in kg) will make no sense. Substituting from (3.71), this yields:

$$y^* = V p = B\hat{\chi} p = B\hat{p}\chi, \tag{3.73}$$

where the last equality is due to the fact that both χ and p are vectors of the same order.

3.3.2 Industry-Based Technology

Assuming the constancy of the $m \times n$ matrix of input-coefficients

$$C^* = (u_{ij}/y_j^*) \tag{3.74}$$

it follows that

$$U = C^*\widehat{y^*} \tag{3.75}$$

Substitution of this expression into (3.69) gives:

$$\chi = C^*\widehat{y^*}\iota_n + f = C^*y^* + f$$
$$= C^* B\hat{p}\chi + f, \tag{3.76}$$

where the last equality is from (3.73). Solving for χ gives:

$$\chi = (I - C^* B \widehat{p})^{-1} f = (I - \acute{A})^{-1} f. \tag{3.77}$$

The matrix $\acute{A}$ is called the industry-based technology matrix [22], as it is assumed (in the definition of B in (3.70)) that the total output of a commodity (product) is provided by industries (sectors) in fixed proportions [22, 29].

From (3.70), (3.75), and (3.72), $\acute{A}$ is expressed in terms of U and V as

$$\acute{A} = U \widehat{y^*}^{-1} V \widehat{\chi}^{-1} \widehat{p} = U (\widehat{V p})^{-1} V \widehat{\chi}^{-1} \widehat{p}, \tag{3.78}$$

which indicates that $\acute{A}$ is always nonnegative, and that it depends on the price of products, that is, on the units of measurement of products. The latter point, not being price-invariant, is a weakness of the model based on industry technology assumption (see [29] for further details on this point). In particular, the model cannot be used for analysis involving prices. Because of this, we abstain from further discussion of $\acute{A}$.

3.3.3 Commodity-Based Technology

For the case of $m = n$, an alternative to industry-based technology, commodity-based technology, can be introduced. In this case, $\chi = x$, and V becomes a square matrix. Assuming further that V is nonsingular, an alternative matrix of input-coefficients, $\grave{A}$, can be derived:

$$\grave{A} = U (V^\top)^{-1}, \tag{3.79}$$

which is called the commodity-based technology matrix, as it assumes that every commodity has its own input structure irrespective of the sector of fabrication ([29], p.88).

Note that in the absence of by-product, this reduces to A. In contrast to the industry-technology matrix $\acute{A}$, the commodity-technology matrix $\grave{A}$ does not depend on product prices, and can have negative elements.

3.3.4 The Relationship Between $\breve{A}$ and $\grave{A}$

Two matrices of input coefficients, $\breve{A}$ and $\grave{A}$, have been introduced to cope with the presence of by-product. They have common characteristics as the absence of product prices and the possible presence of negative elements. It is of interest to see how they are related to each other. In order to do so, decompose V into two parts:

$$V = V_0 + V_1, \tag{3.80}$$

where V_0 refers to a matrix consisting of the diagonal elements of V, that is, $\hat{x}$, and V_1 refers to a matrix consisting of the off-diagonal elements. The following expressions can then be obtained for D and U:

$$D = V^{\top} V_0^{-1}, \tag{3.81}$$

$$U = A V_0. \tag{3.82}$$

Rewriting (3.81) gives:

$$V^{\top} = D V_0. \tag{3.83}$$

Substituting (3.82) and (3.83) into (3.79) gives:

$$\grave{A} = U (V^{\top})^{-1} = A V_0 (V^{\top})^{-1} = A V_0 (D V_0)^{-1}$$
$$= A D^{-1} \tag{3.84}$$

Compared to $\breve{A}$ in (3.60), the derivation of $\grave{A}$ in (3.84) involves the multiplication and inversion of matrices, which is intuitively not as straightforward as (3.60). Given the availability of information on the production processes of each sector including the production of by-product, $\breve{A}$ will be much easier to obtain than $\grave{A}$.

3.4 Extension Towards a Closed Model

In Section 3.1, several cases were considered where the trade flows between two regions are internalized under consideration of the obvious fact that an export from region a to b is an import of b from a, and vice versa. In other words, the model is closed with regard to bilateral trade: the exports between the two regions no longer occur as final demand sectors, because they are reckoned as imports, the level of which are determined by the remaining elements of the final demand such as household consumption and investment in plant and equipment. This section deals with several approaches to closing the IO model with regard to these remaining items of final demand.

3.4.1 Integrating Consumption

The closing of IOA in terms of household consumption was already mentioned in Section 2.1.1.2. For the sake of simplicity, assume the homogeneity of labor, and neglect the presence of exogenous inputs and waste. Denoting the household sector as the $n + 1$st sector, write the $(n + 2) \times 1$ vector of its unit process as

$$P_{n+1} = \begin{pmatrix} \text{inputs (except labor)} \\ \text{input of labor} \\ \dots\dots\dots\dots \\ \text{output of labor (hours)} \end{pmatrix} = \begin{pmatrix} (a)._{n+1} \\ a_{n+1,n+1} \\ \dots\dots \\ 1 \end{pmatrix}, \tag{3.85}$$

where $(a)._{n+1}$ refers to the $n \times 1$ vector of inputs into the household sector, and $a_{n+1,n+1}$ to its own input of labor in the household sector. The ith element of $a._{n+1}$ refers to the input of product i per unit of hours of work. In this simple representation, the household sector is treated just like any production process, where its output (labor hours) is produced out of the input of consumption goods such as food and clothing.

Denote by f^C the $n \times 1$ vector of household consumption, and by x_{n+1} the amount of labor supply. From (3.85), f^C can then be given by

$$f^C = (a)._{n+1} x_{n+1}. \tag{3.86}$$

Writing $(a)_{n+1}.$ for the $1 \times n$ vector of the input of labor, and $\bar{f}$ for the $(n+1) \times 1$ vector of final demand that excludes household consumption, the extended version of the quantity balance (2.139) with household as the $n + 1$st endogenous sector is given by

$$\begin{pmatrix} x \\ x_{n+1} \end{pmatrix} = \begin{pmatrix} A & (a)._{n+1} \\ (a)_{n+1}. & a_{n+1 n+1} \end{pmatrix} \begin{pmatrix} x \\ x_{n+1} \end{pmatrix} + \bar{f}, \tag{3.87}$$

or using obvious matrix notations

$$\bar{x} = \bar{A}\bar{x} + \bar{f}. \tag{3.88}$$

This way of closing the IO model with regard to household consumption was first proposed by Leontief [15]. The unit process of household consumption (3.85) implies that the share of each product in household expenditure is a constant, and does not change when the expenditure changes. This is a strong assumption, and is inconsistent with the Engel's law of consumption, a rare example of an empirical law in economics found by Ernst Engel, a 19th century German statistician, which states that the proportion of expenditure on food (and other items as well) is not proportional to income [11]. Subsequently, there have been substantial extensions of this simple formulation, such as [20] and [1]. See [2] for more recent development.

3.4.2 The Dynamic Model: Closing the IO Model with Regard to Capital Formation

Recall from Section 2.1.1.2 that the inputs in a unit process refer to those that are required for operating an existing process over a unit of time, but do not include those that are required for the construction or expansion of the process itself. For example, in the iron production process, limestone, iron ore, coke, oxygen, and electricity occur as current inputs, but the blast furnace itself does not, while the inputs for its regular repair and maintenance do occur as current inputs. What enters and is used up in the production process is not the furnace itself but the service provided by it: in the production process of iron, current inputs such as limestone, coke, oxygen, and ore are used up, and transformed into outputs of various forms, but the furnace remains as such except for some wear (of bricks constituting the inner wall, for example).

The inputs that are needed for the construction and/or expansion of a production process itself are mostly durables with certain services associated with them, such as design and supervision. Construction and/or expansion of a production process constitutes the investment in equipment and structure, f^K, which is another important component of final demand besides household consumption. This section is concerned with the closing of the IO model with regard to this important component of final demand.

For the sake of simplicity, suppose that the final demand consists of household consumption and investment in plant and equipment only

$$x(t) = Ax(t) + f^C(t) + f^K(t), \qquad (3.89)$$

where $x(t)$ refers to the amount of x in year t (the same holds for $f^C(t)$ and $f^K(t)$ as well).

3.4.2.1 Capital Coefficients

Similar to the representation of a unit process, denote by c_{ij} the input of product i that is needed for a unit of productive capacity of process j. For instance, suppose that j deals with an electrical plant, c_{ij}'s then include transformer boards, transformers, power generating set, control panels, distribution boards, wiring, high-voltage condenser boards etc. The $n \times n$ matrix $C = (c_{ij})$ is called the capital coefficients or stock-flow matrix [15]. Denote by $\tilde{x}_j(t)$ the productive capacity in sector j in time t, and by $\Delta \tilde{x}_j(t)$ the addition to the productive capacity between t and $t + 1$:

$$\Delta \tilde{x}_j(t) = \tilde{x}_j(t+1) - \tilde{x}_j(t) \geq 0. \qquad (3.90)$$

Expansion of productive capacity requires investment in plant and equipment. Writing $k_{ij}(t)$ for the amount of product i that is needed for this expansion, it will be given by:

$$k_{ij}(t) = c_{ij} \Delta \tilde{x}_j(t), \qquad (3.91)$$

and hence

$$f_i^K(t) = \sum_j k_{ij}(t) = \sum_j c_{ij} \Delta \tilde{x}_j(t). \qquad (3.92)$$

This formulation of investment is called the acceleration principle, a theory of investment in macroeconomics, which asserts that the level of investment is accelerated only through the rate of increase in output. Implicit in (3.92) is also the assumption that the gestation period of investment is equal to one year: the investment executed in year t becomes the productive capacity in year $t + 1$.

Table 3.7 The Capital (Investment) Matrix for Japan, 2000.

	PRM	FWP	CHM	MMC	CST	UTL	SRV	Row-sum
PRM	141	0	0	0	0	0	48	189
FWP	50	63	11	128	25	4	751	1,032
CHM	0	0	0	0	0	0	0	0
MMC	900	1,654	1,229	9,071	833	3,455	21,548	38,690
CST	3,025	554	423	1,515	186	6,796	55,832	68,331
UTL	0	0	0	0	0	0	0	0
SRV	517	1,052	1,257	4,647	357	2,783	11,156	21,770
Column-sum	4,633	3,324	2,920	15,362	1,400	13,038	89,335	130,012

Units: 1,000 million Japanese yen. Source [19]. PRM: Primary, FWP: Food, wood and paper, CHM: Chemicals, MMC: Metal, machinery and other manufacturing, CST: Construction, UTL: Utilities, SRV: Service and others.

The matrix $K = (k_{ij})$ is called the capital (or investment) matrix, and gives the sectoral breakdown of the purchase of products for investment for the whole economy, f^K. In Japanese IOTs, the estimate of matrix K is routinely reported as subsidiary information. Table 3.7 gives an aggregated version of the Japanese capital matrix for year 2000 [19]. The row sums give the amount of gross fixed capital formation that occur in the final demand of IO table, while the column sum gives the amount of gross fixed capital formation that was made at each sector. Note that the row sum is equal to the column of capital formation in Table 3.1. Of the seven sectors in the table, the service sector made the largest amount of fixed capital formation, 89 trillion (10^6 million) yen, the largest component of which is the products of the construction sector, followed by those of the metal, machinery, and other manufacturing sector. The capital matrix consists mostly of the products of metals, machinery, and other manufacturing sector, the products of the construction sector, as well as the products of the service sector. No entries occur for the rows referring to chemicals and utilities.

3.4.2.2 The Leontief Dynamic Model

Substitution of (3.92) into (3.89) gives:

$$x(t) = Ax(t) + f^C(t) + C\Delta\tilde{x}(t). \tag{3.93}$$

If the production were *always* kept at the level of full capacity, and hence $x(t) = \tilde{x}(t)$ for all t, (3.93) would become

$$x(t) = Ax(t) + f^C(t) + C\Delta x(t). \tag{3.94}$$

This is known as the Leontief dynamic model [16]. Mathematically, this is a system of first-order linear difference equations, which can be rewritten as:

$$Cx(t+1) = (I - A + C)x(t) - f^C(t) \tag{3.95}$$

If the capital coefficients matrix C were nonsingular, (3.95) would be soluble forward recursively as:

$$x(t+1) = (I + C^{-1}(I-A))x(t) - C^{-1}f^C(t). \tag{3.96}$$

If, furthermore, it were possible to endogenize $f^C(t)$, and close the model as in (3.88), this would reduce to a system of homogeneous difference equations:

$$x(t+1) = (I + C^{-1}(I-A))x(t), \tag{3.97}$$

where it is understood that the dimension of x, A, and C has been adequately altered to accommodate the inclusion of the household sector. Given an initial value of x, say $x(0)$, the growth path of x could then be derived for any $t > 0$. For instance, $x(t+s)$ with $s > 0$ will be given by:

$$x(t+s) = (I + C^{-1}(I-A))^s x(0). \tag{3.98}$$

This system has been the subject of a considerable number of theoretical considerations by economists (see [14] for a recent survey).

3.4.2.3 Issues in the Application of the Leontief Dynamic Model

It is well known, however, that the Leontief dynamic model (3.94) suffers from several fundamental problems [14], such as:

1. The equality of production and productive capacity, or production at full capacity, usually does not hold.
2. The matrix C is in general not invertible because it will contain rows, all the elements of which are zero (see Table 3.7).
3. Even if the matrix C were invertible, the presence of a negative solution would not be excluded.

From the mathematical point of view, one could cope with the second problem by reducing the order of the system "until the condition [of nonsingularity] is satisfied by eliminating variables and equations" ([13], p.109). This, however, comes at the cost of a significant loss in information about the interdependence among different sectors. For Table 3.7, this practice would amount to neglecting the presence of interdependence between the chemical and utilities sectors with the metal and machinery sector. This would certainly not be a practice to be encouraged in Industrial Ecology, LCA, and Life Cycle Management. It suffices to recall that consideration of this sort of interdependence is the main reason IOA is used in these disciplines.

Furthermore, it is not realistic to assume that the values of A and C would stay constant over time, that is, for $t \to \infty$ (a major concern of theories of economic growth consists in the investigation of conditions under which the economy can achieve a steady growth path for $t \to \infty$). It is safe to say that from the point of view of Industrial Ecology, the recursive dynamic model is of limited, if any, practical relevance.

3.4.2.4 An Alternative Approach to Closing the IO Model with Regard to Capital Accumulation

Despite the limitation of its practical applicability, the Leontief dynamic model is an ambitious one that is aimed at describing the future path of economic growth in a simple framework. A less ambitious approach will be to use the capital matrix to make the demand for investment endogenous, but refrain from deriving a model of economic growth. An example of this approach is Lenzen [17], where the following relationship is postulated in place of (3.91):

$$k_{ij} = c_{ij} x_j \tag{3.99}$$

and hence the equivalence to (3.92) becomes

$$f_i^K = \sum_j c_{ij} x_j, \tag{3.100}$$

where the suffix t referring to time is suppressed because the model does not involve different time periods, and is static. Substituting (3.99) into (3.89) gives

$$x = (A + C)x + f^C \tag{3.101}$$

and hence

$$x = \left[I - (A + C) \right]^{-1} f^C. \tag{3.102}$$

Contrary to the dynamic model (3.94), this model does not require the inversion of the matrix C.

A natural question to be raised will be how (3.99) is related to the original acceleration model of investment (3.91). It is obvious that the former reduces to the latter if

$$\Delta x_j(t) = x_j(t+1) - x_j(t) = x_j(t+1) \tag{3.103}$$

that is, when $x_j(t) = 0$. The static model (3.102) thus gives the level of x that is associated with the realization of f^C from scratch, that is, from the situation where no production activity was present. In this case, plants have to be built before production starts. The construction of plants, however, requires inputs, the production of which requires another construction of plants, and so forth. The model (3.102) gives the level of output that is required to satisfy all these requirements. An alternative interpretation of the same result follows when one replaces (3.103) by

$$\Delta x_j(t) = x_j(t+1) - x_j(t) = x_j(t) \tag{3.104}$$

which corresponds to the case where the level of production at $t + 1$ is twice the level of t.

In contrast, the original static model (2.152) does not consider the production activity that is associated with the construction of productive capacity, whilst it considers the activity that is associated with the maintenance of a given productive capacity.

The model (3.102) will thus be useful if one is interested in analyzing the overall effect of realizing a given level of final demand (excluding the final demand for investment) including the construction of associated productive capacity. This type of model was used by Nansai et al. [21] to analyze the environmental burdens associated with alternative household consumption patterns.

Table 3.8 gives an example of the inverse matrix that occurs in (3.102). For comparison, the standard Leontief inverse matrix is also shown in the lower panel of the table. Inclusion of the construction of productive capacity increases the column sum of the inverse matrix by 44% on average. Substantial differences exist among sectors. At 93%, the utilities sector shows the highest rate of increase in the column sum: while the production of a unit of its product requires directly and indirectly 1.8 units of inputs when existing production facilities are used, the amount would increase to 3.5 units when its facilities also have to be constructed from scratch. The construction sector is the main subject of fixed capital formation: it occupies the largest portion (53%) of fixed capital formation, while 88% of its product is devoted to fixed capital formation. It is interesting to find that, with 21%, the overall inputs requirements of the construction sector are least affected by the inclusion of the construction of its productive facilities.

Table 3.8 An Example of $(I - (A + C))^{-1}$ Matrix.

	\multicolumn{7}{c}{$(I - (A+C))^{-1}$ matrix}
	PRM	FWP	CHM	MMC	CST	UTL	SRV
PRM	1.1837	0.1988	0.2004	0.0588	0.0600	0.1440	0.0379
FWP	0.1995	1.3407	0.1054	0.0959	0.1200	0.1129	0.0946
CHM	0.2114	0.1669	1.4643	0.2041	0.2023	0.2108	0.1142
MMC	0.3567	0.2649	0.2772	2.0271	0.4085	0.5630	0.2623
CST	0.3332	0.1417	0.1614	0.1430	1.1037	0.4710	0.2067
UTL	0.0434	0.0509	0.0752	0.0603	0.0365	1.1048	0.0418
SRV	0.6423	0.5711	0.6360	0.7141	0.5682	0.8521	1.5404
Sum	2.9702	2.7350	2.9198	3.3033	2.4993	3.4586	2.2979
	\multicolumn{7}{c}{$(I - A)^{-1}$ matrix}
	PRM	FWP	CHM	MMC	CST	UTL	SRV
PRM	1.1474	0.1847	0.1853	0.0430	0.0506	0.1083	0.0225
FWP	0.1462	1.3147	0.0780	0.0645	0.1015	0.0437	0.0634
CHM	0.1303	0.1273	1.4209	0.1535	0.1731	0.0954	0.0648
MMC	0.0456	0.0735	0.0737	1.7255	0.2602	0.0599	0.0555
CST	0.0131	0.0127	0.0189	0.0158	1.0118	0.0560	0.0180
UTL	0.0254	0.0412	0.0646	0.0469	0.0292	1.0774	0.0305
SRV	0.3357	0.3982	0.4379	0.4736	0.4410	0.3506	1.3483
Sum	1.8437	2.1523	2.2792	2.5228	2.0675	1.7914	1.6030

Source: Japanese IO table for year 2000. PRM: Primary, FWP: Food, wood and paper, CHM: Chemicals, MMC: Metal, machinery and other manufacturing, CST: Construction, UTL: Utilities, SRV: Service and others.

3.4.3 A Fully Closed Model

Consider the case where the level of productive capacity is kept at a constant level, that is, $\Delta x(t) = 0$, and the household consumption is endogenized as in Section 3.4.1. The dynamic model (3.94) then reduces to

$$x = Ax, \text{ or}$$

$$(I - A)x = 0, \quad \text{for} \quad x > 0, \tag{3.105}$$

where it is understood that the dimension of x and A has been adequately altered, say to $\acute{n}$, to accommodate the inclusion of the household sector. This is a fully closed model, which has neither final demand nor value added.

In a fully closed model, the $\acute{n} \times \acute{n}$ matrix $(I - A)$ is no longer invertible: (3.105) implies the linear dependence of the columns of $(I - A)$. Because of this feature, the closed model (3.105) is solvable for relative proportions of x only. In order to see this, partition the left-hand side of (3.105) as:

$$(I - A)x = \begin{pmatrix} I - \check{A} & -(a)_{.n} \\ -(a)_{n.} & 1 - a_{nn} \end{pmatrix} \begin{pmatrix} \check{x} \\ x_n \end{pmatrix}, \tag{3.106}$$

where the choice of n is arbitrary. Substitution of (3.106) into (3.105) gives:;

$$(I - \check{A})\bar{x} - (a)_{.n}x_n = 0 \tag{3.107}$$

Because the $(\acute{n} - 1) \times (\acute{n} - 1)$ matrix $I - \check{A}$ is nonsingular and invertible, provided $x_n > 0$, this can be solved for relative quantities:

$$\check{x}x_n^{-1} = (I - \check{A})^{-1}(a)_{.n}. \tag{3.108}$$

Analogously, the price equation for the closed model is given by

$$p(I - \check{A}) = 0, \tag{3.109}$$

Similar to (3.108), the model can be solved for relative prices

$$\check{p}p_n^{-1} = (a)_{n.}(I - \check{A})^{-1}, \tag{3.110}$$

where $\check{p}$ refers to the vector of p with its nth element excluded.

A comparison of the closed model (3.108) with its open counterpart, (2.152), indicates that while the effects of a change in final demand, f, is the main concern of the open model, this sort of analysis is not applicable to a closed model because f is not an exogenous but endogenous variable, the level of which is determined interdependent with the levels of all the remaining variables. Scenario analysis of the sort (2.169) or (2.177), the easy implementation of which is an important strength of IO models ([10], p.17), are not relevant to a closed model. Instead, a closed model is focused on analyzing the effects of a change in A on the relative quantities. In fact, a substantial portion of Leontief [15] is devoted to this sort of analysis.

It is one thing that the closed IO model is based on a set of strong assumptions. It is another thing that economic variables are intrinsically mutually interdependent, and feedback effects exist: a change in endogenous variables that results from a change in the variables that are assumed to be exogenous in the open model, say final demand, can have effects on some of them. For instance, the saving of expenditure on energy due to an increase in the energy efficiency of home appliances, such as refrigerators or washing machines, can lead to an increase in the expenditure on other consumer products, which is known as a rebound effect. The fact that rebound effects are taken seriously in LCA of consumption indicates the importance of the feedback effects [9, 31].

On the other hand, it seems safe to say that a widely accepted, robust, and reliable modeling of these feedback effects that matches the wide acceptance and robustness of the standardized open IO model does not yet exist. Henceforth, this book takes a rather conservative stance, and limits its focus on the open model, while acknowledging the presence of feedback effects.

3.5 Extension to the System with Inequalities

Up to now in this chapter we have been concerned with the models involving equalities only. Consideration of the limited supply of exogenous inputs calls for the introduction of inequalities to account for the upper bounds of available supplies. The introduction of inequalities will also be useful for considering the possibility of technical substitution from among alternative unit processes.

3.5.1 Limited Supply of Exogenous Inputs

In Section 2.1.2.6, the issue of a limited supply of exogenous inputs was treated, and the upper bound of production (2.30) and final demand (2.31) were obtained. In the case of multiple sectors, however, the situation is not that simple because of the presence of sectoral interdependence. Suppose for instance, that exogenous input k is the limiting factor, the available amount of which for sector i is given by z_{ik}^*. Writing $\widehat{(B)_{k\cdot}}$ for the diagonal matrix of the kth column elements of B, and $(z)_{\cdot k}^*$ for the $n \times 1$ vector with z_{ik}^* as its ith element, the upper level of production that can be achieved for a given z_k^* will be given by

$$x \leq \left(\widehat{(B)_{k\cdot}} \right)^{-1} (z)_{\cdot k}^*, \tag{3.111}$$

the ith element of which is

$$x_i \leq \frac{z_{ik}^*}{b_{ki}}. \tag{3.112}$$

In the case of one endogenous sector, the derivation of the upper bound of final demand is straightforward because it is a simple multiplication by $(1 - a_1)$ of the upper bound of production. In the case of $n \geq 2$, however, this is no longer the case because the final demand obtained by analogy to the one-product case:

$$f = (I - A)(\widehat{B_{k \cdot}})^{-1}(z)^*_{\cdot k}, \tag{3.113}$$

may contain negative elements.

3.5.1.1 An Example

For illustration, consider the matrix of input coefficients given by Table 2.5. Suppose that water is the limiting factor, the available amount of which is given by

$$\begin{pmatrix} z_{11} \ (\text{for rice}) \\ z_{21} \ (\text{for fish}) \end{pmatrix} = \begin{pmatrix} 100 \\ 5 \end{pmatrix} \tag{3.114}$$

From (3.111), the upper bound of production will be given by:

$$x = \begin{pmatrix} 5.7 & 0 \\ 0 & 0.28 \end{pmatrix}^{-1} \begin{pmatrix} 100 \\ 5 \end{pmatrix} = \begin{pmatrix} 17.54 \\ 17.86 \end{pmatrix}. \tag{3.115}$$

Applying (3.113), the corresponding upper bound of f will then be given by

$$f = \begin{pmatrix} 0.993 & -0.08 \\ -0.3 & 0.91 \end{pmatrix} \begin{pmatrix} 5.7 & 0 \\ 0 & 0.28 \end{pmatrix}^{-1} \begin{pmatrix} 100 \\ 5 \end{pmatrix} = \begin{pmatrix} 15.99 \\ 10.99 \end{pmatrix}. \tag{3.116}$$

However, if the allocation pattern of water changes in favor of fish, and is given by

$$\begin{pmatrix} z_{11} \\ z_{21} \end{pmatrix} = \begin{pmatrix} 65 \\ 40 \end{pmatrix} \tag{3.117}$$

calculation by (3.113) yields a negative solution:

$$f = \begin{pmatrix} 0.993 & -0.08 \\ -0.3 & 0.91 \end{pmatrix} \begin{pmatrix} 5.7 & 0 \\ 0 & 0.28 \end{pmatrix}^{-1} \begin{pmatrix} 65 \\ 40 \end{pmatrix} = \begin{pmatrix} -0.10 \\ 126.58 \end{pmatrix}. \tag{3.118}$$

3.5.1.2 Alternative Method Based on Linear Programming

An easy and elegant way to search for the maximum attainable level of f is to formulate the model as an optimization problem subject to constraints, and resort to standard techniques such as linear programming [18]. The set of feasible final demand can be represented by the following set of constraints:

$$(I - A)^{-1} f \geq 0 \tag{3.119}$$

$$\widehat{B}_1 (I - A)^{-1} f \leq z \tag{3.120}$$

$$f \geq 0, \tag{3.121}$$

where the left hand side of the first equation is the usual quantity model, and states that the level of production must be nonnegative, the second inequality refers to the upper bounds of the available exogenous inputs, and the last inequality refers to the nonnegativity of f. The maximum attainable level of f can then be obtained by introducing the objective function

$$c^\top f \tag{3.122}$$

that should be maximized subject to (3.119)–(3.121), where c is a vector of constants referring to the relative importance of individual components. If all the components are of equal importance, $c_i = 1$ can be used for all i. This optimization problem is called linear programming problem because both the objective function and the constraints are linear in variables, that is, in f.

3.5.1.3 An Example Based on Linear Programming

In the above example, the solution for f by (3.113) happened to contain a negative element when z is given by (3.117). Application of linear programming to the system mentioned above yields for $c_i = 1, i = 1, 2$:

$$f = \begin{pmatrix} 0 \\ 125.39 \end{pmatrix}, \tag{3.123}$$

which resembles (3.118), but contains no negative elements (see Section 3.8 for the computation based on Excel). If nonnegativity is the only restriction that is imposed on f, the most efficient use of water will be achieved when the final demand for rice, f_1, becomes zero: water should be allocated to fish production because it can make a more efficient use of water than rice production.

If (3.121) is replaced by

$$f \geq \begin{pmatrix} 10 \\ 0 \end{pmatrix}, \tag{3.124}$$

the optimum solution for f becomes

$$f = \begin{pmatrix} 10 \\ 11.64 \end{pmatrix} \tag{3.125}$$

which indicates the fact that (in this numerical example) a substantial reduction in fish production is required to increase the production of rice. While this fact can be grasped by looking at the data, its quantitative extent cannot be ascertained without resorting to a computation like this.

3.5.2 Issues of Substitution: Programming Model

In the IO models discussed so far in this book, a one-to-one correspondence between products and processes has been assumed, and the possibility of substitution between alternative processes has been ruled out. In reality, however, technical alternatives do exist. Steam for power generation can be obtained from coal, heavy oil, natural gas, nuclear power, or organic waste. Electricity can also be obtained from, among others, hydraulic dams, solar cells, and wind mills. A car can be driven by an internal combustion engine, electric motors, or a hybrid of both of them. Organic waste can be composted, incinerated, gasified, or landfilled. Alternative treatments of waste water include the use of trickling filter beds, biological aerated filters, and membrane biological reactors. This section considers the IO model with alternative unit processes.

For the sake of simplicity, consider the case of two endogenous sectors in Section 2.2. Suppose that two processes, $P_1^{(1)}$ and $P_1^{(2)}$, are available for the production of product 1, and three processes, $P_2^{(1)}$, $P_2^{(2)}$, and $P_2^{(3)}$, are available for the production of product 2. Write P^+ for the matrix of unit process (2.127) extended for the presence of alternative processes:

$$P^+ = \left(P_1^{(1)(2)} \; P_2^{(1)(2)(3)} \right) = \left(\begin{array}{ccc|cc}
a_{11}^{(1)} & a_{11}^{(2)} & a_{12}^{(1)} & a_{12}^{(2)} & a_{12}^{(3)} \\
a_{21}^{(1)} & a_{21}^{(2)} & a_{22}^{(1)} & a_{22}^{(2)} & a_{22}^{(3)} \\ \hline
b_{11}^{(1)} & b_{11}^{(2)} & b_{12}^{(1)} & b_{12}^{(2)} & b_{12}^{(3)} \\
b_{21}^{(1)} & b_{21}^{(2)} & b_{22}^{(1)} & b_{22}^{(2)} & b_{22}^{(3)} \\
b_{31}^{(1)} & b_{31}^{(2)} & b_{32}^{(1)} & b_{32}^{(2)} & b_{32}^{(3)} \\ \hline
1 & 1 & 0 & 0 & 0 \\
0 & 0 & 1 & 1 & 1 \\ \hline
g_{11}^{(1)} & g_{11}^{(2)} & 0 & 0 & 0 \\
0 & 0 & g_{22}^{(1)} & g_{22}^{(2)} & g_{22}^{(3)}
\end{array}\right) = \left(\begin{array}{c} A^+ \\ B^+ \\ D^+ \\ G^+ \end{array}\right), \qquad (3.126)$$

where A^+ is a rectangular matrix of 2×5, with its third-columns elements referring to the input coefficients of $P_2^{(1)}$. The matrices B^+, D^+, and G^+ are also defined in analogous manner.

From among the alternative unit processes given by (3.126), a choice has to be made based on a set of criteria. A straightforward and elegant way of dealing with the choice problem is to formulate it as an optimization problem subject to constraints, and resort to optimization techniques as in Section 3.5.1 above. Write a rectangular matrix H^+ for the matrix of net output coefficients that correspond to P^+:

$$H^+ = D^+ - A^+. \qquad (3.127)$$

Denote by $x^+ = (x_1^{(1)}, \cdots, x_2^{(3)})^\top$ a 5×1 vector of the activity level of five unit processes. When a certain process, say, $P_1^{(2)}$, is not chosen, the corresponding level of activity is zero, that is, $x_1^{(2)} = 0$. The quantity balance equation (2.140) can be extended to the following system of equalities:

$$H^+ x^+ = f, \tag{3.128}$$

which states that whatever choice of unit processes is to be made, it has to satisfy the given final demand f. Another important set of constraints is the nonnegativity condition:

$$x^+ \geq 0. \tag{3.129}$$

Suppose that exogenous input 1 is the scarcest resource for this economy, and hence the choice of processes is made so as to minimize the amount of its overall input. Writing $(B)_1$. for the first row elements of B^+, the choice of processes is then formulated as the following LP:

$$\text{minimize} \quad (B)_1 . x^+ \quad \text{with regard to } x^+ \tag{3.130}$$

subject to (3.128) and (3.129).

3.6 The "Supply-Side Input-Output" Model of Ghosh

As an alternative for the standard Leontief input-output model, which is 'demand-driven' as shown above, Ghosh [7] presented a model that has become known as the 'supply-driven' input-output model. Its main distinguishing feature is the use of fixed output coefficients given by:

$$A^* = (a_{ij}^*) = \left(\frac{x_{ij}}{x_i}\right) = \widehat{x}^{-1} X \tag{3.131}$$

which implies for fixed x and X

$$\widehat{x} A^* = X \tag{3.132}$$

Substituting this into (2.166) gives

$$p\widehat{x} = pX + v$$
$$= p\widehat{x} A^* + v \tag{3.133}$$

It should be recalled that the column sum is applicable to a monetary IO table only, where all the elements are measured in a monetary unit, and hence can be added up.

Solving (3.133) for $p\widehat{x}$ gives

$$p\widehat{x} = v(I - A^*)^{-1}. \tag{3.134}$$

Recalling that v refers to the amount (in value terms) of exogenous inputs, one may be tempted to interpret (3.134) as a supply-side input-output model, where the amount of output is determined by the amount of available supply of exogenous inputs. In fact, there are quite a few studies where the Ghosh model is interpreted this way, including the standard text book by Miller and Blair ([22], pp.317–322).

It was shown by Oosterhaven [24], however, that this interpretation is flawed, who conclude that

> Reviewing the literature we concluded that the model may only be used, if carefully interpreted, in descriptive analyses. Any causal interpretation or application leads to at best meaningless, but probably nonsensical results. ([24], p.215)

Subsequently, it was also shown by Dietzenbacher [4] that if properly interpreted the Ghosh model is in fact identical to the Leontief price model. In other words, the Ghosh model does not represent any alternative model that could be obtained from the same matrix of input-output flows.

In order to see this, note that substitution of (2.160) into (3.131) gives, for a given x and X, the following relationship between the Ghosh matrix and the standard Leontief matrix:

$$A^* = \widehat{x}^{-1}X = \widehat{x}^{-1}A\widehat{x} \tag{3.135}$$

Substituting this into (3.134) gives:

$$p\widehat{x} = v(I - \widehat{x}^{-1}A\widehat{x})^{-1} = v(\widehat{x}^{-1}(I-A)\widehat{x})^{-1}$$
$$= v\widehat{x}^{-1}(I-A)^{-1}\widehat{x} \tag{3.136}$$

Post-multiplying of both sides by $\widehat{x}^{-1}$ gives:

$$p = v\widehat{x}^{-1}(I-A)^{-1} = \pi(I-A)^{-1} \tag{3.137}$$

which is nothing but the Leontief price model (2.168).

It turns out that the so-called "supply driven model" of Ghosh (3.134) is identical with the Leontief price model. It has to be said that the term "supply driven model" is rather misleading, because it gives the impression that the Ghosh model represented an alternative tool that one could obtain from the same set of IO data, while in reality it did not [4, 24].

3.7 The Fundamental Structure of Production

The ordering of sectors in an IO table usually follows historical convention, where the agricultural or primary sectors occur first, followed by the secondary or manufacturing sectors, and ends with the tertiary or service sectors. As such, there are no particular logical grounds for the ordering of sectors.

However, if the sectors are arranged in several groups according to their technical characteristics, say metals, nonmetals, and services, certain patterns may emerge that reflect their technical characteristics: agricultural products will not enter metals as inputs, for example. Furthermore, if within each of these groups the sectors are arranged by degrees of fabrication, a triangular pattern may emerge, which reflects the fact that products of higher degrees of fabrication will not be used for the production of products of lower degrees of fabrication: basic chemicals such as ethylene and propylene are used for the production of plastic resins, but not the other way round.

A proper rearrangement of the ordering of sectors may reveal the pattern of an IO matrix, which is common to different economies, and is rather stable over time.

3.7.1 Identifying the Fundamental Structure of Production

It was Simpson and Tsukui [26] who discovered that a proper rearrangement of the ordering of sectors could reveal the "fundamental structure of production". Using the comparable U.S. and Japanese IO tables that were each aggregated to 38 homogeneous sectors, they found:

1. Decomposability (Figure 3.2 (a)). When sectors were grouped into blocks according to their physical qualities, and blocks were arranged in the following order:

   (a) Metal
   (b) Nonmetal
   (c) Energy
   (d) Services

   then the IO matrix was decomposable in such a way that the blocks followed a triangular order with respect to one another.

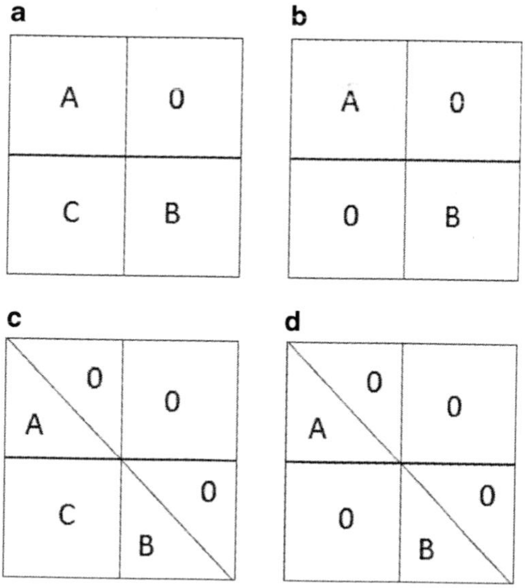

Fig. 3.2 Basic Patterns of the IO Matrix. The Square at the Northeast Corner Refers to a Zero Matrix. (a) Block Decomposability, (b) Block Independence, (c) Block Decomposability and Triangularity, (d) Block Independence and Triangularity. Modified from Figure 1 in [26].

2. Block Independence (Figure 3.2 (b)). A block is said to be independent, when no sector of it is related to another outside its own block. With the exception of the services block, the blocks were found to be almost independent of one another.
3. Triangularity (Figure 3.2 (c)). The rearranged matrix had the property of being almost triangular in all its elements.

Decomposability refers to the broad hierarchical order of input groups based on different degrees of fabrication: a machinery block would depend on the block of metal products, but not vice-versa, because the former is at a higher stage of fabrication. Block independence refers to the fact that there are inputs that are specific to particular groups of products, obvious examples of which are metal products not entering food except for containers, and agricultural products not entering metal products. Triangularity refers to the ordering based on the degree of fabrication inside a given block: stainless steel has a higher degree of fabrication than carbon steel, while the latter has a higher degree of fabrication than pig iron. Because the products would become more specialized with an increase in the degree of fabrication, their application would become limited to narrower ranges of sectors, which leads to the formation of a triangular structure when the sectors are ordered in terms of degrees of fabrication.

If they prevail, fundamental patterns as in Figure 3.2 have strong implications for the nature of interdependence among different producing sectors. For instance, if the blocks are decomposable (Figure 3.2 (a)), a change in the final demand for the products in block A affect the sectors in block B, while the sectors in block A are not affected by any change in the final demand for the products in block B.

3.7.2 Application to the Japanese IO Table

For illustration, an attempt was made to reveal the fundamental structure of production for the Japanese IO table with 104 production sectors for the year 2000 [19]. The 104 sectors were grouped into the following groups

1. Metals: construction, machinery, and metals
2. Nonmetals: food, textiles
3. Chemicals: plastics, chemical products, basic chemicals
4. Energy and mineral ores
5. Services

The ordering within each group was determined based on the degree of fabrication. For the ease of presentation, the service sectors were aggregated to a single sector.

Figure 3.3 shows the result. The significance of rearranging the ordering of sectors in revealing a structural pattern is evident from its comparison with the original form in Figure 3.4. Several observations can be made. First, a clear triangularity is visible for the metal block (see Table 3.9 for its components). Secondly, the metal block is independent of the block consisting of sectors 41–56, which mostly consist of products based on primary industries, and of chemical final products.

Fig. 3.3 The A Matrix in a Triangulated Form: The Elements Smaller Than $1/104 = 0.096$ Set Equal to 0.

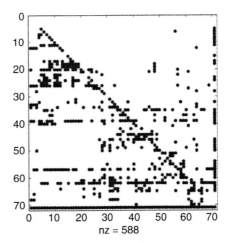

Fig. 3.4 The A Matrix in Original Form: The Elements Smaller Than $1/104$ Set Equal to 0.

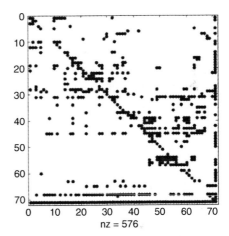

A numerical measure of triangularity is the degree of linearity:

$$\lambda = \frac{\sum_{i,j,i>j} a_{ij}}{\sum_{i,j,i\neq j} a_{ij}} \tag{3.138}$$

which takes unity when the matrix is completely triangulated as in Figure 3.2 (c), and takes 0.5 in the absence of any triangularity. For the original matrix (Figure 3.4) $\lambda = 0.5601$ while for the rearranged matrix (Figure 3.3) $\lambda = 0.7923$. The effectiveness of the reordering of sectors is thus confirmed not only graphically but numerically as well.

Table 3.9 The Order of Metal Related Sectors in Figure 3.3.

1	Civil engineering & construction
2	Public construction
3	Construction
4	Passenger cars
5	Other transportation equipment & repairing
6	Ship building & repair
7	Other cars
8	Household electronic & electric equipment
9	Office machinery
10	Special industrial machinery
11	General machinery for industrial use
12	Metal products for construction & architecture
13	Applied electronic apparatus, electrical measurement devices
14	Communication equipment
15	Precision machinery
16	Electronic computing equipment & accessories
17	Heavy electrical equipment
18	Electronic parts
19	Semiconductors, chips
20	Nonferrous metals products
21	Nonferrous metals refining
22	Other steel products
23	Other general machinery
24	Other electrical equipment
25	Forged iron products
26	Steel products
27	Iron and steel

3.8 Exercise with Excel

3.8.1 Accounting for Competitive Imports

This exercise is continued from Section 2.4.

1. Open sec013.xls.
2. Make Sheet1 the active sheet.

3.8.1.1 Calculating the Leontief Inverse Matrix of the Type $(I - (I - \hat{\mu})A)^{-1}$

1. Enter =-AD6/W6 into AI6 to calculate the share of competitive imports in the total domestic demand by (3.5) (the negative sign is necessary because the amounts
of import are in negative values). Then, copy AI6, and paste it into AI7:AI18.

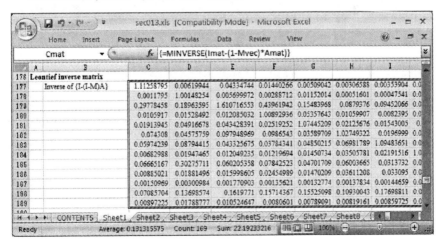

2. Assign the name Mvec to AI6:AI18.
3. Assign the name Amat to C65:O77, where the input coefficients matrix A is stored.
4. Assign the name Imat to C97:O109, where the identity matrix I is stored.
5. Enter the following as an array formula into C177:O189 to calculate the Leontief inverse matrix by (3.9):

```
=MINVERSE(Imat-(1-Mvec)*Amat)
```

3.8.1.2 Calculating the Amounts of Imports Induced by Final Demand Categories (3.11)

1. Assign the name Cmat to C177:O189, where the Leontief inverse matrix is stored.
2. Assign the name FDmat to Q6:U18, where the matrix of five domestic final demand categories are stored.

	O	P	Q	R	S	T	U	
4	13	33	35	36	37	38	40	
5	Activities not elsewhere classified	Total of intermediate sectors	Consumption expenditure outside households (column)	Consumption expenditure (private)	Consumption expenditure of general government	Gross domestic fixed capital formation	Increase in stocks	dome de
6	0	11483153	91221	3874706	0	193481	773717	
7	988	10059677	-437	-6669	0	-4720	-11065	
8	383742	191370775	3337634	61596633	459179	39721886	-635406	10
9	0	8979216	0	0	0	68331313	0	6
10	66691	18102356	5018	8082874	785162	0	0	
11	84146	34568771	1920454	45862001	4485	10660234	117370	5
12	955849	27638188	250	10486047	0	0	0	1
13	49316	9127828	0	56704783	17785	0	0	5
14	208141	30525482	542956	14733328	-41860	738784	32056	1
15	117956	14198691	221898	7793058	0	0	0	
16	708777	708777	0	735152	34781965	0	0	3
17	349908	78237164	13052191	71091948	49699501	10371088	0	14
18	0	4404490	0	36351	0	0	0	
19	2925514	439404568	19171185	280990212	85706217	130012066	276672	51

3. Assign the name EXvec to X6:X18, where the vector of exports is stored.
4. Entering the following into C197:G209 as an array formula:
   =MMULT(Mvec*Amat,MMULT(Cmat,(1-Mvec)*FDmat))+Mvec*FDmat
   gives the amounts of imports induced by each category of domestic final demand.

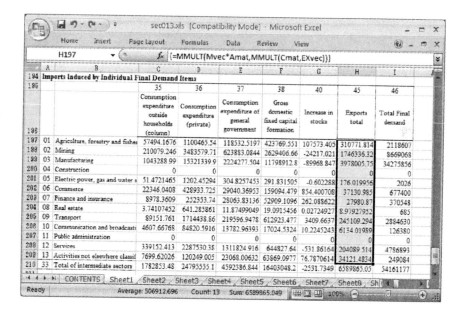

5. Similarly, entering the following into **H197:H209** as an array formula:
=MMULT(Mvec*Amat,MMULT(Cmat,EXvec))
gives the amounts of imports induced by export.

'Consumption expenditure (private)' and 'Export total' amount to 281 trillion yen and 57 trillion yen, which account for 49% and 10% of 'Total final demand', respectively. The imports induced by these two final demand items amount to 24.8 trillion yen and 6.6 trillion yen, which account for 45.8% and 12.2% of total induced imports, respectively. It follows that meeting the demand for exports required larger amounts of imports per unit than meeting the demand for household consumption.

3.8.2 The Upper Bounds of Final Demand When the Supply of Exogenous Inputs Is Limited

In the following, the results of computation by LP in Section 3.5.1.3 are reproduced. To use Excel to solve LP problems the Solver add-in must be included. It is assumed that this add-in has already been installed on your computer.

Suppose that $(I - A)^{-1}$ occurs in F10:G11, $\widehat{B}_1$ in C13:D14, and z in B24:B25 of the active sheet.

1. Enter some numbers as the initial values of f in B35:B36 (you can use, for instance, the results in (3.118)). Enter the expressions on the left-hand side of (3.119) in B24:B25

	A	B	C	D	E
24	Positivity condition for sector 1	=MMULT(F10:G11,B35:B36)			
25	Positivity condition for sector 2	142	0		

The expressions on the left-hand side of (3.120) can then be entered in B27:B28 as

	A	B	C	D	E
27	Constraints on water for sector 1	=MMULT(C13:D14,MMULT(F10:G11,B35:B36))			
28	Constraints on water for sector 2	40	40		

2. Enter the objective function (3.122) in B38

	A	B	C	D
33	c	1	1	
34				
35	Solution for f1	-0.1		
36	Solution for f2	127		
37				
38	Objective function	=MMULT(B33:C33,B35:B36)		

where c's are set equal to unity.

3. On selecting the menu option Data → Analysis → Solver reveals the dialogue box

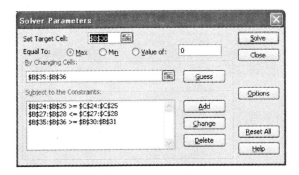

Set the target cell to **B38**, where the objective function occurs, and select Max. Enter the range of cells where the decision variables occur, that is, B35:B36, to By Changing Cells.

4. Click [**Add**] button to enter the constraints (3.119)–(3.121). By default <= is selected. Clicking on the drop-down arrow reveals a list of other constraint types. Go into [**Options**] and check the 'Assume Linear Model' box

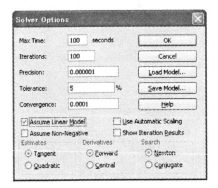

This can speed up the length of time taken for Solver to find a solution to the problem. Finally, clicking [**Solve**] gives the optimal values.

3.8.3 The Choice of Technology

We now turn to the illustration of the choice of technology in Section 3.5.2. Use is made of the Japan–US international IO table (Section 3.1.4.1), which can be obtained from the Internet:

```
http://www.meti.go.jp/english/statistics/tyo/kokusio/
excel/wio2000-27sec.xls
```

3.8.3.1 Setting Up the Problem

Following (3.33), write the balancing equations as follows

$$
\begin{aligned}
x^J &= A^{JJ}x^J + A^{JU}x^U + f^{JJ} + f^{JU} + x^{JR} \\
x^U &= A^{UJ}x^J + A^{UU}x^U + f^{UJ} + f^{UU} + x^{UR} \\
x^R &= A^{RJ}x^J + A^{RU}x^U + f^{RJ} + f^{RU},
\end{aligned}
\tag{3.139}
$$

where

$$
A^{rs} = X^{rs}(\hat{x}^s)^{-1} \quad (r = J, U, R, \ s = J, U)
\tag{3.140}
$$

Assuming that a U.S. sector and the corresponding Japanese sector produce a homogeneous product, we obtain the 27×27 matrices of input coefficients for Japan and the U.S., A^J and A^U, as

$$
A^J = A^{JJ} + A^{UJ}, \quad A^U = A^{JU} + A^{UU}.
\tag{3.141}
$$

Abstracting away from issues of location and transport for simplification, we assume that for each sector producing jth product two processes, P_j^J and P_j^U, are available for choice:

$$
P_j^J = \begin{pmatrix} (A^J)_{\cdot j} \\ (A^{RJ})_{\cdot j} \end{pmatrix}, \quad
P_j^U = \begin{pmatrix} (A^U)_{\cdot j} \\ (A^{RU})_{\cdot j} \end{pmatrix}.
\tag{3.142}
$$

The extended form of (the first two equations of) (3.139) allowing for the possibility of substitution, that is, (3.128), then becomes

$$
x^J + x^U - A^J x^J - A^U x^U = f^J + f^U,
\tag{3.143}
$$

where

$$
f^J = f^{JJ} + f^{JU} + x^{JR}, \quad f^U = f^{UJ} + f^{UU} + x^{UR}.
\tag{3.144}
$$

As the objective to be optimized, we consider minimization of the input of 'Petroleum and coal products' (the product of sector 10). The optimization problem (LP) then becomes

$$
\begin{vmatrix}
\text{minimize} & c^J x^J + c^U x^U \\
\text{subject to} & x^J + x^U - A^J x^J - A^U x^U = f^J + f^U, \\
& x^J \geq 0, \ x^U \geq 0
\end{vmatrix}
\tag{3.145}
$$

where

$$c^J = (A^J)_{10.} + (A^{RJ})_{10.}, \quad c^U = (A^U)_{10.} + (A^{RU})_{10.} \tag{3.146}$$

3.8.3.2 Applying Solver to the Problem

1. Assign the names Xjj, Xju, Xuj, Xuu, Xrj, Xru, Fj, Fu, Xj, and Xu to the ranges where X^{JJ}, X^{JU}, X^{UJ}, X^{UU}, X^{RJ}, X^{RU}, f^J, f^U, x^J, and x^U are stored.
2. Enter $=(Xjj+Xuj)/TRANSPOSE(Xj)$ into D104:AD130 as an array formula to compute A^J, and assign the name Aj to the result.

	A	C	D	E
103	Input coefficient matrix		Aj	
104	1 Agriculture		=(Xjj+Xuj)/TRANSPOSE(Xj)	
105	2 Forestry		0.0018	0.0000
106	3 Fishing		0.0000	0.0000
107	4 Mining		0.0001	0.0002

Similarly, compute A^U and assign the name Au to the result.
3. Compute c^J in D134:AD134 by entering
   $=INDEX(Xjj+Xuj+Xrj,10,0)/TRANSPOSE(Xj)$
   as an array formula, and assign the name Cj to the result.

	A	C	D	E	F	G	H
133	Objective function		Cj				
134	10 Petroleum and coal products		=INDEX(Xjj+Xuj+Xrj,10,0)/TRANSPOSE(Xj)				0.0050
135							
136							

Similarly, compute c^U and assign the name Cu to the result.
4. Choose BH104:BI130 as the area for the decision variables (of order 27×2), assign the names Xj_var and Xu_var to the two columns, and enter arbitrary initial values into them.
5. Enter the following as an array formula
   $=Xj\_var+Xu\_var-MMULT(Aj,Xj\_var)-MMULT(Au,Xu\_var)$
   into BK104:BK130 to obtain the left-hand side of (3.143).

	BH	BI	BJ	BK	BL	BM	BN
103	Xj_var	Xu_var		LHS	RHS		
104	1	1		=Xj_var+Xu_var-MMULT(Aj,Xj_var)-MMULT(Au,Xu_var)			
105	1	1		2			
106	1	1		2			
107	1	1		1			
108	1	1		2			
109	1	1		1			
110	1	1		1			
111	1	1		2			

6. Enter =Fj+Fu into BL104:BL130 as an array formula to obtain the right-hand side of (3.143).

	BH	BI	BJ	BK	BL
103	Xj_var	Xu_var		LHS	RHS
104	1	1		1	=Fj+Fu
105	1	1		2	68,673
106	1	1		2	167,639
107	1	1		1	274,210
108	1	1		2	5,977,653
109	1	1		1	1,149,859
110	1	1		1	1,155,903
111	1	1		2	649,067

7. Enter the objective function =MMULT(Cj,Xj_var)+MMULT(Cu,Xu_var) into BN104:

	BK	BL	BM	BN	BO	BP
103	LHS	RHS		Obj val		
104	1	754,938		=MMULT(Cj,Xj_var)+MMULT(Cu,Xu_var)		
105	2	68,673				
106	2	167,639				
107	1	274,210				

8. Arrange the parameters of Solver as follows

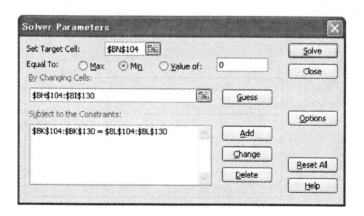

9. Check the 'Assume Non-Negative' option because all the decision variables are nonnegative.

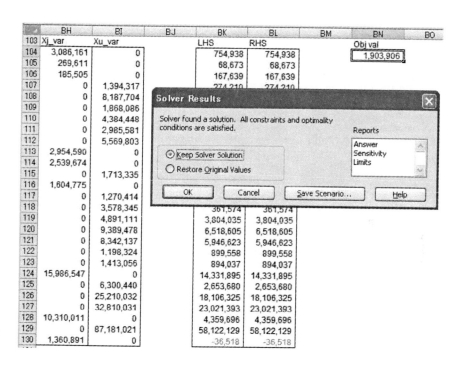

The following shows the results

The results indicate that in all the producing sectors either the US or the Japanese process was chosen, a typical result when there are no constraints on available productive capacity. The Japanese process was chosen for 'Agriculture', 'Forestry', 'Fishing', 'Petroleum and coal products', 'Plastic, rubber and leather products', 'Steel and steel products', 'Construction', 'Transport', and 'Unclassified, etc.', while the U.S. process was chosen for the remaining 18 sectors. It was found that the optimization reduces the industrial demand for 'Petroleum and coal products' to 80% of the original level.

References

1. Batey, P. W., & Rose, A. Z. (1990). Extended input-output models: Progress and potential. *International Regional Science Review, 13*(1), 27–49.
2. Bazzazan, F., & Batey, P. (2003). The development and empirical testing of extended input-output price models. *Economic Systems Research, 15*(1), 69–86.
3. Chenery, H. (1953). Regional analysis. In H. Chenery, P. Clark, & V. Pinna (Eds.), *The structure and growth of the Italian economy* (pp. 91–129). Rome: US Mutual Security Agency.
4. Dietzenbacher, E. (1997). In vindication of the Ghosh model: A reinterpretation as a price model. *Journal of Regional Science, 37*(4), 629–651.
5. Duchin, F. (2005). A world trade model based on comparative advantage with m regions, n goods, and k factors. *Economic Systems Research, 17*(2), 141–162.
6. European Topic Center on Resource and Waste Management. http://waste.eionet.europa.eu/definitions/waste
7. Ghosh, A. (1959). Input-output approach in an allocation system. *Economica, 25,* 58–64.
8. Heijungs, R., & Suh, S. (2002). *The computational structure of Life Cycle Assessment.* Dordrecht: Kluwer.
9. Hertwich, E. (2006). Consumption and the rebound effect: An industrial ecology perspective. *Journal of Industrial Ecology, 9*(1–2), 85–98.
10. Hoekstra, R. (2005). *Economic growth, material flows and the environment.* Cheltenham: Edward Elgar.
11. Houthakker, H. S. (1957). An international comparison of household expenditure patterns, commemorating the centenary of Engel's law. *Econometrica, 25*(4), 532–551.
12. Isard, W. (1951). Interregional and regional input-output analysis: A model of a space-economy. *Review of Economics and Statistics, 33*(4), 318–328.
13. Jorgenson, D. W. (1961). Stability of a dynamic input-output system. *Review of Economic Studies, 28,* 105–116.
14. Kurz, H., Dietzenbacher, E., & Lager, C. (Eds.). (1998). *Input-output analysis* (Vol. 1). Cheltenham: Edward Elgar.
15. Leontief, W. (1951). *The structure of American economy 1919–1939.* New York: Oxford University Press.
16. Leontief, W. (1953). *Dynamic analysis, in studies in the structure of the American economy* (Chapter 3, pp. 53–90). New York: Oxford University Press.
17. Lenzen, M. (1998). Primary energy and greenhouse gases embodied in Australian final consumption: An input-output analysis. *Energy Policy, 26*(6), 495–506.
18. Luenberger, D. (2003). *Linear and nonlinear programming* (2nd ed.). Boston: Kluwer.
19. Ministry of Economy, Trade and Industry, The Government of Japan http://www.meti.go.jp/english/statistics/index.html
20. Miyazawa, K. (1968). Input-output analysis and interrelational income multipliers as a matrix. *Hitotsubashi Journal of Economics, 18,* 39–58 (English translation of the original Japanese publication in 1963).
21. Nansai, K., Inaba, R., Kagawa, S., & Moriguchi, Y. (2008). Identifying common features among household consumption patterns optimized to minimize specific environmental burdens. *Journal of Cleaner Production, 16,* 538–548.
22. Miller, R., & Blair, P. (1985). *Input-output analysis.* Englewood Cliffs, NJ: Prentice-Hall.
23. Moses, L. (1955). The stability of interregional trading patterns and input-output analysis. *American Economic Review, 45*(5), 803–826.

24. Oosterhaven, J. (1988). On the plausibility of the supply driven input-output model. *Journal of Regional Science, 28*(2), 203–217.
25. Rogoff, K. (1996). The purchasing power puzzle. *Journal of Economic Literature, 34*, 647–668.
26. Simpson, D., & Tsukui, J. (1965). The fundamental structure of input-output tables, an international comparison. *Review of Economics and Statistics, 47*(4), 434–446.
27. Statistics Directorate, OECD OECD statistics on Purchasing Power Parities (PPP). http://www.oecd.org/department/0,3355,en_2649_34357_1_1_1_1_1,00.html
28. Stone, R. (1961). *Input-output and national accounts.* Paris: The Organization for European Economic Development.
29. ten Raa, T. (2005). *Input-output analysis.* Cambridge: Cambridge University Press.
30. The Institute of Developing Economies (IDE) http://www.ide.go.jp/English/Publish/Books/Sds/
31. Thiesen, J., Christensen, T.S., Kristensen, T.G., Andersen, R.D., Brunoe, B., Gregersen, T.K., Thrane, M., & Weidema, B.P. (2008). Rebound effects of price differences. International Journal of Life Cycle Assessment, *13*(2), 104–114.
32. Weisz, H., & Duchin, F. (2006). Physical and monetary input-output analysis: What makes the difference? *Ecological Economics, 57,* 534–541.
33. Wonnacott, R. J. (1961). *Canadian-American dependence: An interindustry analysis of production and prices.* Amsterdam: North-Holland Press.

Chapter 4
Microeconomic Foundations

Abstract This chapter provides some basics of microeconomics that are closely related to IOA. First, some basic concepts of great relevance to IOA in microeconomics, such as the concept of production- and cost functions, will be discussed. When it comes to practical application, general concepts need to be specified in order to be quantitatively implemented to real data. Issues related to the specification of technology in microeconomics constitute the second topic of this chapter. Some remarks on CGE (the Computable General Equilibrium model) with regard to its specification of technology will also be made. Finally, the microeconomic characteristics of IOA will be derived.

4.1 Introduction

Nowadays, IOA is widely used in areas, such as LCA and IE, the origins of which are only remotely, if at all, connected to the traditional area of economics. IOA, on the other hand, does have its origins in economics, and is deeply rooted in microeconomics in particular. Basic knowledge of the microeconomic foundations of IOA will therefore be useful for a better understanding of its underlying features. This will particularly be the case when one tries to generalize the basic framework of IOA to accommodate topics of interest which may not be adequately dealt with within the existing framework. Among others, the assumption of fixed coefficients may appear unduly strong. Whilst a straightforward extension of the standard IOA to accommodate this issue was discussed in Section 3.5.2, an alternative approach, known as applied or computable general equilibrium (CGE) modeling, is also available, one that is more closely related to the standard settings of neoclassical economics.

This chapter is devoted to providing some basics of microeconomics that are closely related to IOA. First, some basic concepts of great relevance to IOA in microeconomics, such as the concept of production- and cost functions, will be discussed. When it comes to practical application, general concepts need to be specified in order to be quantitatively implemented to real data. Issues related to the specification of technology in microeconomics constitute the second topic of this chapter. Some remarks on CGE with regard to its specification of technology will also be made. Finally, as a summary of the whole discussion, the microeconomic characteristics of IOA will be derived.

Readers who are interested in the treatment of IOA in more general settings of economic theory are advised to consult [27]. On the other hand, readers with little interest in economics can skip this chapter without losing the integrity of the content.

4.2 Representation of Technology

In economics, the relationship between inputs and outputs, such as (2.6) and (2.160), is called the factor (input) demand function. Technical relationships underlie the input-output relationships depicted by the demand function. This section deals with basic approaches in economics to representing technical input-output relationships.

4.2.1 Technology and Production Function

Fundamental to the microeconomics of production (or the theory of the firm) is the mathematical representation of technical input-output relationships of production processes. In an LCA, this representation is usually given by a unit process. The microeconomics of production provides a more general representation, to which we now turn.

4.2.1.1 Production Possibility Sets

The starting point of the microeconomics of production is the notion of production possibility set T which represents the totality of all possible combinations between inputs and outputs that are available for a producer. For the sake of simplicity, first consider the case of two inputs and one output, the quantities of which are respectively denoted by $z = (z_1, z_2)$ and y:

$$T = [y, z \,|\, z \text{ can produce } y > 0].$$ (4.1)

This set is assumed, among others, to satisfy the following conditions (for more rigorous treatments, see [9])

- [Exclusion of lotus land or mass balance]
  No output is obtained from no input:

$$(0, 0, 0) \in T$$ (4.2)

- [Free disposal]
  An increase in input does not decrease output. Frequently, this condition is also
  termed the monotonicity condition or free disposal:

$$\text{If } (y', z') \in T, \text{ then } (y', z'') \in T \text{ for } z'' \geq z'. \tag{4.3}$$

Each element of T, say, $T^i = (y^i, z^i_1, z^i_2)$, represents a particular process, an example
of which is the process in Table 2.1.

Note that T consists of processes with different efficiency. Let $T^0 = (y^0, z^0_1, z^0_2)$
and $T^1 = (y^0, z^1_1, z^1_2)$ be its two elements. Process T^0 is then said to be more efficient
than the process T^1 if the former can produce the same amount of output by using
smaller amount of inputs, that is, $z^1_i \geq z^0_i$, $i = 1, 2$, with the strict inequality holding
for at least one of the inputs.

4.2.1.2 Production Functions

The subset of T that refers to its efficient combinations of inputs and output corresponds to the concept of production function. A production function, say f, gives
the maximum amount of output that can be obtained under a given technology from
a given bundle of z:

$$f(z) = \max[y \,|\, (y, z) \in T]. \tag{4.4}$$

The function f has the following properties:

- [No input implies no output]

$$f(0) = 0. \tag{4.5}$$

  This follows from (4.2).
- [Monotonicity] Increase in input does not decrease output.

$$f(z') \geq f(z'') \text{ for } z' \geq z''. \tag{4.6}$$

  This follows from (4.3).
- [Quasi-concavity] Let $V(y)$ be the set of all possible combinations of z that can
  produce y, that is, $V(y) = \{z \,|\, f(z) \geq y\}$. Then

$$V(y) \text{ is convex for all } y > 0 \tag{4.7}$$

If z' and z'' belong to $V(y)$, their linear combination $tz' + (1-t)z''$ with $0 \leq t \leq 1$
also belongs to $V(y)$.

Simply stated, the last condition states that if two processes can produce a given output, their linear combinations can produce it as well. This is depicted in Figure 4.1
where the shaded area refers to $V(y)$. The southwest frontier of the set represents
the set of the most efficient input bundles to produce a given level of output, and is
called the isoquant. Formally, this is a subset of the boundary of the set $V(y)$ defined
by ([25], p.17)

Fig. 4.1 The Set of Technical Possibilities, and the Isoquant: the Case of an Infinite Number of Processes. The Shaded Area Represents all the Possible Combinations of z_1 and z_2 to Produce y, That Is, $V(y)$ in (4.7). The Southwest Frontier of the Combinations Represents the Set of Most Efficient Combinations, the Isoquant Curve.

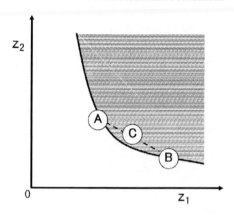

Fig. 4.2 The Set of Technical Possibility, $V(y)$, and the Isoquant: The Case of a Finite Number of Processes.

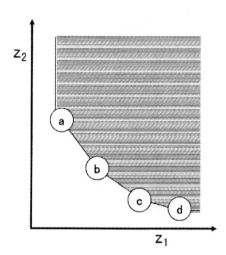

$$\{z|z \geq 0, z \in V(y), \lambda z \notin V(y), \text{ for } 0 \leq \lambda < 1\}. \tag{4.8}$$

Because $V(y)$ is convex, the isoquant curve is convex to the origin. If the points A and B happen to lie on the same isoquant, then their linear combination C must also be able to produce y, and hence must lie in the shaded area. A function f with the property (4.7) is called a quasi-concave function. This explains the naming of the last property (for more details of quasi-concavity, see [25], especially p.297).

Figure 4.1 refers to the case of a continuous isoquant curve where there are an infinite number of alternative processes to choose from, a case usually supposed in economics for the ease of mathematical convenience. In LCA and IE, however, it is usual to consider more realistic cases where there are only a finite number of processes, and hence the isoquant has kinks, and is not continuous (Figure 4.2). This is more realistic because each finite point on the isoquant can be associated with a specific production process.

4.2.2 Cost and Input Demand Functions

An isoquant such as in Figures 4.1 and 4.2 represents the set of equally efficient alternative processes from which a choice should be made. In order to make a choice, criteria of the choice or underlying behavioral assumptions are needed. The standard assumption in economics that is widely used as the criterion is the minimization of production costs:

$$\text{Cost of production} = \sum_i p_i z_i \qquad (4.9)$$

to achieve a given level of output $y > 0$ for a given set of input prices, $p = (p_i)$, and technology, T or f. Several important implications, which are of relevance for empirical application, follow from this assumption.

4.2.2.1 Cost Functions

Minimization of (4.9) with respect to z under given y, p and f gives:

$$\min_z \{ \sum_i p_i z_i \, | \, f(z) \geq y \} = \sum_i p_i z_i^*(p,y) = C(p,y), \qquad (4.10)$$

where $z_i^*(p,y)$ is the demand function for input i, which gives the cost-minimizing quantity of input i, and C is the minimum cost function. Henceforth, C is simply called the cost function. The form of C depends on the form of f.

Major properties of C that follow from (4.10) include:

- [Homogeneity in p]: The cost is proportional to input prices,

$$C(\lambda p, y) = \lambda C(p,y), \quad \lambda > 0$$

This is obvious from (4.10). If C is differentiable with regard to p, this implies:

$$\sum_i \frac{\partial C}{\partial p_i} p_i = C(p,y) \qquad (4.11)$$

- [Increasing in y]: An increase in production does not decrease the cost

$$C(p, y') \geq C(p, y'') \text{ for } y' \geq y''. \qquad (4.12)$$

This follows from the condition that f is an increasing function.
- [Concavity]: For given $p' > 0$, $p'' > 0$, and $0 \leq t \leq 1$:

$$C(tp' + (1-t)p'', y) \leq tC(p', y) + (1-t)C(p'', y) \qquad (4.13)$$
$$= C(tp', y) + C((1-t)p'', y) \qquad (4.14)$$

The inequality in (4.13) follows from the condition of minimization. Its left-hand side gives the minimum cost for the price level $p^* = tp' + (1-t)p''$. Cost minimization implies that this is not larger than the sum of the minimum costs at tp' and at $(1-t)p''$ given by the right-hand side: the minimum cost for a linear combination of prices is not larger than the linear combination of minimum costs at respective prices. If C is twice differentiable, this implies the negative definiteness of its Hessian $H = [\partial^2 C/\partial p_i \partial p_j]$:

$$\theta^\top H \theta \le 0, \quad \text{where } \theta \text{ is a vector.} \tag{4.15}$$

The equality in (4.14) follows from the homogeneity condition (4.11).

4.2.2.2 Shephard's Lemma and Input Demand Functions

When the cost function C is differentiable with respect to p, the following holds

Theorem 4.1 (Shephard's lemma [24]).

$$\frac{\partial C}{\partial p_i} = z_i^*(p, y). \tag{4.16}$$

Proof. The partial derivative of C with respect to p_i gives

$$\frac{\partial C}{\partial p_i} = \frac{\partial (\sum_i p_i z_i^*)}{\partial p_i} = z_i^* + \sum_j p_j \frac{\partial z_j^*(p, y)}{\partial p_i} \tag{4.17}$$

Multiplying both sides by p_i and summing up, we obtain;

$$\sum_i \frac{\partial C}{\partial p_i} p_i = \sum_i p_i z_i^* + \sum_i p_i \sum_j p_j \frac{\partial z_j^*(p, y)}{\partial p_i} = \sum_i p_i z_i^* \tag{4.18}$$

where the last equality results from (4.11). Because this has to hold for any p, the second term of the expression in the middle has to be zero:

$$\sum_j p_j \frac{\partial z_j^*(p, y)}{\partial p_i} = 0 \tag{4.19}$$

and hence the last term on the far right of (4.17) diminishes, and reduces to (4.16).

This is an extremely useful result, especially for empirical application, which states that for a given cost function its derivative with regard to an input price gives its cost minimizing demand function.

4.2.2.3 Homothetic Technology

For empirical application, it is usual to restrict the class of f or C to a smaller class by imposing further properties. Homotheticity is one such property, and is widely used in economics.

The technology is said to be homothetic if the production function f can be written as follows:

$$f(z) = g(h(z)), \tag{4.20}$$

where h is a homogeneous function of degree 1:

$$h(\lambda z) = \lambda h(z), \quad \lambda > 0 \tag{4.21}$$

and g is a monotone increasing function. Under homothetic technology, the slope of the isoquant will be the same along rays coming from the origin: given an isoquant for a certain level of output, the isoquant curves for different levels of output can simply be obtained by its proportional enlargement or reduction (Figures 4.3 and 4.4). When g is linear, f reduces to a function homogeneous of degree 1, and the technology is subject to constant returns to scale.

In order to derive implications of homothetic technology on the demand for inputs, it is helpful to first consider its implications for the cost function C. If f is

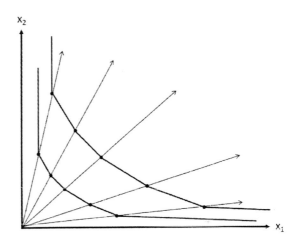

Fig. 4.3 Isoquants for Homothetic Production Function.

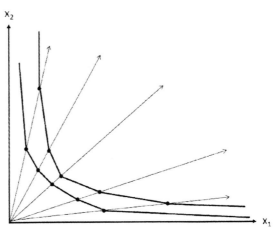

Fig. 4.4 Isoquants for Nonhomothetic Production Function.

homothetic, the associated cost function can be written as the product of a function that is independent of the level of output, and a function that is independent of input prices:

$$
\begin{aligned}
C(p,y) &= \min_z \left[\sum_i p_i z_i \,\middle|\, g(h(z)) \geq y \right] \\
&= \min_z \left[\sum_i p_i z_i \,\middle|\, h(z) \geq g^{-1}(y) \right] \\
&= g^{-1}(y) \min_z \left[\sum_i p_i z_i / g^{-1}(y) \,\middle|\, h(z/g^{-1}(y)) \geq 1 \right] \\
&= g^{-1}(y) \min_w \left[\sum_i p_i w_i \,\middle|\, h(w) \geq 1 \right] \\
&= g^{-1}(y) C(p,1) = g^{-1}(y) c(p)
\end{aligned}
\tag{4.22}
$$

where the second equality follows from the monotonicity of g (the inverse of g exists and is positive), the third equation follows from (4.21), and $w_i = z_i/g^{-1}(y)$ (divide both sides of the constraint by $g^{-1}(y)$). The function c is independent of y, and is called the unit cost function.

Application of (4.16) to (4.22) yields:

$$
z_i^* = g^{-1}(y) \frac{\partial c(p)}{\partial p_i}
\tag{4.23}
$$

and hence the cost minimizing proportion of inputs

$$
\frac{z_i^*}{z_j^*} = \frac{\partial c(p)}{\partial p_i} \left(\frac{\partial c(p)}{\partial p_j} \right)^{-1}
\tag{4.24}
$$

becomes independent of the level of output.

In the special case of f being homogeneous of degree 1 (constant returns to scale), we obtain

$$
C(p,y) = c(p)y.
\tag{4.25}
$$

Application of (4.16) then yields:

$$
\frac{z_i^*}{y} = \frac{\partial c(p)}{\partial p_i}
\tag{4.26}
$$

which states that the amount of input per output, or input coefficients, becomes independent of the level of output.

4.2.3 Duality Between Cost and Production Functions

Above, the cost function was derived from a given production function f via cost minimization (4.10). It is interesting to note that under fairly general conditions the opposite is also the case, that is, from a given C it is possible to derive the corresponding f via the following operation ([2], Theorem A.3):

$$f(z) = \max_{y} \left\{ y \mid \sum_i p_i z_i \geq C(p,y), \forall p > 0 \right\}. \tag{4.27}$$

This is known as the duality between cost and production functions.

This is a very useful result, because it enables one to derive input demand functions from any cost function which is consistent with the definition (4.10) by use of (4.16), that is, by simply computing the first derivatives. In contrast, derivation of the input demand functions from f via (4.10) requires solving a constrained optimization problem, which can be tedious.

For instance, suppose that the technology is subject to constant returns to scale, and that the unit cost function c is given by:

$$c(p) = \sum_i p_i a_i \tag{4.28}$$

where $a_i > 0$ are constants. From (4.16), it immediately follows:

$$\frac{z_i^*}{y} = a_i, \tag{4.29}$$

that is, the fixed coefficients input demand function (2.6) of the IOA is obtained. This implies that the technology underlying the IOA is given by the linear cost function (4.28). It is noteworthy that in fact the same functional form has already occurred above (see (2.43), (2.94), (2.95)) without resorting to cost minimization.

4.3 Specification of Technology

In economics, theoretical models are formulated in a highly general and abstract manner. This applies to the model of production as well, which implies that in order to conduct quantitative analysis one has to specify the functional form of the production or cost function. This section deals with issues related to the specification of production and/or cost functions.

4.3.1 Top-Down and Bottom-Up Approaches

For quantitative analysis, one has to specify T, f, or C. Two approaches can be distinguished: top-down and the bottom-up approaches. Under a top-down approach, one starts from the specification of f or C, without paying special attention to details of individual processes, that is, to individual elements of T. The major concern in this approach is to find out an appropriate functional form of f or C which is analytically convenient and does not place too much constraints on substitution possibilities, that is, the shape of the isoquant.

Exactly the opposite is the case for the bottom-up approach. Its main concern consists in the identification of individual processes, that is, individual elements of T, based on detailed technical information. Almost no attention is paid to the identification of a functional form of f or C that encompasses all the processes that lie above the isoquant curve.

The difference of the two approaches can best be characterized by comparing the continuous isoquant in Figure 4.1 with the noncontinuous isoquant in Figure 4.2. In the top-down approach, one is interested in the specification of a particular f or C that best approximates the isoquant, but is not interested in the technical details of processes A, B and C. On the other hand, in the bottom-up approach, one is concerned with the identification of individual processes, a, b, c, and d, but has little interest in the functional form of the isoquant itself: once the former is available, the latter can be constructed ex-post, while it is not certain if the resulting isoquant can be represented in terms of a simple functional form.

The top-down approach is one that is usually adopted in economics, whereas in LCA and IE the bottom-up approach is mostly used. The use of different approaches between the different disciplines can partly be attributed to historical difference in the degree of resolution that is required. In economics, one was typically concerned with highly aggregated issues like distribution of income between capital and labor at the national level, where the bottom-up approach cannot be effectively applied. On the other hand, the very nature of LCA and IE requires the use of a much higher degree of resolution, because a meaningful consideration of environmental issues needs a certain level of detail. With increased interests in applying economic models to environmental issues, however, this "division of labor" between the two disciplines has increasingly become blurred, an important example of which is the use of a CGE model such as the global trade analysis project (GTAP) model [12] or the Asia-Pacific integrated (AIM) model [17] to issues of global warming.

4.3.2 The Elasticity of Substitution Between Inputs

For a given production process, two inputs are said to be perfect substitutes if production is possible by using only one of them: Portland cement produced in Canada and the US will be perfect substitutes for construction purposes. In this case, the isoquant curve becomes a straight line (isoquant A in Figure 4.5). The opposite of this is the case of perfect complements, where no production is possible without using a fixed proportion of each of them: for a given process of iron production, coke and iron ore are perfect complements. In this case, the isoquant curve assumes a rectangular shape (isoquant B in Figure 4.5).

The elasticity of substitution is a measure of the degree of substitution: its value gives information about the curvature of the isoquant. Assuming constant return to scale, the elasticity of substitution, say σ, is given by ([14], p.86)

Fig. 4.5 Isoquant Curves
Corresponding to Different
Degree of Substitution.

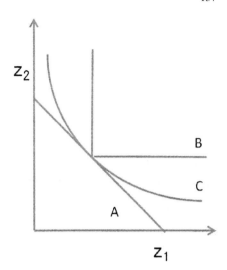

$$\sigma = \left(\frac{\partial f}{\partial z_1}\frac{\partial f}{\partial z_2)}\right) / \left(\frac{\partial^2 f}{\partial z_1 \partial z_2} y\right). \tag{4.30}$$

The inputs are perfect substitutes when $\sigma = \infty$, and perfect complements when $\sigma = 0$. In the case of perfect substitutes, the isoquant is a straight line, and $\sigma = \infty$ (the isoquant A in Figure 4.5). With a decrease in the degree of substitution, the curvature of the isoquant becomes sharper, and eventually becomes a rectangular with $\sigma = 0$ (the isoquant B in Figure 4.5). The isoquant C in Figure 4.5 refers to the situation in between with $0 < \sigma < \infty$.

Because of the duality between f and C, an alternative representation of σ in terms of C is also possible [9]:

$$\sigma = \frac{C}{\partial C/\partial p_1\, \partial C/\partial p_1} \cdot \frac{\partial^2 C}{\partial p_1 \partial p_2} \tag{4.31}$$

4.3.3 CES Functions

We now turn to some examples of functional forms which are frequently used in empirical application. The constant elasticity of substitution (CES) function [1] is presumably the most widely used functional form in the top-down approach.

4.3.3.1 The CES Production Function

For the case of constant returns to scale with two inputs and one output, the CES production function is given by

$$y = A\left[\theta_1(a_1z_1)^\rho + \theta_2(a_2z_2)^\rho\right]^{1/\rho}, \tag{4.32}$$

where $0 < \theta_i < 1, i = 1, 2$ are the share parameters with $\theta_1 + \theta_2 = 1$. The parameters A, a_1, and a_2 depend upon the units in which the output and inputs are measured and play no important role. The parameter $\rho \leq 1$ determines the degree of substitutability of the inputs. From (4.30) it follows:

$$\sigma = 1/(1-\rho). \tag{4.33}$$

The naming of the CES function originates from the fact that σ is a constant. Its popularity comes from the fact that no a-priori restriction is imposed on the value of σ. The first-order condition of the constrained optimization problem (4.10) with f given by (4.32) yields

$$(1/\rho)y^{1-\rho}\theta_i\rho a_i(a_iz_i^*)^{\rho-1} = \lambda p_i \tag{4.34}$$

where λ refers to the Lagrange multiplier. Division of the expression for input 1 with that for input 2 yields the following expression, which does not include λ:

$$\frac{\theta_1 a_1}{\theta_2 a_2}\left(\frac{a_1z_1^*}{a_2z_2^*}\right)^{\rho-1} = \frac{p_1}{p_2} \tag{4.35}$$

Solving for the proportion of the inputs results in:

$$\frac{z_1^*}{z_2^*} = \frac{a_2}{a_1}\left(\frac{\theta_2 a_2 p_1}{\theta_1 a_1 p_2}\right)^{\frac{1}{\rho-1}} \tag{4.36}$$

The two extreme cases are when $\rho = 1$ or $\rho \to -\infty$. If $\rho \to 1$, $\sigma \to \infty$, and (4.32) reduces to

$$y = \theta_1 a_1 z_1 + \theta_2 a_2 z_2. \tag{4.37}$$

In this case, a given amount of y can be obtained using only one of the inputs provided enough amount of it is available, which implies that these inputs are perfect substitutes.

On the other hand, if $\rho \to -\infty$, then $\sigma \to 0$, and the inputs are complements, which cannot be substituted for each other. In this case, f can be given by

$$y = A\min_z\{a_1z_1, a_2z_2\}. \tag{4.38}$$

The level of output is then determined or limited by the least level of a_iz_i. This resembles Liebig's Law of the Minimum in agricultural science, which states that growth is controlled not by the total of resources available, but by the scarcest resource. For the cost minimizing combination of the inputs, we would have (assuming $A = 1$ without loss of generality)

$$y = a_1z_1^* = a_2z_2^* \tag{4.39}$$

and hence

$$z_i^* = \frac{y}{a_i} \qquad (4.40)$$

which corresponds to the fixed coefficient input demand function of IOA (2.55). The production function underlying the IOA is thus given by (4.38).

If $\rho = 0$, $\sigma = 1$, and the CES function reduces to the Cobb–Douglas function (this result makes use of L'Hôpital's rule; see [14] for details)

$$y = Az_1^{a_1}z_2^{a_2} \qquad (4.41)$$

The Cobb–Douglas production function has been used in economics since its introduction by Cobb and Douglas [4] over 80 years ago, and is still today one of the most frequently used forms.

4.3.3.2 The CES Cost Function

The cost function that is dual to the CES production function (4.32) is again a CES function:

$$c(p_1, p_2) = A^{-1} \left(t_1^\sigma (p_1/a_1)^{1-\sigma} + t_2^\sigma (p_2/a_2)^{1-\sigma} \right)^{1/(1-\sigma)} \qquad (4.42)$$

The fact that this is dual to (4.32) and represents the same technology, can easily be confirmed by deriving the associated input demand function, and comparing it with the result obtained from the production function (4.36). Applying (4.16) to (4.42), the input demand is given by

$$z_i^* = \frac{\partial c}{\partial p_i} = A^{-1}c^\sigma (t_i/a_i)(p_i/a_i)^{-\sigma} \qquad (4.43)$$

Hence, the proportion of the inputs is given by

$$\frac{z_1^*}{z_2^*} = \frac{t_1 a_2}{t_2 a_1} \left(\frac{p_1 a_2}{p_2 a_1} \right)^{-\sigma} \qquad (4.44)$$

which is (4.36) with $t_i = \theta_i^\sigma$. The duality between (4.32) and (4.42) has thus been confirmed.

4.3.3.3 CES Function with $n > 2$ Inputs

Up to now, the number of inputs has been limited to two. In reality, the number of inputs is considerable. Recall that even the simple example in Table 2.5 involves five inputs. It turns out, however, that a simple extension of the CES function to the case of more than two inputs is subject to some difficulty. A straightforward extension to the case of n inputs (A is omitted for the sake of simplicity)

$$c(p) = \left[\sum_i t_i (p_i/a_i)^\delta \right]^{1/\delta} \tag{4.45}$$

imposes the equality of σ among all the inputs, which is a strong condition, and may be unrealistic.

Its more general version

$$c(p) = \left[\sum_i t_i (p_i/a_i)^{\delta_i} \right]^{1/\delta} \tag{4.46}$$

may be free of this shortcoming, but fails to satisfy the homogeneity condition (4.11) unless $\delta_i = \delta$, and hence cannot be regarded as a proper cost function. Its production function counterpart:

$$y = \left[\sum_i \theta_i (a_i z_i)^{\rho_i} \right]^{1/\rho}, \tag{4.47}$$

cannot be homogeneous of degree 1 in z unless all ρs are equal, and hence cannot represent a constant returns to scale technology.

4.3.4 Flexible Functional Forms

The sudden and significant rise in petroleum prices in the 1970s caused by oil embargoes had profound effects on wide areas of society. One of its noticeable impacts on economics was an acknowledgment of the need for the incorporation of energy as an additional factor of production besides the traditional ones such as capital and labor. With this background, a number of so-called flexible functional forms (FFF) were developed, with the aim of finding a functional form that overcomes the restrictive properties of the CES function. Another important factor that contributed to this development was an advancement in the knowledge of duality theory between production and cost functions. In particular, the application of Shephard's lemma (4.16) to a flexible cost function was instrumental in that it has made possible an easy derivation of input demand functions without bothering about the elimination of the Lagrange multiplier, which is frequently the case when one tries to solve the constrained minimization problem (4.10) for a given form of production function directly.

4.3.4.1 Flexible Functional Forms

FFFs are characterized by the property that they can provide a second-order approximation to an arbitrary twice differentiable function satisfying, at least locally, the appropriate regularity conditions [2]. They place no a-priori restriction on the value of elasticity of substitution, and can accommodate any number of inputs. Widely used FFFs are the generalized Leontief function due to Diewert [5] and the translog function due to Christensen et al. [3].

The Generalized Leontief Function

For the case of constant returns to scale technology, the generalized Leontief (GL) unit cost function is given by:

$$c(p) = \sum_{i,j} b_{ij} \sqrt{p_i} \sqrt{p_j}, \quad b_{ij} = b_{ji} \tag{4.48}$$

The term "generalized Leontief" originates from the fact that when $b_{ij} = 0, \forall i \neq j$, this form reduces to

$$c(p) = \sum_i b_{ii} p_i \tag{4.49}$$

that is, to the linear cost function (4.28). Application of (4.16) yields:

$$z_i^*/y = b_{ii} + \sum_j b_{ij} \sqrt{p_j/p_i}. \tag{4.50}$$

The (i, j) element of the Hessian (4.15) of the GL cost function is

$$(H)_{ij} = \begin{cases} -\frac{1}{2} b_{ii} p_i^{1/2} p_j^{-3/2}, & \text{for } i = j \\ \frac{1}{2} b_{ij} p_i^{-1/2} p_j^{-1/2}, & \text{for } i \neq j \end{cases} \tag{4.51}$$

A sufficient condition for global concavity (concavity at any $p \geq 0$) of GL is $b_{ij} \geq 0$, $\forall i \neq j$ [21].

The Translog Function

The translog (TL) unit cost function is given by

$$\ln c(p) = a_0 + \sum_i a_i \ln p_i + \frac{1}{2} \sum_{i,j} b_{ij} \ln p_i \ln p_j, \quad b_{ij} = b_{ji}, i \neq j, \tag{4.52}$$

where a_i and b_{ij} are subject to the following homogeneity conditions:

$$\sum_i a_i = 1, \quad \sum_i b_{ij} = 0 \tag{4.53}$$

This functional was used by Hudson and Jorgenson [15] with the explicit purpose of determining the input coefficients endogenously as functions of the prices of products of all sectors, the prices of labor and capital services, and the prices of competing imports. Applying the logarithmic version of (4.16):

$$\frac{\partial c}{\partial p_i}\frac{p_i}{c} = \frac{\partial \ln c}{\partial \ln p_i} = \frac{p_i z_i^*}{cy} \tag{4.54}$$

to (4.52) results in the input demand functions in the form of cost share equations:

$$s_i(p) = \frac{p_i z_i^*}{\sum_j p_j z_j^*} = a_i + \sum_i b_{ij} \ln p_j, \quad i = 1, \cdots, n \tag{4.55}$$

where $s_i(p)$ is the ith input's share in cost.

The (i, j) element of the Hessian (4.15) of the TL cost function is

$$(H)_{ij} = \begin{cases} (b_{ii} - s_i + s_i^2)\dfrac{c}{p_i^2}, & \text{for } i = j \\[2ex] (b_{ij} + s_i s_j)\dfrac{c}{p_i p_j}, & \text{for } i \neq j \end{cases} \tag{4.56}$$

The TL cost function is locally concave if, assuming $s_i \geq 0$ for all i, the matrix $G = (g_{ij})$ with g_{ij} referring to the (i, j) element inside the parenthesis on the right of (4.56) is negative semidefinite, while it is global concave if the matrix $B = (b_{ij})$ itself is negative semidefinite [6].

Implementation of Flexible Functional Forms

Because of their characteristic as a second-order approximation to an arbitrary production or cost function, FFFs such as (4.48) or (4.52) may appear as the ideal functional form to be used in the top-down approach to the specification of technology. When it comes to practical application, however, the full exploitation of this characteristic is hampered by the very fact that the parameters of FFFs are in general unknown, and have to be estimated from whatever data are available by use of statistical regression (econometric) techniques. Econometric estimation of this sort of model is characterized by notorious shortages of the degree of freedom due to the small number of time series observations, a mechanical extension of which is at risk of ignoring possible change in the parameters over time. The presence in FFFs of the huge number of parameters to be estimated, which originates from their very nature of being a second-order approximation, further deteriorates this problem.

To circumvent this problem, therefore, it is necessary to substantially reduce the number of unknown parameters by introducing a-priori restrictions (information). For FFFs, the option of reducing unknown parameters by use of process knowledge is almost nonexistent because by its very nature it is almost impossible to establish relationships between their parameters and concrete technical knowledge on processes: otherwise, one would not resort to a second order approximation of an unknown isoquant. The only applicable option then will be to impose zero restrictions on some parameters, which is usually done by introducing a set of nested structures, under which individual inputs can be grouped into several groups of inputs like material or energy [15] (see Section 4.3.5 for further details on this point).

It follows that when implemented to real data, FFFs cannot be as flexible as they claim to be. In determining the form of a particular nested structure to be imposed, it is usual to pay very little, if any, attention to process information. It is mostly determined from the point of view of analytical convenience. Furthermore, as will be shown in the next section, the FFFs such as the generalized Leontief and translog functions are not free of theoretical drawbacks, which make them less ideal than they appear. In this regard, the following observation once made by Leontief more than 50 years ago still seems valid:

> Theoretical economists deal with production functions in their quite general form. More specific characteristics, if introduced at all, take the form of hypothetical assumption rather than systematically observed and measured facts. ([20], p.37)

4.3.5 Tree Structure of Technology: Separability

Replacement of the fixed coefficients model with more general production functions allowing for the possibility of substitution has been one of the central concerns of CGE models [11]. The CES function with two inputs is the functional form that is most widely used in CGE models.

It is noteworthy that in spite of the superiority of FFF's over CES functions in terms of generality, the FFFs are seldom used in CGE models. The above-mentioned difficulty associated with obtaining reliable estimates of their unknown parameters is one reason for this. Another more fundamental reason is the fact the FFFs, while capable of providing a second order *local* approximation to an arbitrary production or cost function, are not able to satisfy the theoretical conditions of a production or cost function *globally* without imposing further restrictions, which can deprive them of the very nature of flexibility.

For instance, the translog cost function cannot satisfy the concavity condition (4.15) globally, unless the matrix of its second order parameters is negative semidefinite (see (4.56)), which implies nonpositiveness of b_{ii} for all i. Accordingly, if initial econometric estimation of the model yields a positive estimate of b_{ii}, it has to be reestimated subject to a nonpositiveness constraint, say be imposing a zero to its value. Because of the homogeneity condition (4.53), however, putting zeros to some b_{ii}'s can result in imposing zero restrictions on b_{ij}'s as well. For instance, if all b_{ii}'s are put to zero, all the b_{ij}'s will also be put to zero, which reduces the TL function to the Cobb–Douglas function (4.41) (for an example of studies where zero restrictions are widely imposed, see [16]). The GL cost function can be globally concave if all the off-diagonal second-order parameters are nonnegative, which excludes the presence of complementarity among inputs.

In CGE models, it is usual to consider dozens of producing sectors. The latest version of GTAP, for instance, deals with about 50 production sectors. In addition to these are other inputs such as labor and capital services. It is obvious that a CES function with two inputs is not applicable to CGE models in its original form. Recall from Section 4.3.3.3 that a simple extension of the CES function to the case of more than two inputs is not a meaningful option. In CGE models, this problem is coped with by introducing a tree or nested structure of technology ([10], p.59).

4.3.5.1 Nested Structure

For illustration, consider the case of rice production with four inputs (2.1), and write the production function as:

$$\underbrace{y}_{\text{rice}} = f(\underbrace{x_1}_{\text{rice seeds}}, \underbrace{z_1}_{\text{water}}, \underbrace{z_2}_{\text{labor}}, \underbrace{z_3}_{\text{land}}). \tag{4.57}$$

Suppose that there are functions F, g_A, and g_B, which are subject to (4.5), (4.6), (4.7), and constant returns to scale, such that (4.57) can be written as

$$y = F[g_A(x_1, z_1), g_B(z_2, z_3)] = F(A, B). \tag{4.58}$$

where g_A and g_B are subproduction (or aggregation) functions that respectively give the aggregate of rice and water, and the aggregate of labor and land, and F is the production function in terms of these aggregates. In this case, the production function f is said to be (homothetic) separable in (x_1, z_1) and (z_2, z_3). (The concept of separability was independently introduced into economics by Sono [26] and Leontief [19] in the 1940s. See Blackorby et al. [2] for details of this concept).

An important implication of the separability in (4.58) is that the composition of the components of the aggregate A and the composition of the components of the aggregate B are mutually independent: the composition of rice and water is independent of the composition of labor and land, and vice versa. Apart from whether this condition is consistent with the reality of technology, it is necessary for the technology to be represented in terms of aggregates alone as in (4.58). Otherwise, a change in the composition of A, which keeps the quantity of A unchanged, can change the quantity of B via its effects on the composition of B. In that case, F will no longer be applicable for analytical purposes (see [2] for further details).

Because F, g_A, and g_B contain only two inputs for each, they can be specified by CES functions as follows:

$$F[g_A(x_1, z_1), g_B(z_2, z_3)] =$$
$$\left(c\left(ax_1^{\rho_1} + (1-a)z_1^{\rho_1}\right)^{\frac{\rho_3}{\rho_1}} + (1-c)\left(bz_2^{\rho_2} + (1-b)z_3^{\rho_2}\right)^{\frac{\rho_3}{\rho_2}}\right)^{\frac{1}{\rho_3}} \tag{4.59}$$

where a, b, and c are constants lying between zero and unity. This form is called the nested or two-stage CES function [22].

Imposition of the separability assumption reduces the number of unknown elasticity of substitution between input i and input j, $\sigma_{i,j}$. For (4.57) there are six unknown $\sigma_{i,j}$'s, but for (4.59) there are only three: σ_{x_1,z_1}, σ_{z_2,z_3}, and $\sigma_{A,B}$. What about the elasticity for the remaining combinations of inputs, say σ_{x_1,z_3}? Application of (4.30) to (4.59) gives the following expression for σ_{x_1,z_3}:

$$
\begin{aligned}
\sigma_{x_1,z_3} &= \left(\frac{\partial f}{\partial x_1}\frac{\partial f}{\partial z_3}\right) \Big/ \left(\frac{\partial f^2}{\partial x_1 \partial z_3}y\right) \\
&= \left(\frac{\partial A}{\partial x_1}\frac{\partial F}{\partial A}\frac{\partial B}{\partial z_3}\frac{\partial F}{\partial B}\right) \Big/ \left(\frac{\partial A}{\partial x_1}\frac{\partial F^2}{\partial A \partial B}\frac{\partial B}{\partial z_3}y\right) \\
&= \left(\frac{\partial F}{\partial A}\frac{\partial F}{\partial B}\right) \Big/ \left(\frac{\partial F^2}{\partial A \partial B}y\right) = \sigma_{A,B},
\end{aligned}
\tag{4.60}
$$

which implies that σ_{x_1,z_3} depends on the aggregates A and B, but not on its individual components. Accordingly, the following holds

$$
\sigma_{x_1,z_3} = \sigma_{z_1,z_3} = \sigma_{x_1,z_2} = \sigma_{z_1,z_2} = \sigma_{A,B}.
\tag{4.61}
$$

An alternative way of representing the nested structure is to use a tree diagram like Figure 4.6. The nested structure can be generalized to involve a larger number of inputs, and multiple stages. For instance, the AIM model [17], a CGE model developed by the National Institute of Environmental Studies (NIES), Japan, for the estimation of greenhouse gas emission and the assessment of policy options for its reduction, involves multiple stages, a schematic diagram of which is given in Figure 4.7. Similar tree diagrams are also found in GTAP models [11].

Empirical validity of the separability assumption depends upon the extent to which a condition like (4.60) holds in reality. In Figure 4.7, the aggregates 'capital' and 'energy' are separable from each other, which implies that the composition of 'energy' can be changed independently of the composition of 'capital'. The use of alternative fossil fuels, however, is likely to require the use of different types of equipment: LNG or petroleum cannot be applied to a coal-firing boiler, and vice versa. To the extent that this is the case, doubt is cast on the empirical validity of the separability in Figure 4.7.

The separability assumption has strong implications for the structure of technology, the empirical validity of which should be properly assessed. Uncritical use of

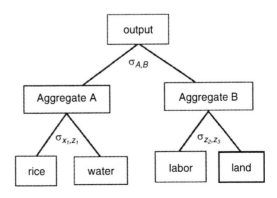

Fig. 4.6 A Tree Structure of Technology.

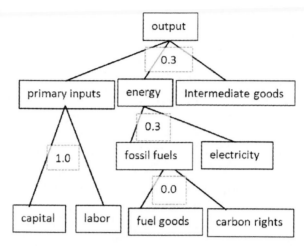

Fig. 4.7 The Tree Structure of the AIM Model. The Numbers in the Boxes with Broken Lines Indicate the Values of Elasticity of Substitution. Source: Own Drawing After Figure 2 in [17].

separability based on analytical convenience alone risks excluding important factors prior to conducting any empirical investigation. In this connection, the following remark about the practice in CGE models by Scrieciu [23] is worth noting:

> The CGE model builder tends to be satisfied with the choice of some specific functional forms and closure rules, and modifies the available representation of the real world instead of rejecting the model. In other words, specific functional forms or restrictions need to be employed in order to ensure a unique and stable equilibrium, though the economic realism of these restrictions has often been overlooked. ([23], p.660)

4.4 Technology in the Leontief IO Model

Based on results in the previous sections of this chapter, this section first characterizes the IOA in terms of its underlying microeconomic assumptions. This section also addresses the so-called substitution theorem, and some issues related to the implicit form of the supply and demand curves in IOA. A remark on the relevance of bottom-up and top-down approaches to IOA closes the section.

4.4.1 Characteristics of Technology in IOA

When $\sigma = 0$, the CES cost function (4.42) reduces to the linear cost function (4.28), with the demand function given by (4.29), which corresponds to the case of $\rho \rightarrow -\infty$ in the CES production function (4.32). The fixed coefficient input demand function used in IOA corresponds to the case where the technology is characterized by

Fig. 4.8 The Isoquant Curve
of the Production Function
Underlying IOA.

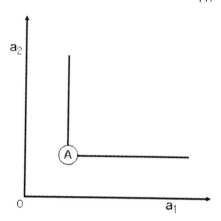

1. Constant return to scale and
2. The absence of substitution possibilities between inputs

In terms of the cost function, it is given by (4.28), while in terms of the production function it is given by (4.38).

From $\sigma = 0$, the input coefficients are independent of input prices. From the assumption of constant returns to scale, the input coefficients are independent of the level of production as well (see (4.26)). Because the level of input prices and production are the only arguments of the input demand function, it follows that the input coefficients are constants. Figure 4.8 shows the corresponding isoquant curve. At whatever level of input prices, the process at point **A** will be chosen.

4.4.2 Substitution Theorems

Exclusion of the possibility of substitution between alternative processes is at odds with the fundamental ideas of neoclassical (mainstream) economics where great emphasis is placed on the working of market mechanisms which manifest themselves through changes in prices. With the possibility of substitution among inputs excluded, a change in the prices of inputs is not able to invoke any change in the demand for inputs that could result in further changes in the relative prices of diverse goods and services. Given the practical importance of IOA, which can be traced back to the 1940s when the US Bureau of Labor Statistics initiated the compilation of an IO table, it is no surprise that substantial efforts have been spent by theoretically oriented (mainstream) economists to reconcile this odd situation with mainstream ideas.

One such endeavor that has attracted quite a few number of prominent economists, including several Nobel laureates, centers around the notion of a "substitution theorem", the purpose of which is to search for theoretical conditions under which "Substitution possibilities have not been explicitly introduced in his models by Leontief but are not incompatible with them" ([18], p.4).

Substitution theorems give conditions under which the input coefficients remain fixed in the presence of alternative processes. A standard version of the theorem [18] states that

Theorem 4.2 (nonsubstitution theorem).
If the following conditions are met, the cost minimizing technology is represented by given fixed coefficients that are independent of the size and composition of final demand f:

(a)   *The technology is subject to constant return to scale.*
(b)   *There is only one exogenous input with positive price, all other exogenous inputs have zero price.*
(c)   *There is no by-product.*

Proof. See [8] for a general proof.

For the purpose of illustrating the basic idea behind the substitution theorem, a proof for the simple case of one endogenous input in Section 2.1 is shown. Suppose that homogeneous labor corresponds to the exogenous input in the above conditions. Because there is only one product in this case, no consideration is needed of the presence of a by-product.

Let there be $m > 1$ types of alternative unit processes to choose from:

$$P = \left(P^{(1)} \; P^{(2)} \; \cdots \; P^{(m)} \right) \tag{4.62}$$

where each process is represented by a set of fixed input coefficients:

$$P^{(i)} = \begin{pmatrix} a_1^{(i)} \\ b_1^{(i)} \end{pmatrix} \tag{4.63}$$

From (2.19) the amount of labor required for process i is given by:

$$b_1^{(i)} (1 - a_1^{(i)})^{-1} \tag{4.64}$$

Because labor is the only exogenous input, one chooses from among the set of alternative processes the one that uses the minimum (direct and indirect) amount of it. Let process $P^{(1)}$ be the one with this property, that is, the most profitable process:

$$b_1^{(1)} (1 - a_1^{(1)})^{-1} < b_1^{(i)} (1 - a_1^{(i)})^{-1}, \quad i \neq 1 \tag{4.65}$$

Because the input coefficients are fixed, the profitability of processes does not depend on f.

It is important to note that the substitution theorem(s) do(es) not provide any theoretical justification to the use of fixed input coefficients in a practical sense. The extent to which the coefficients remain constant is a matter of empirical question but not of a theoretical one. What the theorems do is to provide theoretical conditions under which the coefficients can remain invariant to changes in the final demand in spite of the presence of alternative processes. In other words, to the extent that these conditions are not likely to be met in reality, there are no theoretical grounds to assume that the coefficients are constant. Of the three conditions mentioned, the one that appears least acceptable will be the second one where there is only one "scarce" exogenous input whose use is to be minimized. This is so because even in the simple case of one producing sector, land occurs besides labor as exogenous input with positive price (rent) (see Section 2.1.3).

When a second exogenous input with a positive price is present, a simple ordering of processes like (4.65) will no longer be possible. Because processes will differ in their requirements of labor and land, a change in the composition of final demand would have effects of different magnitudes on these inputs, which would result in changes in the relative prices thereof, and hence changes in the profitability of processes as well ([7], p.252). If the condition of constant returns to scale (condition (a)) does not hold, the processes to be chosen will depend on the size of final demand because the amount of inputs per unit of production, and hence the profitability, depends on the size of production. In the presence of by-products (a violation of condition (c)), certain processes might be ruled out because they produce products in proportions too different from the composition of final demand. With a change in the composition, these processes might become profitable ([7], p.252).

4.4.3 Supply and Demand Curves in IOA

In economics, one usually talks about the determination of the price and quantity of a product based on a downward-sloping demand curve and an upward-sloping supply curve, with the price and quantity of the product being simultaneously determined at the point of intersection of both the curves. The way prices and quantities of products are determined in IOA as introduced above is significantly different from this conventional view. The prices and quantities are determined not simultaneously but independently, the former by (2.45), where no quantities of products occur, and the latter by (2.15), where no prices of inputs occur.

A short introduction to the theoretical background of supply and demand curves seems useful to make clear the specific features of IOA that give rise to this independent determination of price and quantity. The slope of the demand function for an input refers to the effect of a change in its price on the demand for it. The magnitude of this effect can be obtained by further differentiation of (4.16) with respect to p_i:

$$\partial z_i^* / \partial p_i = \partial^2 C / \partial p_i^2 \leq 0 \tag{4.66}$$

where inequality follows from the concavity of C. It follows that the demand function for an input is in general downward sloping in its price. Under the assumption of fixed coefficients technology or linear cost function, however,

$$\partial z_i^* / \partial p_i = \partial^2 C / \partial p_i^2 = 0 \tag{4.67}$$

Hence, the demand function for an input is independent of input prices. Furthermore, the level of final demand f is exogenously determined from outside the model. Consequently, the demand curves for the inputs become independent of input prices.

As for the supply curve of a product, it is helpful to start from the underlying assumptions of economics which are responsible for making it upward-sloping or increasing in the price. The supply curve refers to the behavior of a supplier of a product to the level of the price of the product over which he or she has no control. Essential to the derivation of a supply curve are the following:

1. Behavioral assumption: the producers (suppliers) maximize profits, which are given by the difference between the revenue from the sale of the product (at the price of p_y) and the cost of production: $p_y y - C(p,y)$.
2. Competitive structure of the market: for individual suppliers the prices of products are given. In particular, p_y is independent of y. It then follows that the level of production y is the only variable with which suppliers can change the level of their profits.
3. Characteristics of technology: the production function is subject to decreasing returns to scale, that is, the marginal cost is increasing in y, $\partial^2 C(p,y)/\partial y^2 > 0$.

From the first two assumptions, the following well-known condition in economics is derived

$$p_y = \partial C(p,y)/\partial y, \qquad (4.68)$$

which states that the level of output is determined at the point where the marginal cost becomes equal to the output price the level of which is exogenously given. The supply function is obtained by solving this equation for y. For a solution to exist, the marginal cost has to be a function of y, and for it to be upward sloping in p_y, the marginal cost has to be increasing in y, which implies that the technology has to be subject to decreasing returns to scale.

It is now clear that an upward-sloping supply curve does not occur in IOA because of the assumption of constant returns to scale, under which the right-hand side of (4.68) becomes a constant. The standard assumption of decreasing returns to scale in economics is based on the presence of exogenous inputs, the level of which is of limited supply (see [25], especially Section 2.7). Implicit in IOA is the assumption that there is no shortage in the supply of exogenous inputs at least within the range of production level under consideration. In other words, the IOA is a demand-driven model, where the level of production is not determined by supply constraints but by demand conditions alone.

Figure 4.9 depicts the supply and demand curves that are implicit in IOA. In fact, these curves occur as mutually orthogonal straight lines. An increase in the final demand shifts the demand line to the right, and increases the supply by the amount equal to the increase in demand without causing any change in price level.

Fig. 4.9 The Supply and Demand Curves in IOA. The Supply Curve Does Not Depend on the Level of Output Because of Constant Returns to Scale. The Demand Curve Does Not Depend on Any Price, Because of the Absence of Substitution Among Inputs and of the Fact That the Final Demand Is Exogenous.

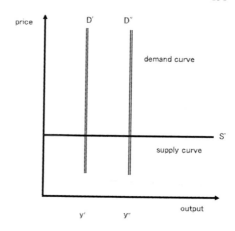

4.4.4 IOA: Bottom-Up or Top-Down?

The top-down approach to the specification of technology, as characterized by the use of FFFs or CES functions as mentioned above, is notable for the fact that it makes little use, if any, of technical process knowledge. Accordingly, it is almost impossible to establish relationships between their parameters and technical process knowledge. An important implication of this is that the extent to which FFFs or CES can be used to analyze concrete technical processes will be of a very limited nature, if any.

In the sense that the IOA also makes use of a particular form of production function, it might appear to have a feature common to the top-down approach mentioned above. The same, however, applies to a process-based LCA too, because of the use of linear technology just like IOA [13]. From the point of view of the incorporation of technical process knowledge, however, the IOA stands closer to the bottom-up approach. In fact, there is nothing in the features of IOA that prevents the incorporation of process knowledge whenever it is available at whatever level of details.

References

1. Arrow, K., Chenery, H., Minhas, B., & Solow, R. (1961). Capital-labor substitution and economic efficiency. *Review of Economics and Statistics, 43*, 225–250.
2. Blackorby, C., Primont, D., & Russell, R. (1979). *Duality, separability, and the functional structure; theory and economic applications.* New York: Elsevier.
3. Christensen, R., Jorgenson, W., & Lau, L. (1971). Conjugate duality and the transcendental logarithmic utility function. *Econometrica, 39*(4), 255–256.
4. Cobb, C. W., & Douglas, P. H. (1928). A theory of production. *American Economic Review, 18*(1), 139–165.
5. Diewert, E. (1971). An application of the Shephard duality theorem: A Generalized Leontief production function. *Journal of Political Economy, 79*, 481–507.

6. Diewert, E., & Wales, T. (1987). Flexible functional forms and global curvature conditions. *Econometrica, 55,* 43–68.

7. Dorfman, R., Samuelson, P., & Solow, R. (1958). *Linear programming and economic analysis.* New York: McGraw-Hill.

8. Fujimoto, T., Silva, J., & Villar, A. (2005). A nonsubstitution theorem with non-constant returns to scale and externalities. *Metroeconomica, 56,* 25–56.

9. Fuss, M., & McFadden, D. (1978). *Production economics: A dual approach to theory and applications I.* Amsterdam: North Holland.

10. Ginsburgh, V., & Keyzer, M. (1997). *The structure of applied general equilibrium models.* Cambridge, MA: MIT Press.

11. Gohin, A., & Hertel, T.W. *A note on the CES functional form and its use in the GTAP model.* GTAP Research Memorandum No. 2. https://www.gtap.agecon.purdue.edu/resources/download/1610.pdf

12. https://www.gtap.agecon.purdue.edu/

13. Heijungs. R., & Suh, S. (2002). *The computational structure of Life Cycle Assessment.* Dordrecht: Kluwer.

14. Henderson, J., & Quandt, R. (1971). *Microeconomic theory: A mathematical approach.* New York: McGraw-Hill.

15. Hudson, E., & Jorgenson, D. (1974). U.S. energy policy and economic growth, 1975–2000. *Bell Journal of Economics and Management Science, 5*(2), 461–514.

16. Jorgenson, D., & Fraumeni, B. (1981). Relative prices and technical change. In E. Berndt & B. Field (Eds.), *Modeling and measuring natural resources substitution.* Cambridge, MA: MIT Press.

17. Kainuma, M., Matsuoka, Y., & Morita, T. (1999). Analysis of Post-Kyoto scenarios: The Asian-Pacific integrated model. *The Energy Journal,* (Special Issue, The Cost of the Kyoto Protocol: A Multi-Model Evaluation), *20,* 207–220.

18. Koopmans, T. (1951). *Activity analysis of production and allocation.* New York: Wiley.

19. Leontief, W. (1947). A note on the interrelation of subsets of independent variables of a continuous function with continuous first derivatives. *Bulletin of the American Mathematical Society, 53,* 343–350.

20. Leontief, W. (1951). *The structure of American economy 1919–1939.* New York: Oxford University Press.

21. Ryan, D., & Wales, T. (2000). Imposing local concavity in the translog and generalized Leontief cost functions. *Economics Letters, 67,* 253–260.

22. Sato, K. (1967). A two-level constant-elasticity-of-substitution production function. *Review of Economic Studies, 34*(2), 201–218.

23. Scrieciu, S. (2007). The inherent dangers of using computable general equilibrium models as a single integrated modeling framework for sustainability impact assessment. A critical note on Böhringer and Löschel. *Ecological Economics, 60*(4), 678–684.

24. Shephard, R. (1953). *Cost and production functions.* Princeton, NJ: Princeton University Press.

25. Shephard, R. (1970). *Theory of cost and production functions.* Princeton, NJ: Princeton University Press.

26. Sono, M. (1961). The effect of price changes on the demand and supply of separable goods. *International Economic Review, 2*(3), 239–271 (English translation of the original Japanese publication in *Kokumin keizai zassi* 74-3, 1943, 261–311).

27. ten Raa, T. (2005). *The economics of input-output analysis.* Cambridge: Cambridge University Press.

Part II
Waste Input-Output Analysis

Chapter 5
Basics of WIO

Abstract The purpose of this chapter is to introduce the basic concepts of Waste Input-Output (WIO) tables and the model (Nakamura [37, 38]). The WIO is a variant of environmental Input-Output (EIO) models, and focuses on issues of waste management (waste treatment and recycling). Familiarity with major EIO models is useful for a proper understanding of the main features of WIO. The first section of this chapter gives a brief review of EIO models from rather broad areas, which include economy–environment (ecology) IO tables and models, IO-based energy analysis, and IO models of emissions. The review in the second section is focused on EIO models that deal with pollution (waste) and its abatement (treatment), which include the seminal work of Leontief [28], and its extension by Duchin, as well as the development of IOTs incorporating waste flows (the Dutch NAMEA and German PIOT). Section 5.3 introduces the concepts of WIO table and its modeling for analytical purposes (further details of WIO model will be dealt with in the next chapter).

5.1 Environmental IO (EIO)

Just like any living organism, the human economic system can be regarded as a metabolic system, the working of which is maintained by a steady inflow of useful inputs (resources) from the ecosystem ecosystem that surrounds it and the corresponding outflow of unwanted outputs (waste and waste heat) into the ecosystem. In other words, the economy is an open subsystem embedded in the ecosystem (Figure 5.1).

In Figure 5.1, the circled arrows denoted by "α" inside the circle termed "economy" refers to the internal flow of goods and services in the economy, the modeling of which by IOA has been dealt with in preceding chapters. The inflow of useful inputs and the outflow of unwanted outputs are respectively denoted by "γ" and "β". So far as the size of economic activities remains "small" relative to the size of the supporting ecosystem, these inflows and outflows do not exert any noticeable effects

S. Nakamura, Y. Kondo, *Waste Input-Output Analysis*, Eco-Efficiency in Industry and Science 26, © Springer Science+Business Media B.V. 2009

Fig. 5.1 The Economy and
the Ecosystem: The Former
is an Open Subsystem of the
Latter.

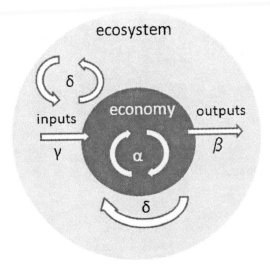

on the ecosystem. As has been already noticed in the famous report of the Club of Rome in 1972 by Meadows et al. [33], however, the size of human economic activities has long surpassed this threshold. In fact, at the level of a limited geographical area (in particular, on an isolated island), human history is not short of examples where extensive deforestation due to intensive agriculture and/or cattle breeding has resulted in the collapse of its economic activity, that is, its civilization [9].

In Figure 5.1, the possible effects of human economic activities on the supporting ecosystem are indicated by the round arrows denoted by "δ": an excessive consumption of useful inputs such as fish and timber can result in resource depletion and the loss of biodiversity. The emission of CO_2 from fossil fuels is a widely acknowledged example of the discharge of unwanted outputs with significant global consequences.

In Chapter 2, the inflow of resources from and the outflow of waste into the environment were incorporated into the framework of IOA in a straightforward manner without resorting to related studies. This section gives a review of environmental IO models and tables which deal with the integration of economic activities in the ecosystem as depicted in Figure 5.1.

5.1.1 Linking the Economy and the Environment in IOA

This subsection gives a brief review of EIO models concerned with the link between economies and ecologies as in Figure 5.1, the roots of which can be traced back to Daly [4] and Isard [22]. While Daly introduced an economy–ecology account, it was Isard who implemented it for the first time for a marine ecosystem. Characteristic of the Isard model is the incorporation of a food web into IOA, which was further elaborated by Hannon into a general IO model of ecological systems. The economy-ecology interactions conceived by the Isard model are illustrated by its simple but full implementation by Jin et al. [24]. Summarizing the review, the focus of the subsequent analysis is shown.

5.1.1.1 The Economy–Environment IO Models of Isard and Daly

Isard et al. [22] and Daly [4] can be mentioned as the first economists who noticed the usefulness of IOA to consider the link between economies and ecologies as in Figure 5.1. For instance, noting

> How does one integrate the world of commodities into the larger economy of nature? [Leontief's input-output model] may offer the most promising analytical framework within which to consider the above question. ([4], p.400)

Daly proposed a basic economy–environment account (Table 5.1) with four quadrants. The standard IO table occupies quadrant (α) with both the rows and columns referring to human economic sectors. Quadrant (γ) refers to the inputs from non-human sectors consisting of living sectors (such as animal, plant, and bacteria) and non-living sectors (such as atmosphere, hydrosphere, and lithosphere) into the human sectors, an example of which is the input of water that is required to produce a given quantity of alfalfa ([4], p.403). Quadrant (β) refers to the outputs from the human sectors to the nonhuman sectors, and represent all the environmental interventions caused by human activities such as river, lake and ocean pollution, soil erosion, genetic diversity losses, and so forth. Finally, quadrant (δ) refers to the interrelationships among ecological sectors such as the web of food chains. These quadrants thus give an IO-based representation of the flows in Figure 5.1.

The work of Daly remained at a conceptual level. The work of Isard is distinguished by its attempt to empirically implement the model to regional data (the Plymouth Bay area in Philadelphia, US) based on a US IO table at the level of four-digit SIC (standard industrial classification). Of particular interest is the implementation of quadrant (δ) based on a marine food web, a schematic representation of which is given by Figure 5.2. In the figure, the sectors (species) are arranged to represent the hierarchy of the food chain with top carnivores (say cod) at the top of the hierarchy (or trophic levels), and producers (say marine plants and phytoplankton) at its bottom. This ordering results in a triangular structure of the matrix discussed in Section 3.7, because the hierarchy of a food chain represents a kind of degree of fabrication: for instance, the production of cod requires the inputs of carnivorous invertebrates and small fish, among others, the production of which in turn requires the inputs of herbivores, the production of which requires detritus, and so forth.

Table 5.1 An Input-Output Representation of the Interaction Between Economy and Environment by Daly [4].

From\to	Human sectors	Nonhuman sectors
Human sectors	α	β
Nonhuman sectors	γ	δ

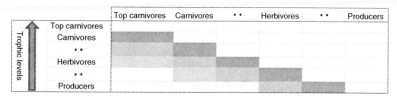

Fig. 5.2 A Schematic Representation of the Food Web in the Form of an IO Table: The Color Density Shows the Relative Size of Values. See [22], for Numerical Examples of the Marine Food Web.

As for the elements in quadrant β, the discharge of waste water from different sectors of the economy (household, laundry services, leather manufacturing, etc.) was considered. It could disturb the ecological production via its negative effects on water quality: waste water discharges can reduce the transparency of water, which in turn can interfere negatively with photosynthesis ([23], p.73). However, the lack of established data for the ecological system prevented Isard from actually implementing the linkage between economy and environment, and putting his ambitious model to work (see Victor [48], p.47).

5.1.1.2 The IO Model of an Ecological System by Hannon

Hannon [17] accomplished transferring the input-output concept into ecology with the aim of describing the structure of ecosystems. His framework features production and respiration energy flows between individuals, species or trophic levels, or in general, compartments, and is based on the principle of energy conservation:

$$q_i - \sum_j q_{ij} - r_i = 0, \quad i = 1, \cdots, n, \text{ or}$$

$$q = Q\iota + r \tag{5.1}$$

where q_{ij} refers to the energy output of ecological component i used by ecological compartment j in an ecosystem in the given time period, an example of which could be the daily amount of algal biomass carbon consumed by a particular herbivore (j), r_i refers to the ith element of respiration flows or outflows that leave the system (for example, by energy dissipation into heat, mass flow out of the system, or extraction by humans), and ι to an $n \times 1$ vector, all the elements of which are unity. The matrix $Q = (q_{ij})$, which is called the production matrix, is made square by including only the substances or energy that are identical to some of the substances produced by the system; the other substances or forms of energy occur as exogenous inputs.

Defining the activities matrix $G = (g_{ij})$ with $g_{ij} = q_{ij}/q_j$, the constancy of which can be accepted for "not very large changes in output" ([18], p.1921), (5.1) is solved as

$$q = (I - G)^{-1}r \tag{5.2}$$

which is identical to the Leontief quantity model. The interpretation of the inverse matrix $(I - G)^{-1}$ is also identical with its ith row and jth column element referring, for instance, to the quantity of algae indirectly needed to produce 1 kg of top carnivore. If one of the elements of r is changed by a small amount (say, a small increase in fishing success), (5.2) gives the amount of change in the level outputs q that is required for it to happen.

Hannon introduced also the "special energy matrix":

$$(I - G)^{-1}\hat{r} \tag{5.3}$$

the ith row and jth column element of which refers to the amount of total direct and indirect energy i consumed by the jth component so that the jth component can carry out its respiration at the given rate.

The concept of the Hannon IO model has been applied widely ever since including the measurement of ecosystem structure and function, sensitivity, cycling, dependency and contribution effects between compartments, loop analysis, and so forth (see [27] for further details).

5.1.1.3 The Jin et al. Model: A Simplified Version of the Isard Model

Actual implementation of the Isard model was hampered, among others, by its huge data requirements in particular on ecological systems, which are not well established even today. Turning to a simpler model that is less demanding in terms of data, Jin et al. [24] developed an integrated IO model that considers both the economic and ecological systems for a region with a restricted set of industry sectors and food web trophic levels. The model was applied to a numerical example for the New England region. It thus represents a full empirical implementation of all the four quadrants of Table 5.1, albeit for a simpler case, which was made possible by an explicit consideration and modeling of the impact of economic activity on the productiveness of the ecological sectors, which were not explicit in the Isard model.

Their starting point is the $(n + m) \times (n + m)$ matrix of input coefficients with n economic sectors, and m ecological compartments:

$$\begin{pmatrix} A & E \\ B & G \end{pmatrix} \tag{5.4}$$

where A is the standard matrix of economic input coefficients, G is the activities matrix of Hannon, B is the input coefficients matrix of ecological commodities into economic sectors (the supply of ecosystem resources, goods, and services to economic sectors), and E is the matrix describing the impacts of economic sectors on the ecosystem (e.g. marine pollution or destruction of habitat) ([24], p.370).

The demand for ecological outputs, s, that are necessary to satisfy a given final demand f is

$$s = B(I - A)^{-1}f \tag{5.5}$$

In the matrix E, negative elements only occur to represent the negative impact of economic activity on ecological sectors, say, a reduction in fish stock due to fishing. Writing d for the (negative) environmental impact of a given final demand f, we have:

$$d = E^\top (I - A)^{-1} f \tag{5.6}$$

Write w for the biomass (measured in energy flow as in the Hannon model) available for consumption by the economy. When the economy affects the ecosystem, the ecosystem needs to provide enough resource stocks to support the economy as described in (5.5) and, in addition, to offset the negative impact of the economy on the stocks as described in (5.6). Accordingly, w is given by

$$w = s - d = (B - E^\top)(I - A)^{-1} f \tag{5.7}$$

Compared to (5.5), this captures the effects of economic activity on the environment via three channels, A (technology), B (dependence of the economy on the environment), and E (the negative impact of the economy on the environment).

The relationship between energy flow and respiration at the regional level, w and r^*, is obtained by multiplying both sides of (5.2) by the area of the study. Assuming that the impact of the economy on the ecosystem d and ecological commodity inputs to the economy s are also modeled explicitly with respiration ([24], p.374), a variant of the Hannon model at the regional level is derived

$$w = (I - G)^{-1} (r^* + s - d) \tag{5.8}$$

where r^* and $r^* + s - d$ can respectively be considered as respiration at the regional level without and with fishing. Substituting from (5.7) finally yields

$$w = (I - G)^{-1} \left(r^* + (B - E^\top)(I - A)^{-1} f \right) \tag{5.9}$$

which represents the impact of f on ecological stocks through all the four channels, A, B, E, and G (the ecosystem). From (5.9), the resource multiplier that captures the direct and indirect effects of an increase in the unit of f on the energy flow (i.e. resource stock) is given by

$$\frac{\partial w}{\partial f} = (I - G)^{-1} (B - E^\top)(I - A)^{-1} \tag{5.10}$$

The above model was implemented for marine data on the New England region for $n = 4$ (agriculture, fishery, manufacturing, and others) and $m = 4$ (phytoplankton, zooplankton, fish, and the sun). As for B and E, a rather simple structure was assumed, where only one nonnegative element occurs in both: for B it is the element that refers to the amount of fish harvested per unit of fishing activity, and for E it is the element that refers to the decline of the natural fish stock by the fishing activity (see ([24] for further details).

5.1.1.4 Limiting the Scope

The crucial question in an integrated economic and ecological model has been whether the nonhuman, ecological quadrant δ in Table 5.1 can be determined with sufficient accuracy and detail in order to provide a useful decision-making tool. While a limited range of nonhuman indicators has been incorporated into economic analysis, it seems safe to say that uncovering natural interdependencies in quadrant δ has been "anything but successful !" [3] (see [27] for a recent review).

Henceforth, we will be concerned with only flows of ecological commodities from the environment into the economy and of waste (of solid, liquid, and/or gaseous forms) from the economy into the environment. Quadrant (δ) will be excluded from our focus. With quadrant (δ) excluded, Table 5.1 can be written as in Table 5.2. Notice that the physical IO table in Table 2.7 corresponds to the format of Table 5.2.

5.1.2 Energy Analysis

Energy analysis in the early 1970s by Wright [50] and Bullard and Herendeen [2] belong to the first examples of IOA that explicitly considered quadrant (γ) in Table 5.1. Responding to the increasing awareness of energy issues caused by the 1973 oil crisis, the main focus of these studies was quantitative identification of the dependence of the economy on primary energy inputs.

Write $B°$ for a $k \times n$ matrix, with $(B°)_{ij}$ referring to the input of ith primary energy (in Joules) per unit of production of product j. Following (2.162), the total quantity of primary energy that is required to satisfy a given final demand f is then given by

$$\text{primary energy requirements} = B°(I - A)^{-1}f \qquad (5.11)$$

Bullard and Herendeen [2] applied this to the US IO able for 1967 with $n = 357$ producing sectors and $k = 5$ types of primary energy that consisted of coal, crude oil, and natural gas, as well as electricity from hydro- and nuclear power. Secondary energy such as refined petroleum and fossil electricity were not included in $B°$ to avoid double counting.

The items of primary energy do not occur in A, while the items of secondary energy do. In terms of Table 5.2, the items of primary energy occur as row elements

Table 5.2 A Simplified Version of the Daly Diagram with the Quadrant δ Excluded.

	Human sectors (economy)
Human sectors	α
Input from nonhuman sectors (environment)	γ
Output to nonhuman sectors: waste (solid, liquid, gas)	$\beta^\top$

$\beta^\top$ indicates the transpose of the quadrant β.

of quadrant γ, while the items of secondary energy occur in quadrant α as both row and column elements. In other words, primary energy was counted as an exogenous input, whilst secondary energy was counted as an endogenous input. No attention was paid to issues related to emissions resulting from the use of energy: quadrant (β) was not considered.

5.1.3 Emission IO Model

Combustion of fossil fuel is associated with the emission of airborne waste such as CO, CO_2, NO_x, N_2O, SO_x, and SPM (small particle matters), and with the emission of solid waste such as bottom ash and fly ash (dust caught by exhaust gas filters). Victor [48] is one of the earliest IO-based studies that explicitly considered most of these emissions (with the exception of CO_2, which was not widely considered as a factor of environmental load in the 1970s), and hence, quadrant β.

Given the linear relationship between the amount of a particular type of fuel consumed and the amount of emissions, the amount of emissions can easily be incorporated into the IO model by augmenting the input-coefficients matrix with a set of emission generation coefficients. In this case, the emission associated with f will be given by

$$\text{emission} = \varepsilon B^*(I-A)^{-1}f \tag{5.12}$$

where ε refers to the $m \times k$ matrix of fuel specific emission factors with $(\varepsilon)_{ij}$ referring to the quantity of emission i per unit of consumption of fuel j, and B^* refers to the $k \times n$ matrix of fuel inputs coefficients. Note that in contrast to (5.11) where the elements of $B°$ must not occur in A to avoid double counting, in (5.12) fuel producing sectors occur in A because it is the emission associated with the combustion of fuels that matters. Recall that $B°$ refers to the use of primary energy, but not to the emissions associated with it.

The linearity between consumption and emission will in particular be the case for CO_2 that originates from the incineration of fossil fuels, and from the use of limestone in iron and steel, cement, and glass. To be specific, the following hold for the combustion of gasoline (octane) and the transformation of limestone into lime:

$$C_8H_{18} + 12.5O_2 \rightarrow 8CO_2 + 9H_2O \tag{5.13}$$

$$CaCO_3 \rightarrow CaO + CO_2 \tag{5.14}$$

Stoichiometrically, combustion of 1 mol of C_8H_{18} and the transformation of 1 mol of $CaCO_3$ into CaO generate 8 mol and 1 mol of CO_2, respectively. Assuming perfect combustion, the CO_2 emission factors for each fossil fuel can thus be estimated based on its carbon content and molecular weight.

While NO_x, SO_x, and SPM are also generated by the combustion of fossil fuels, their emission factors cannot be estimated by the same procedure as was applied to CO_2 [39]. These emissions have been, at least in advanced countries, the subject of pollution control over decades, and various measures have been taken to reduce them, say by installing (in the case of stationary sources) desulfurization equipment, denitrification equipment, and low NO_x burners. Consequently, the emission from stationary sources depends on the types and efficiency of pollutant-removal technology that are specific to industry and combustion facilities. For the emission of NO_x and SPM from mobile sources (mostly automobiles), the situation is further complicated because of additional dependence on traffic densities and traveling conditions [39].

5.1.3.1 The IO Model of CO_2 Emission for Germany and the UK

Proops et al. [43] is one of the earliest IO studies that analyzed the interdependence between economic activity and the emission of CO_2 that results from the combustion of gas (CH_4), coal, and liquid fuels. As the origin of emission, they considered both fuel consumption (5.13) and nonfuel consumption (5.14). Furthermore, the direct emission from the final-demand sector that results from such purposes as personal transport, home heating, and government activities was also taken into account.

Without loss of generality, let fuel sectors be the first k sectors, and denote the first k elements of f by f^E and the remaining elements by f^R:

$$f = \begin{pmatrix} f^E : \text{for fuel} \\ f^R : \text{for others} \end{pmatrix} \tag{5.15}$$

Extending (5.12) to allow for the emission from nonfuel sources and from the final demand, the total emission is given by

$$\text{Total } CO_2 \text{ emission} = (\varepsilon B^* + \mu)(I - A)^{-1} \begin{pmatrix} f^E \\ f^R \end{pmatrix} + \varepsilon f^E \tag{5.16}$$

where μ is the $1 \times n$ vector of CO_2 emission factors that refer to nonfuel sources. The first term on the right-hand side of (5.16) refers to the indirect emission invoked by f, and the second term to the emission directly invoked by f. Of the first term, the portion involving f^E refers to the emission that results from the production of fuels including the delivery to the final-demand sectors, but excludes the emission from their consumption. It is the second term that refers to the emission from the consumption in the final-demand sectors. In [43], (5.16) was applied to German and UK IO tables taken from several different years from the late 1960s to the middle to late 1980s, which were aggregated to mutually comparable 47 sectors.

5.1.3.2 Large Scale Emission IO Models

For a serious application to issues of industrial ecology, the degree of resolution provided by IO tables with less than 100 sectors will not be adequate (according to [42], a hybrid IO model with 2,000–3,000 variables is not uncommon). With more than 400 sectors, the US (432 industry sectors for the 2002 benchmark IO accounts [47]) and Japanese (517 rows × 405 columns for the 2000 table [31]) IO tables can be mentioned as the largest ones that are currently publicly available (another example is the Multi-Regional-Input-Output (MRIO) system for Australia, which covers 344 industrial sectors over eight Australian regions, see [15]). For both the US and Japanese tables, intensive work has been done to integrate them with extensive data on emissions based on the above-mentioned modeling framework. The eiolca.net at Carnegie Melon University [11,20] for the US, and the 3EID (Embodied Energy and Emission Intensity Data for Japan Using Input-Output Tables) database at the National Institute for Environmental Studies (NIES) [39] for Japan can be mentioned as the most representative studies.

eiolca.net at Carnegie Melon

Characteristic of the eiolca.net is its extensive coverage of environmental categories which include energy use, GWP emissions, air pollutants (the US EPA's 1996 AIR-Data NET SIC Report and emission estimates of the National Air Quality and Emissions Trends Report for 1996), toxics emissions (the US EPA Toxics Release Inventory (TRI) data), and dollar estimates of external air pollution costs. The web site of eiolca.net provides a unique opportunity for its visitors to estimate the overall environmental impact from producing a certain dollar amount of any of 500 commodities or services in the US. The estimation is based on

$$R(I - A)^{-1} f \qquad (5.17)$$

where R is the matrix representing the emissions per dollar of output ([20], p.15). In general terms, R is called the intervention matrix [19].

The 3EID Data Base of NIES

The 3EID database deals with the emissions (CO_2, NO_x, SO_x, SPM) that result from the combustion of 8 types of coal-based fuel, 12 types of petroleum-based fuel, 3 types of natural gas-based fuels, and 5 types of other fuels including municipal and industrial waste [39]. Considerable attention has been paid to excluding the nonenergy use of fossil fuels, naphtha, and LPG used as raw materials for chemical products such as plastics and ammonia. The nonfuel source of emissions such as the use of limestone in cement production, and the incineration of waste are also taken into consideration.

Due to these features, the underlying IO model is similar to (5.16), and is given by

$$\text{Total emissions} = \left(\varUpsilon(H \odot B^*) + \varLambda \right)(I - A)^{-1} f + \varUpsilon f^E \qquad (5.18)$$

where $\varUpsilon$ is the 4×28 matrix of emission coefficients from the combustion of 28 types of fuels, $H = (h_{ij})$ is the $28 \times n$ matrix with

$$h_{ij} = \begin{cases} 1: & \text{if fuel } i \text{ is used for combustion} \\ 0: & \text{if fuel } i \text{ is used for nonfuel use} \end{cases} \qquad (5.19)$$

Λ is the $4 \times n$ matrix of emission coefficients from nonfuel sources, and $\odot$ is the Hadamard product which refers to the element-wise product of two matrices

$$H \odot B^* = (h_{ij} b_{ij}^*) \qquad (5.20)$$

For NO_x, SO_x, and SPM, both stationary and mobile sources are considered, including the emission of SPM resulting from tire wear. See [41] for further details. Remarkable for 3EID is that the whole set of emissions data is available free of charge on the Internet [40].

5.2 The IO Models of Pollution Abatement and Their Relevance to Waste Management

In the IO model of emissions mentioned above, the level of emissions is assumed to be proportional to the level of production via given coefficients of emissions per unit of production. Consequently, for a given set of technical input and emissions coefficients and a given vector of final demand f, the level of emissions is uniquely determined. In other words, in these models the pollution abatement processes such as desulfurization, denitrification, and the filtering of flue gases do not occur separately, but occur integrated into production processes. This can be understood to reflect the reality of the majority of advanced economies where the installation of abatement facilities/devices is mandatory by regulation.

In terms of Table 5.1, the operation of a pollution abatement process requires inputs from quadrants α and γ, and produces outputs to quadrant β. For instance, a flue gas desulfurization (SO_x removal) process requires, among others, slaked lime ($Ca(OH)_2$) and water as inputs, and generates gypsum ($CaSO_4 2H_2O$), the chemical reaction of which is given by ([16])

$$Ca(OH)_2 + CO_2 \rightarrow CaCO_3 + H_2O \qquad (5.21)$$

$$CaCO_3 + SO_2 + \frac{1}{2}H_2O \rightarrow CaSO_3 \cdot \frac{1}{2}H_2O + CO_2 \qquad (5.22)$$

$$CaSO_3 \cdot \frac{1}{2}H_2O + \frac{1}{2}O_2 + \frac{3}{2}H_2O \rightarrow CaSO_4 2H_2O \qquad (5.23)$$

The treatment of 1 mol of sulferdioxide requires 1 mol of $Ca(OH)_2$, and produces 1 mol of gypsum. In the removal of NO_x from a flue gas (denitrification), on the other hand, ammonia (NH_3) is required, the underlying chemical equation of which is given by ([16])

$$4NO + 4NH_3 + O_2 \rightarrow 4N_2 + 6H_2O \qquad (5.24)$$

Removing 1 mol of NO thus requires 1 mol of ammonia.

If pollution abatement processes occurred separately from production processes in an IO framework, it would enable one to analyze the effects on economic production and emissions of different levels of pollution abatement. Leontief [28] is a seminal work that proposed an IO model for this kind of analysis by explicitly considering the interdependence between pollution abatement processes and the remaining sectors of the economy. This subsection is concerned with this model.

5.2.1 The Leontief Model of Pollution Abatement

A good way to introduce the basic features of the Leontief model of pollution abatement is to resort to a numerical example. For the sake of ease of illustration, consider an economy that consists of two conventional production sectors and a pollution-abatement sector, the unit processes of which are given in Table 5.3. It is assumed that the operation of sectors 1 and 2 emits a single pollutant, say SO_2, while the operation of the abatement process emits none. The third row of the table gives the emission coefficients. It is also assumed that the final-demand sector emits no pollutant.

Neglecting the issue of the way in which the pollutant occurs in the final-demand vector f, for a moment, suppose that the first two elements of f are given by $f_1 = 55$ and $f_2 = 30$. Analogous to (5.17), the resulting emission is then given by

$$
\begin{aligned}
&\left(r_1\ r_2\right) \begin{pmatrix} 1-a_{11} & -a_{12} \\ -a_{21} & 1-a_{22} \end{pmatrix}^{-1} \begin{pmatrix} f_1 \\ f_2 \end{pmatrix} \\
&= \left(0.5\ 0.2\right) \begin{pmatrix} .75 & -.04 \\ -.014 & 0.88 \end{pmatrix}^{-1} \begin{pmatrix} 55 \\ 30 \end{pmatrix} = 60
\end{aligned}
\tag{5.25}
$$

Without any pollution abatement activity, 60 units of the pollutant is emitted into the environment.

5.2.1.1 The Leontief EIO Model Without Abatement Activity

One may find nothing substantially new in Table 5.3: it resembles the matrix of unit processes with waste and an exogenous input as in (2.128). What makes the Leontief pollution abatement model innovative and unique is the way the amount

Table 5.3 A Matrix of Unit Processes with Pollution Abatement.

Input\output	Sector 1	Sector 2	Pollution abatement
Sector 1	0.25	0.40	0.00
Sector 2	0.14	0.12	0.20
Pollutant	0.50	0.20	0.00
Labor	0.80	3.60	2.00

of pollutant occurs in the final demand. In the Leontief pollution abatement model, the final demand for pollutant refers to the negative of its amount that is *tolerated* or *accepted* by the final-demand sectors, i.e., by the households.

It was obtained above that in the absence of abatement activity 60 units of pollutant is emitted into the environment. In the Leontief abatement model, this corresponds to a situation where society is ready to tolerate that level of emissions. Denoting by f_3 the tolerated quantity of pollutant, f now becomes

$$f = \begin{pmatrix} 55 \\ 30 \\ -60 \end{pmatrix} \tag{5.26}$$

The level of production of the three sectors including the abatement sector is then given by

$$x = (I - A)^{-1} f = \begin{pmatrix} 0.75 & -0.4 & 0 \\ -0.14 & 0.88 & -0.2 \\ -0.5 & -0.2 & 1 \end{pmatrix}^{-1} \begin{pmatrix} 55 \\ 30 \\ -60 \end{pmatrix} = \begin{pmatrix} 100 \\ 50 \\ 0 \end{pmatrix} \tag{5.27}$$

Because society accepts all the emissions, there is no operation of the abatement process, and hence $x_3 = 0$.

Table 5.4 gives the IO table that is obtained by use of (2.160) for the level of x given by (5.27). All the elements of the column referring to the abatement sector are zero, because the activity level of the abatement process is zero. Notice that the activity level of the abatement process is measured by the amount of pollutant treated. The third row of the table refers to the condition of mass conservation with regard to the pollutant:

$$\underbrace{r_1 x_1 + r_2 x_2}_{\text{emission}} - \underbrace{(-f_3)}_{\text{tolerated}} = \underbrace{x_3}_{\text{abated}} \tag{5.28}$$

5.2.1.2 The Leontief EIO Model with Abatement Activity

Consider now the situation where society is no longer ready to accept all the emission of pollutants, but only half, i.e., $f_3 = -30$. With the value of f_3 in (5.27) reduced to half, application of the same procedure as was used for obtaining the IO table in Table 5.4 gives the result in Table 5.5.

Table 5.4 The Leontief EIO Table with No Pollution Abatement.

Input\output	Sector 1	Sector 2	Pollution abatement	Final demand	Production
Sector 1	25	20	0	55	100
Sector 2	14	6	0	30	50
Pollutant	50	10	0	-60	0
Labor	80	180	0		
Production	100	50			

Table 5.5 The Leontief EIO Table with Pollution Abatement: 50% Reduction of Emission into the Environment.

Input\output	Sector 1	Sector 2	Pollution abatement	Final demand	Total
Sector 1	26	23	0	55	104
Sector 2	15	7	7	30	58
Pollutant	52	12	0	−30	34
Labor	84	210	68	0	362

The sum of numbers does not match exactly because they are round-up.

Table 5.6 The Leontief EIO Table with Pollution Abatement: 100% Reduction of Emission into the Environment.

Input\output	Sector 1	Sector 2	Pollution Abatement	Final demand	Total
Sector 1	27	27	0	55	109
Sector 2	15	8	14	30	67
Pollutant	54	13	0	0	68
Labor	87	241	136	0	464

The sum of numbers does not match exactly because they are round-up.

Comparison of Table 5.5 with Table 5.4 reveals several remarkable results. First, the production levels of sectors 1 and 2 increase because of the increase in the direct and indirect demand for their products that is invoked by the operation of the abatement sector. Secondly, the total amount of emission from sectors 1 and 2 increases due to the increase in their production. It is interesting that the very operation of the pollution-abatement sector contributes to an increase in the amount of pollutant because of its positive effects on economic activity. Thirdly, as a consequence of the increased amount of total emission of pollutant, a larger amount of pollutant than was perceived previously, i.e., $x_3 = 30$, has to be abated to keep the amount of emission into the environment at 30.

Finally, Table 5.6 shows the case where emission is abated by 100%, i.e., $f_3 = 0$. The total amount of emission now becomes 68, and the amount of labor input increases about 80% relative to the case of no abatement at all in Table 5.4.

5.2.1.3 Pollution Abatement Cost and Product Prices

In the above hypothetical numerical example, an increase in pollution abatement calls for a sizable increase in the demand for labor. Since labor is the only exogenous input, this will imply a nonnegligible increase in the cost of production and abatement, and hence in product prices. By use of the Leontief price model (2.168) it is possible to estimate the effects of pollution abatement on product prices.

Consider first the case where the cost of abatement is not borne by the producers who emit the pollutant but by the final-demand sector, say by tax money. Because the cost of production then does not include the cost of abatement, the product prices of sectors 1 and 2 are determined without considering p_3, and are given by

$$(p_1 \ p_2) = (b_1 \ b_2) \begin{pmatrix} 1-a_{11} & -a_{12} \\ -a_{21} & 1-a_{22} \end{pmatrix}^{-1} = (2.0 \ 5.0) \qquad (5.29)$$

The price of the abatement sector is then given by

$$p_3 = p_1 a_{13} + p_2 a_{23} + b_3 = 3.0 \qquad (5.30)$$

Consider next the case where the cost of abatement is borne by the producers who emit the pollutant at a rate proportional to the amount of emission. In this case, the abatement cost becomes a component of production cost in both the sectors 1 and 2, and hence all the prices p_1, p_2, and p_3 need to be simultaneously determined:

$$(p_1 \ p_2 \ p_3) = (b_1 \ b_2 \ b_3) \begin{pmatrix} 1-a_{11} & -a_{12} & -a_{13} \\ -a_{21} & 1-a_{22} & -a_{23} \\ -a_{31} & -a_{32} & 1-a_{33} \end{pmatrix}^{-1} = (4.6 \ 7.0 \ 3.4) \quad (5.31)$$

Internalization of the abatement cost increases the product price of sector 1 by a factor of 2.3, and that of sector 2 by a factor of 1.4.

It is important to notice that in both cases the total expenditure of the final-demand sector, that is the household sector, remains exactly the same. In Table 5.6, for instance, the final expenditure of the household is

$$\underbrace{(2.0 \ 5.0 \ 3.0) \begin{pmatrix} 55 \\ 30 \\ 68 \end{pmatrix}}_{\text{no internalization in product prices}} = 463.6 = \underbrace{(4.6 \ 7.0) \begin{pmatrix} 55 \\ 30 \end{pmatrix}}_{\text{full internalization in product prices}} \qquad (5.32)$$

This is so because the cost of abatement has to be finally borne by the final-demand sector independent of who bears the cost initially. In the case of full internalization, the producers bear the cost initially, but transfer it to product prices, and it is the final-demand sector that finally bears the cost by purchasing the products at the prices that include the cost of abatement.

5.2.1.4 Extension of EIO to the Case of Multiple Pollutants

Leontief [29, 30] extended the above simple model by considering the presence of multiple pollutants and the generation of pollutants from the final-demand sectors (the emission of pollutants from passenger cars etc.). The presence of multiple pollutants was coped with by adding for each pollutant a new row referring to its generation and a corresponding new column referring to its abatement.

Table 5.7 gives the extended matrix of input and pollution generation coefficients for the general case of n goods-producing sectors and m pollutants, augmented with the column of final demand. In the table, the subscripts "I" and "II" on matrices respectively refer to goods/goods-producing sectors and pollutants/pollution-abatement sectors: the matrix $A_{I,I}$ refers to the $n \times n$ matrix of standard input

Table 5.7 The A and R Matrices of EIO with Multiple Pollutants.

	Goods (n)	Abatement (m)	Final demand
Goods-producing sectors (n)	$A_{I,I}$	$A_{I,II}$	f_I
Pollutants (m)	$A_{II,I}$	$A_{II,II}$	$A_{II,f}f_I$

The numbers in () refer to the numbers of rows or columns of the relevant matrices.

coefficients, and the $n \times m$ matrix $A_{I,II}$ refers to the inputs from goods-producing sectors (electricity, water, lime, ammonia, and so forth) to pollution-abatement sectors.

Three novel features of the extended EIO model can be mentioned (they are taken from Mathematical Appendix of [30]). First, the generation of pollutants from the final-demand sector is taken into account by introducing the $m \times n$ matrix $A_{II,f}$, with $(A_{II,f})_{ij}$ referring to the generation of pollutant i in the final-demand sector in the process of consuming a unit of good j. The product $A_{II,f}f_I$ then gives the generation of m pollutants from the final-demand sector. Secondly, the generation of pollutants from pollution abatement processes is taken into account by the $m \times m$ matrix $A_{II,II}$ with $(A_{II,II})_{km}$ referring to the generation of pollutant k per unit of treatment of pollutant m. Finally, the recycling or reuse of pollutants is taken into account by allowing the matrices $A_{II,i}$, $i = I, II, f$, to include negative elements. For instance, $(A_{II,I})_{ij} < 0$ refers to the use of pollutant i in goods producing sector j, and $(A_{II,II})_{kl} < 0$ refers to the use of pollutant k in pollution-abatement sector l.

Denoting the amounts of pollutants that are tolerated by society by an $m \times 1$ vector f_{II}, the output-balancing equations for goods production x_I and pollution abatement activity x_{II} are given by

$$A_{I,I}x_I + A_{I,II}x_{II} + f_I = x_I \tag{5.33}$$

$$A_{II,I}x_I + A_{II,II}x_{II} + A_{II,f}f_I - f_{II} = x_{II} \tag{5.34}$$

where the first three terms on the left of (5.34) respectively refer to the emission of pollutants from goods production sectors, pollution-abatement sectors, and final demand (household) sectors. Rearranging this system of equations in a matrix form

$$\begin{pmatrix} I - A_{I,I} & -A_{I,II} \\ -A_{II,I} & I - A_{II,II} \end{pmatrix} \begin{pmatrix} x_I \\ x_{II} \end{pmatrix} = \begin{pmatrix} f_I \\ A_{II,f}f_I - f_{II} \end{pmatrix} \tag{5.35}$$

and noting that the right-hand side variables are exogenous, this system can be solved for x:

$$\begin{pmatrix} x_I \\ x_{II} \end{pmatrix} = \begin{pmatrix} I - A_{I,I} & -A_{I,II} \\ -A_{II,I} & I - A_{II,II} \end{pmatrix}^{-1} \begin{pmatrix} f_I \\ A_{II,f}f_I - f_{II} \end{pmatrix} \tag{5.36}$$

If each production/pollution-abatement sector is to bear (and include in its product price) the costs of eliminating all the pollution it emitted directly, the unit cost equations will become

$$\begin{pmatrix} p_I & p_{II} \end{pmatrix} = \begin{pmatrix} p_I & p_{II} \end{pmatrix} \begin{pmatrix} A_{I,I} & A_{I,II} \\ A_{II,I} & A_{II,II} \end{pmatrix} + \begin{pmatrix} \pi_I & \pi_{II} \end{pmatrix} \tag{5.37}$$

where p_I and p_{II} respectively refer to an $n \times 1$ vector of the prices of goods & services and an $m \times 1$ vector of the prices of pollution abatement services, while π_I and π_{II} refer to the corresponding row vectors of value added ratios.

Solving for the prices gives

$$\begin{pmatrix} p_I & p_{II} \end{pmatrix} = \begin{pmatrix} \pi_I & \pi_{II} \end{pmatrix} \begin{pmatrix} I - A_{I,I} & -A_{I,II} \\ -A_{II,I} & I - A_{II,II} \end{pmatrix}^{-1} \tag{5.38}$$

which is a generalized version of (5.31) for the case of n producing sectors, and m pollutants (abatement sectors), and represents the Leontief EIO price model (Equation (5) of Leontief and Ford [29]).

Leontief and Ford [29] attempted to develop an EIO table as in Table 5.7 for the US economy with $n = 90$ and $m = 5$ (particulates, SO_x, HC, CO, and NO_x). The lack of data on the elements of the matrix $A_{I,II}$ prevented them from its full implementation.

The $(n + m) \times (n + m)$ matrix in Table 5.7 may appear similar to the integrated matrix of economic activities and ecological system as in (5.4), with the matrices $A_{I,I}$, $A_{II,I}$, $A_{I,II}$, and $A_{II,II}$ respectively corresponding to A, B, E, and G. Except for $A_{I,I}$ and a part of $A_{II,I}$, however, the analogy does not hold. First of all, an equivalence of B does not occur in Table 5.7, because inputs from the ecological system (into the economic system) are not considered. The matrix $A_{II,I}$ is similar to E to the extent that it refers to the generation of pollutants. Except for the case where all the pollutants are directly emitted into the ecological system without any treatment, however, $A_{II,I}$ are not an element of E.

5.2.2 Further Extensions of the Leontief EIO Model

The Leontief EIO model was developed around 1970 against the background of growing concern about environmental pollution caused by the uncontrolled emission of gaseous substances such as SO_x and NO_x from factories and automobiles, and the discharge of toxic waste water from factories into water systems. In particular, Japan in the 1960s had faced serious pollution problems that caused widely known pollution diseases such as Minamata disease (mercury poisoning due to the discharge of waste water from a chemical plant), Itai-itai disease (cadmium poisoning due to mining), and Yokkaichi asthma (caused by SO_x emitted from petrochemical plants). It is interesting to note that the seminal paper of Leontief [28] was first presented at a symposium on pollution in Tokyo. Since then, these types of pollution have come under ever-tighter regulations and have been mostly resolved, at least in developed countries.

With classic types of environmental pollution having been subjected to tighter regulation, environmental issues have shifted to GWP and resource depletion.

Closely related to issues of resource depletion is the issue of waste management. Since the seminal work of Leontief [28], quite a few studies have evolved to extend his model to issues of waste management, to which we now turn.

5.2.2.1 The Duchin EIO Model: Accounting for the Generation of Residues

As pointed out by Duchin [10], a distinguishing feature of the Leontief EIO model consists in that it is closed for pollution in the sense that there is a one-to-one correspondence between pollutant (waste) and its abatement (treatment): each unit of pollutant (waste) generated is either 'eliminated' or tolerated by the households ([10], p.254). In mathematical terms, this feature is represented by the matrix $A_{II,II}$ being square. The Leontief EIO model thus excludes the presence of treatment residues produced by pollution abatement processes which cannot be further processed, and are emitted to the environment:

> Leontief's model produces no residuals because it is completely closed for pollutants. ([10], p.254)

In reality, however, any treatment of pollutants and/or waste produces residues which are not subjected to further treatment but are emitted to the natural environment. For instance, incineration of waste generates ash (bottom- and fly ash) and flue gases. While bottom ash could be recycled as construction material, fly ash usually ends up in landfill. This feature of the Leontief EIO model thus represents a conceptual weakness. To cope with this, Duchin extended the original Leontief EIO model by introducing new rows referring to the generation of treatment residues of this type, which have no corresponding columns.

Duchin presented an extended model in the form of a numerical example with two producing sectors, food products and machinery, two types of waste water with different (high and low) levels of BOD (biological oxygen demand), and two types of waste water treatment sectors specialized to each type of waste water (Table 5.8). Her model differs from the Leontief EIO model by explicit consideration of the generation of sludge (a treatment residue) in each of the four sectors, which is not subjected to any treatment, but is emitted to the environment. In Table 5.8, this feature is shown by the presence of the row of sludge generation coefficients at the bottom, and the absence of a corresponding column.

Table 5.8 The Input-Output Coefficients Matrix of the Duchin Model.

	Goods production		Waste water treatment	
	Foods	Machinery	Waste water (H)	Waste water (L)
Foods	0.4	0.3	0	0
Machinery	0.2	0.3	0.2	0.1
Waste water (H)	0.4	0	0	0
Waste water (L)	0.4	0.1	0.3	0.2
Sludge	0.05	0	0.05	0.02

Source: [10].

Denoting by R the row at the bottom of Table 5.8 that refers to sludge generation coefficients, and using notations similar to Table 5.7, the coefficients of Table 5.8 are represented by

$$\begin{pmatrix} A_{I,I} & A_{I,II} \\ A_{II,I} & A_{II,II} \\ R_I & R_{II} \end{pmatrix} \tag{5.39}$$

The level of activity in this economy invoked by a given final demand will be obtained as (5.36), and the associated discharge of sludge into the environment by (for the sake of simplicity, Duchin considered the case of $A_{II,f} = 0$)

$$x_R = \begin{pmatrix} R_I & R_{II} \end{pmatrix} \begin{pmatrix} I - A_{I,I} & -A_{I,II} \\ -A_{II,I} & I - A_{II,II} \end{pmatrix}^{-1} \begin{pmatrix} f_I \\ A_{II,f} f_I - f_{II} \end{pmatrix} \tag{5.40}$$

Formally, this is identical to (5.17).

Note that in the Duchin model dumping into the environment is the only way sludge is disposed of, or treated in the broadest sense of the word. If one broadens the definition of waste treatment to include untreated dumping into the environment as well, one can establish a one-to-one correspondence between sludge and its treatment, that is, dumping into the environment, and close the model for all types of waste, including sludge. In other words, by extending waste treatment to include "untreated disposal", the Duchin model can be made formally identical with the Leontief EIO model. In the case of Table 5.8, for instance, this can be facilitated by introducing a 4×1 column of zeros that refers to 'dumping into the environment', which results in Table 5.9.

The quantity model then becomes formally identical with the Leontief model (5.36), and given by

$$\begin{pmatrix} x_I \\ x_{II} \\ x_R \end{pmatrix} = \begin{pmatrix} I - A_{I,I} & -A_{I,II} & O \\ -A_{II,I} & I - A_{II,II} & O \\ -R_I & -R_{II} & I \end{pmatrix}^{-1} \begin{pmatrix} f_I \\ A_{II,f} f_I - f_{II} \\ O \end{pmatrix} \tag{5.41}$$

where O refers to a vector of zeros with appropriate dimensions.

Table 5.9 The Input-Output Coefficients Matrix of the Duchin Model Made Square by Introducing a Column Referring to Dumping.

	Goods production		Waste water treatment		
	Foods	Machinery	Waste water (H)	Waste water (L)	Dumping
Foods	0.4	0.3	0	0	0
Machinery	0.2	0.3	0.2	0.1	0
Waste water (H)	0.4	0	0	0	0
Waste water (L)	0.4	0.1	0.3	0.2	0
Sludge	0.05	0	0.05	0.02	0

Source: [10].

The fact that (5.40) and (5.41) produce identical results can easily be confirmed. From (5.36) and (5.40), the following result is obtained by use of the original Duchin model characterized by the nonsquare inputs and outputs matrix

$$
\begin{pmatrix} x_F \\ x_M \\ x_H \\ x_L \end{pmatrix} = \begin{pmatrix} 0.6 & -0.3 & 0 & 0 \\ -0.2 & 0.7 & 0 & 0 \\ -0.4 & 0 & 1 & 0 \\ -0.4 & 0.1 & -0.3 & 0.80 \end{pmatrix}^{-1} \begin{pmatrix} 10 \\ 3 \\ 0 \\ 0 \end{pmatrix} = \begin{pmatrix} 25.2 \\ 17.0 \\ 10.1 \\ 18.5 \end{pmatrix} \tag{5.42}
$$

$$
x_R = \begin{pmatrix} 0.05 & 0 & 0.05 & 0.02 & 0 \end{pmatrix} \begin{pmatrix} 25.2 \\ 17.0 \\ 10.1 \\ 18.5 \end{pmatrix} = 2.1 \tag{5.43}
$$

where $x_i, i = F, M, H, L$ refer to the activity level of foods, manufacturing, waste water (H), and waste water (L), and x_R refers to the generation of sludge.

The same result can be obtained by use of the revised Duchin model characterized by the square inputs and outputs matrix (5.41)

$$
\begin{pmatrix} x_F \\ x_M \\ x_H \\ x_L \\ x_R \end{pmatrix} = \begin{pmatrix} 0.6 & -0.3 & 0 & 0 & 0 \\ -0.2 & 0.7 & 0 & 0 & 0 \\ -0.4 & 0 & 1 & 0 & 0 \\ -0.4 & 0.1 & -0.3 & 0.80 & 0 \\ -0.05 & 0 & -0.05 & -0.02 & 1 \end{pmatrix}^{-1} \begin{pmatrix} 10 \\ 3 \\ 0 \\ 0 \\ 0 \end{pmatrix} = \begin{pmatrix} 25.2 \\ 17.0 \\ 10.1 \\ 18.5 \\ 2.1 \end{pmatrix} \tag{5.44}
$$

where the 5×1 vector before the second equality refers to the vector of final demand extended for one zero element that refers to the final demand for sludge or dumping into the environment.

In the reality of developed countries, at least, sludge from waste water treatment is not dumped into the environment directly, but is subjected to further treatment including dehydration, incineration, and controlled (sanitary) landfill. Subjected to dehydration, sludge is transformed into a dehydrated cake, which can be incinerated, and/or landfilled. Subjected to incineration, it is transformed into ash. When sludge is incinerated after dehydration, and incineration residues go to a landfill, the final discharge into the environment takes the form of land use and the discharge of leachate from a landfill. It is important to note the presence of alternative processes that can be applied to a given waste, sludge in this particular example. In other words, in general there is no one-to-one corresponded between wastes and waste treatment. This point is not taken into account in the Leontief–Duchin EIO-model.

5.2.2.2 Other Modeling Approaches Based on Leontief EIO

Huang et al. [21] proposed another extension of the Leontief EIO model, which can be distinguished by its explicit focus on "3R" activities that consist of waste reduction, waste reuse, and waste recycling. So far as the formal mathematical structure is concerned, however, their model is the same as the Leontief–Duchin EIO model with a symmetric $A_{II,II}$ matrix, where 3R activities occur as both row and column elements. While a hypothetical numerical example is presented and some scenario analysis are conducted, they give no explicit account of the way waste are associated with the 3R activities. This leaves some uncertainty about the way their model is to be implemented to real data on waste flows.

Characteristic of both Duchin [10] and Huang et al. [21] is the fact that they did not manage to implement their theoretical models to real data, but presented a simple numerical example only, which is purely hypothetical.

In sharp contrast to these is the EIO based LCA studies conducted by Hendrickson et al. [20], where the major concern consists in the analysis of real data, but not in exercise based on purely hypothetical numbers. In their extensive application of EIO to LCA of various case studies, they considered the end of life (EoL) phase as well for construction debris (p.142), fuel tank systems (p.84), and automobile catalysts (p.91). As far as the EoL phase is concerned, however, no IO model of the Leontief–Duchin type was used for its description. Instead, the analysis was done base on a rather ad-hoc and case specific way without resorting to any general analytical form. With this regard it seems noteworthy that in their LCA of passenger cars the EoL phase was not considered:

Since the EIO-LCA model does not have a sector that appropriately represents vehicle end of life, this phase is not included in this study. ([20] p.67)

5.2.3 IO Tables with Waste and Waste Management

Physical waste flows do not occur in the standard IOTs, which are measured in monetary units. However, in large scale IOTs, like those of the US and Japan, at least, it is usual that waste treatment occurs as a separate sector. It is then of interest to see how, if any, the way in which these IOTs record waste treatment is related to the Leontief–Duchin EIO model. The Dutch NAMEA (national accounting matrix including environmental accounts) represents a hybrid IOT where the standard monetary IOT was combined with physical flows of substances and waste. Closely related to Dutch NAMEA is the German PIOT (Physical Input-Output Table), which also includes physical flows of waste. We start with the way waste treatment is recorded in a standard IOT.

5.2.3.1 Waste and Waste Treatment in Standard IO Tables

Japanese IO tables include three sectors concerned with waste management consisting of waste water treatment, public waste treatment, and private waste treatment. Waste treatment is not distinguished by treatment processes, but by institutional characteristics. These sectors occur as both row and column elements, with the columns referring to the inputs from other sectors, and the rows referring to the inputs into other sectors of waste treatment services (the payment for waste treatment services). All the elements are measured in a monetary term.

The matrix composed of the columns of waste treatment sectors can be regarded as the matrix $A_{I,II}$ of the Leontief–Duchin EIO model expressed in monetary terms. The highly aggregated nature of waste treatment sectors, which are distinguished by institutional differences only, precludes any identification of individual waste treatment processes. The rows of waste treatment sectors, however, do not contain any information about the physical flow of waste, and hence do not have any correspondence to the matrix $A_{II,I}$ of the Leontief–Duchin EIO model. It is concluded that the way Japanese IO tables account for waste treatment does not correspond to the Leontief–Duchin EIO model. The same applies to the US IO table, where "Waste management and remediation services" occurs as the single sector related to waste treatment.

Because the IO model obtained from these IOTs is closed for waste treatment, the demand for waste treatment services induced by a given final demand can be obtained by use of the standard Leontief quantity model just like any endogenous sector. Due to the above mentioned nature, however, physical information about the content of these services, such as the mass and composition of wastes, cannot be obtained.

5.2.3.2 The Dutch NAMEA

With regard to explicit consideration of waste flows within the framework of national economic accounts, the Dutch NAMEA (national accounting matrix including environmental accounts) compiled by Statistics Netherlands is a pioneering work [8, 25]. In the Dutch NAMEA, the physical flow of waste and waste water occur as row elements corresponding to the matrix $A_{II,I}$ of the Leontief–Duchin EIO model in what is called "substances account", and measured in physical units:

> Both the pollution generated by economic activities and the accumulation of hazardous agents in the Dutch environment are incorporated in these supplementary accounts. The accumulation is equal to the domestic generation of pollutants, minus their absorption by environmental cleansing (e.g. waste water treatment) and minus the balance of transboundary pollutant flows to and from other countries. ([8], p.17)

This is in sharp contrast to the standard IO table mentioned above, where no consideration is made of physical information about waste flows. It is also noteworthy that the physical transformation of waste by waste treatment (reduction in volume by incineration, and of weight by dehydration) is also taken into account.

On the other hand, the column and row of "Environmental cleansing and sanitary services", the single sector dealing with waste and waste water treatment in Dutch NAMEA, share the same characteristic as the standard practice in US and Japanese IO tables mentioned above: it is an aggregate of all the waste treatment processes with all the column and row elements measured in monetary terms.

Table 5.10 A Prototype of Dutch NAMEA.

Input\output	(P)	(E)	Final demand
(P) Producing sectors[a]	$X_{I,I}$	$X_{I,II}$	f_I
(E) Environmental cleansing			
& sanitary services[a]	$X_{II,I}$	$X_{II,II}$	f_{II}
		Absorption (input)[b]	
CO_2			
N_2O			
CH_4			
CFCs and Halons			
NOx			
SO_2	W_I^{in}	W_{II}^{in}	W_f^{in}
NH_3			
P			
N			
Waste			
Waste water			
		Generation (output)[b]	
CO_2			
N_2O			
CH_4			
CFCs			
NOx			
SO_2	W_I^{out}	W_{II}^{out}	W_f^{out}
NH_3			
P			
N			
Waste			
Waste water			

[a]Measured in monetary units. [b]Measured in physical units. In [8], all the elements of W^{in} matrices are zero, except for the elements in the last two rows of W_{II}^{in}.

Table 5.10 gives a schematic view of the Dutch NAMEA with regard to the way it accounts for waste flows within the framework of an IOT (the feature of NAMEA as an extended national account is omitted). In Table 5.10 the matrix of intermediate flows and the vector of final demand (including waste treatment sectors) are indicated by X and f, all the elements of which are measured in monetary terms. The subscripts "I", "II", and "f" attached on them respectively refer to producing sectors, waste treatment sectors, and final demand.

The elements of substances account are indicated by W, where the super scripts "out" and "in" respectively refer to the generation and use of waste. For instance, the tenth rows of W_I^{out} and W_{II}^{in} respectively refer the amounts of waste generated by producing sectors, and the amounts of waste absorbed by waste treatment sectors. In the event of waste recycling in producing sectors, this will be represented as positive elements of W_I^{in}.

In line with (5.36), the levels of production of industry sectors and environmental cleansing services are given by

$$\begin{pmatrix} x_I \\ x_{II} \end{pmatrix} = \begin{pmatrix} I - A_{I,I} & -A_{I,II} \\ -A_{II,I} & I - A_{II,II} \end{pmatrix}^{-1} \begin{pmatrix} f_I \\ f_{II} \end{pmatrix} \tag{5.45}$$

with the (i, j) element of input coefficients matrices given by

$$(A_{k,l})_{ij} = (X_{k,l})_{ij}/(x_l)_j, \quad k, l = I, II \tag{5.46}$$

The substance account table enables one to compute the amounts of net generation of substances that are induced by a given final demand. Writing the matrix of net generation coefficients G_I and G_{II} as

$$\begin{pmatrix} G_I & 0 \\ 0 & G_{II} \end{pmatrix} = \begin{pmatrix} W_I^{out} - W_I^{in} & 0 \\ 0 & W_{II}^{out} - W_{II}^{in} \end{pmatrix} \begin{pmatrix} \hat{x}_I & 0 \\ 0 & \hat{x}_{II} \end{pmatrix}^{-1} \tag{5.47}$$

the net generation of substances or waste is given by

$$\begin{pmatrix} w_I \\ w_{II} \end{pmatrix} = (G_I \ G_{II}) \begin{pmatrix} I - A_{I,I} & -A_{I,II} \\ -A_{II,I} & I - A_{II,II} \end{pmatrix}^{-1} \begin{pmatrix} f_I \\ f_{II} \end{pmatrix} \tag{5.48}$$

or

$$w = G(I - A)^{-1}f \tag{5.49}$$

The above model formally resembles (5.17), but is fundamentally different from it by the presence of absorption activities, that is, some elements of G_{II} may be negative. When G_{II} contains negative elements, an increase in the final demand for waste treatment services may reduce the amount of net discharge of waste.

5.2.3.3 The German PIOT

The physical input-output table compiled at German Statistical Office (German PIOT) by Stahmer and others [44] can be mentioned as another pioneering study concerned with the incorporation of physical flows of waste into the framework of IOT (see [14] for a recent state of German PIOT). Table 5.11 shows the accounting structure of German PIOT with regard to the flow of waste.

It is similar to the Dutch NAMEA in several respects with regard to the incorporation of waste flows, but differs in some important ways. First, as the name suggests, all the elements of it are measured in physical units, even including services, so far as they involve material flows ([44], p.17). Secondly, waste is distinguished into:

- Waste for recycling
- Waste for treatment (say incineration) and
- Waste for landfill

to "make the transition from waste production to disposal easier" ([44], p.22). Individual waste types, say sludge, slag etc, are not identified.

Table 5.11 A Schematic Representation of Waste Flows in German PIOT.

		I	r	II	Final demand	ΔA_{npn}
I:	Producing sectors (n)	$X_{I,I}$	$X_{I,r}$	$X_{I,II}$	f_I	
r:	Secondary materials (1)	$X_{r,I}$	$X_{r,r}$	$X_{r,II}$	f_r	
II:	Waste treatment (1)	$X_{II,I}$	$X_{II,r}$	$X_{II,II}$	f_{II}	
		Input				
r:	Waste for recycling (1)	0	$W_{r,r}$	0	0	0
t:	Waste for treatment (1)	0	0	$W_{r,II}$	0	0
L:	Waste for landfill (1)	0	0	0	0	$W_L \iota$
		Output				
r:	Waste for recycling (1)	0	0	0	0	$W_{r,r} \iota$
t:	Waste for treatment (1)	0	0	0	0	$W_{t,II} \iota$
L:	Waste for landfill (1)	W_L	0	0	0	0

ΔA_{npn}: Changes in nonproduced natural assets (consist of several column elements). The numbers in parenthesis refer to the number of respective sectors or items. ι refers to a column vector with all the elements equal to unity, which is used to obtain the row sums of a matrix.

The third difference refers to the way in which the generation of waste is accounted for. While the generation of waste for landfill is recorded by its origin in the same way as in the Dutch NAMEA, a different procedure is employed with "waste for recycling and treatment". Instead of recording their generation by sectors of origin, their total amount occurs as an element of "Changes in nonproduced natural assets" (see the column denoted ΔA_{npn} in Table 5.11). Furthermore, the generation of residue (say ash) in the waste treatment sector is not recorded as an output of waste for landfill from that sector.

Finally, waste recycling is set to be carried out exclusively by a single "Secondary materials production sector" (Herstellung von Sekundärrohstoffen), which processes the entire amount of "waste for recycling" into secondary materials of all sorts. In the Dutch NAMEA, waste recycling occurs as the direct input of waste into the sectors using it, and there is no sector the primary product of which is secondary materials. In reality, a significant portion of waste recycling is being carried out by established industrial sectors (waste paper is recycled in paper mills, iron scrap is recycled in steel mills, waste tires are recycled in cement kilns, and so forth), and one like "Secondary materials production sector" will be difficult to find. This way of accounting for recycling should be seen as a conceptual bookkeeping device which is required for an integrated environmental and economic accounting system.

For analytical purposes, the third feature of the German PIOT is a rather inconvenient one, because it allows the computation like (5.49) for waste for landfill only, but not for waste for recycling and for treatment. If information about the origin of waste is available, however, the accounting system can be changed accordingly.

Before closing this review of waste accounts based on IO, it is important to note that the German PIOT (and the Dutch NAMEA as well) is an integrated environmental and economic accounting system, the contents of which can be extracted in many different formats depending on one's interests.

5.3 Waste IO: Concepts and Modeling

We now turn to the derivation of Waste IO, the main topic of this book, which explicitly focuses on the extension of Leontief–Duchin EIO model to account for the generation, recycling, and treatment of waste. The Dutch NAMEA provides a useful framework for incorporating the flow of waste into an IOT, while the Leontief–Duchin EIO gives a basic modeling framework for analytical purposes. Integration of the former accounting system into the modeling framework of the latter offers a good starting point for deriving Waste IO, to which we now turn.

5.3.1 The Leontief–Duchin EIO and the Dutch NAMEA

The fundamental characteristic of Leontief–Duchin EIO model consists in its assumption that waste and waste treatment correspond to each other in a one-to-one way. To the extent that this assumption holds, the row elements of matrices $X_{II,I}$ and $X_{II,II}$ represent the flows of individual waste types. To illustrate this point, suppose that there are k types of waste, to which correspond k treatment processes in a mutually exclusive one-to-one fashion. Table 5.12 shows a variant of Dutch NAMEA (Table 5.10) for this case, which differs from the original one by the absence of the rows $X_{II,\cdot}$ corresponding to the inputs of waste treatment services into other sectors of the economy.

Define the net waste generation matrices by subtracting waste absorption matrices from the corresponding waste generation matrices:

Table 5.12 An Input-Output Account with Waste Flow (I): the Case of One-to-One Correspondence Between Waste and Treatment.

From\to	Producing sectors				Waste treatment sectors				Final demand
	1	2	$\cdots$	n	1	2	$\cdots$	k	
Production sector 1									
Production sector 2			$X_{I,I}$				$X_{I,II}$		f_I
...									
Production sector n									
				Absorption					
Waste 1									
Waste 2			W_I^{in}				W_{II}^{in}		w_f^{in}
...									
Waste k									
				Generation					
Waste 1									
Waste 2			W_I^{out}				W_{II}^{out}		w_f^{out}
...									
Waste k									

Table 5.13 An Input-Output Account with Waste Flow (II): the Case of One-to-One Correspondence Between Waste and Treatment with Net Waste Generation.

From\to	Producing sectors				Waste treatment sectors				Final demand
	1	2	$\cdots$	n	1	2	$\cdots$	k	
Production sector 1									
Production sector 2		$X_{I,I}$				$X_{I,II}$			f_I
...									
Production sector n									
				Net waste generation					
Waste 1									
Waste 2		W_I				W_{II}			w_f
...									
Waste k									

$$W_i = W_i^{out} - W_i^{in}, \quad i = I, II \tag{5.50}$$

$$w_f = w_f^{out} - w_f^{in} \tag{5.51}$$

Using the net waste generation matrices thus defined, Table 5.12 can be simplified to Table 5.13. Note that by the assumption of one-to-one correspondence between wastes and treatment processes the matrix W_{II} is square, and hence the whole matrix of endogenous sectors is square as well.

Denote by x_{II} a $k \times 1$ vector of the activity level of waste treatment sectors, with $(x_{II})_i$ as its ith element. Denoting by x_I an $n \times 1$ vector of the activity levels of conventional producing sector, with $(x_I)_j$ as its jth element, the following balancing equations can be obtained from Table 5.13:

$$\sum_{j=1}^{n}(X_{I,I})_{i,j} + \sum_{j=1}^{k}(X_{I,II})_{ij} + (f_I)_i = (x_I)_i, \quad i = 1, \cdots, n \tag{5.52}$$

$$\sum_{j=1}^{n}(W_I)_{ij} + \sum_{j=1}^{k}(W_{II})_{ij} + (w_f)_i = (x_{II})_i, \quad i = 1, \cdots, k \tag{5.53}$$

Note that (5.53) holds because of the one-to-one correspondence between wastes and treatments: each waste is exclusively processed by a particular treatment sector that treats no other waste.

By use of input and waste generation coefficients, the system of equations (5.52) and (5.53) can be rewritten as:

$$\sum_{j=1}^{n}(A_{I,I})_{ij}(x_I)_j + \sum_{j=1}^{k}(A_{I,II})_{ij}(x_{II})_j + (f_I)_i = (x_I)_i, \quad i = 1, n \tag{5.54}$$

$$\sum_{j=1}^{n}(G_{I,I})_{ij}(x_I)_j + \sum_{j=1}^{k}(A_{I,II})_{ij}(x_{II})_j + (w_f)_i = (x_{II})_i, \quad i = 1, m \tag{5.55}$$

which can be represented by use of obvious matrix notations as

$$\begin{pmatrix} A_{I,I} & A_{I,II} \\ G_I & G_{II} \end{pmatrix} \begin{pmatrix} x_I \\ x_{II} \end{pmatrix} + \begin{pmatrix} f_I \\ w_f \end{pmatrix} = \begin{pmatrix} x_I \\ x_{II} \end{pmatrix} \tag{5.56}$$

Solving for x_I and x_{II} gives

$$\begin{pmatrix} x_I \\ x_{II} \end{pmatrix} = \begin{pmatrix} I - A_{I,I} & -A_{I,II} \\ -G_I & I - G_{II} \end{pmatrix}^{-1} \begin{pmatrix} f_I \\ w_f \end{pmatrix} \tag{5.57}$$

Note that (5.57) is identical to the Leontief EIO model (5.36) (with the amount of waste untreated, f_{II}, set equal to zero), which implies:

$$A_{II,I} = G_I, \quad A_{II,II} = G_{II}, \quad w_f = A_{II,f} f_I \tag{5.58}$$

The solution for waste treatment given by (5.57) differs from the corresponding expression for Dutch NAMEA given by (5.49) in several important respects.

1. The activity levels of treatment sectors correspond to the physical amounts of wastes treated.
2. A change in the composition of final demand can have different effects on the activity levels of different treatment sectors because of the difference in the amounts and types of wastes generated in different sectors.

In the Dutch NAMEA model (5.49) both waste and waste treatment occur separately, with the former expressed in physical terms and the latter expressed in monetary terms. To the extent that it is possible to obtain the monetary value of the activity of a waste treatment sector by multiplying the amount of waste it treated by the price it charges per unit of treatment, the monetary value can easily be retrieved from the physical value: separate information on the monetary value will then be redundant.

The presence of one-to-one correspondence between waste types and treatment processes is fundamental to the derivation of the above model with waste (5.57). In reality, however, waste types and treatment processes will not correspond to each other in a one-to-one fashion, nor will the correspondence be exclusive:

1. A given treatment process can be applied to different types of waste: any solid waste can be landfilled.
2. A given waste can be subjected to different types of treatment: garbage can be, among others, incinerated, gasified, composted, or landfilled.
3. The number of types of waste will exceed the number of types of treatment processes.

The final point, although relevant in practice, is conceptually inessential: satisfaction of the equality condition alone does not imply the presence of a one-to-one correspondence between waste and treatment. A mere equality of the number of waste types and that of treatment types may make the matrix W_{II} square. The solution (5.57) will then be of no economic meaning because in this case x_{II} is not defined ((5.53) does not hold).

5.3.2 The Waste IO

The case where the one-to-one correspondence between waste and waste treatment does not hold is now considered.

Consider first the case of n producing sectors and two waste types given in Table 5.14, where waste 1 is combustible (organic waste or plastics), and waste 2 is not combustible (inorganic waste). While Table 5.14 represents the generation of waste, it does not give a closed system of the flow of waste because of the lack of its treatment processes. The rest of this section is concerned with the closing of the system with regard to the flow of waste.

5.3.2.1 The Case Where Landfill Is the Only Treatment Process

To begin with, consider a simple case where all wastes are landfilled. Write a_{n+1} for an $n \times 1$ vector of input coefficients of the landfill process, and assume that the landfill process generates no waste. Augmenting Table 5.14 with the landfill process as its $n + 1$st column elements, Table 5.15 is obtained.

Writing x_{n+1} for the activity level of the landfill process, that is, the amount of waste going to a landfill, the balancing equations for production and waste treatment are given by

Table 5.14 Unit Processes with Waste Generation and Final Demand.

	Producing sectors			Final demand
Input\output	1	2 $\cdots$	n	
1 Sector 1				
2 Sector 2		$A_{I,I}$		f_I
$\cdots$				
n Sector n				
		Waste generation		
Waste 1 (combustible) Waste 2 (not combustible)		G_I^{out}		w_f

Table 5.15 Unit Processes with Waste Generation and Final Demand with Landfill.

	Producing sectors			Waste treatment	Final demand
Input\output	1	2 $\cdots$	n	Landfill	
1 Sector 1					
2 Sector 2		$A_{I,I}$		a_{n+1}	f_I
$\cdots$					
n Sector n					
		Waste generation			
Waste 1 Waste 2		G_I^{out}		0	w_f

Table 5.16 Unit Processes with Waste Generation and Final Demand with Landfill.

		Producing sectors	Waste treatment	Final demand
		1 2 $\cdots$ n	Landfill	
1	Sector 1			
2	Sector 2		a_{n+1}	f_{I}
	...	$A_{\mathrm{I},\mathrm{I}}$		
n	Sector n			
$n+1$	Landfill	$\iota^{\top} G_{\mathrm{I}}^{\mathrm{out}}$	0	$\iota^{\top} w_f$

$$
\begin{aligned}
A_{\mathrm{I},\mathrm{I}} x_{\mathrm{I}} + a_{n+1} x_{n+1} + f_{\mathrm{I}} &= x_{\mathrm{I}} \\
\iota^{\top} G_{\mathrm{I}}^{\mathrm{out}} x_{\mathrm{I}} + 0 x_{n+1} + \iota^{\top} w_f &= x_{n+1}
\end{aligned}
\tag{5.59}
$$

or in a matrix form

$$
\begin{pmatrix} A_{\mathrm{I},\mathrm{I}} & a_{n+1} \\ \iota^{\top} G_{\mathrm{I}}^{\mathrm{out}} & 0 \end{pmatrix} \begin{pmatrix} x_{\mathrm{I}} \\ x_{n+1} \end{pmatrix} + \begin{pmatrix} f_{\mathrm{I}} \\ \iota^{\top} w_f \end{pmatrix} = \begin{pmatrix} x_{\mathrm{I}} \\ x_{n+1} \end{pmatrix}
\tag{5.60}
$$

The left-hand side of the second equation of (5.59) (or its counterpart in (5.60)) refers to the generation of waste for landfill, while the right-hand side refers to the total amount of waste disposed to a landfill. The column elements of $G_{\mathrm{I}}^{\mathrm{out}}$ and w_f are added up because of the assumption that all wastes are landfilled. In other words, in deriving the second equation of (5.60), the rectangular input coefficients matrix in Table 5.15 has been transformed to a square one in Table 5.16 where landfill occurs as both column and row elements. Solving (5.60) for x, it follows that

$$
\begin{pmatrix} x_{\mathrm{I}} \\ x_{n+1} \end{pmatrix} = \begin{pmatrix} I - A_{\mathrm{I},\mathrm{I}} & -a_{n+1} \\ -\iota^{\top} G_{\mathrm{I}}^{\mathrm{out}} & 1 \end{pmatrix}^{-1} \begin{pmatrix} f_{\mathrm{I}} \\ \iota^{\top} w_f \end{pmatrix}
\tag{5.61}
$$

which corresponds to a special case of the Leontief EIO model (5.57), where a single waste treatment process is applied to two waste types.

5.3.2.2 The Case Where Waste Can Also Be Incinerated

Next, consider the case where incineration occurs as an additional waste treatment process. Because waste 1 is combustible, it can be incinerated instead of being landfilled. For waste 2 landfill remains the only treatment option that is applicable because it is not combustible. Denote by a_{n+2} a vector of input coefficients that is associated with the incineration of a unit of waste 1, and by g_{n+2}^{out} the associated generation of residues (bottom- and fly ash). The introduction of incineration increases the types of waste from two to three due to the additional generation of residues (waste 3). Suppose that residues are landfilled. Augmenting Table 5.15 with the incineration process as its $n+2$nd column results in Table 5.17.

Table 5.17 Unit Processes with Waste Generation and Final Demand with Landfill and Incineration.

	Producing sectors				Waste treatment		Final demand
	1	2 $\cdots$	n		Landfill	Incineration	
Sector 1							
Sector 2		$A_{I,I}$			a_{n+1}	a_{n+2}	f_1
...							
Sector n							
	Waste generation						
Waste 1 (combustible)		$G_{1,I}^{out}$			0	0	w_{f1}
Waste 2 (noncombustible)		$G_{2,I}^{out}$			0	0	w_{f2}
Waste 3 (residue)		0			0	g_{n+2}^{out}	0

Denote by x_{n+2} the activity level of the incineration sector, that is, the amount of waste incinerated. Noting that waste 1 is incinerated, while waste 2 and waste 3 are landfilled, (5.60) is extended as follows:

$$\begin{pmatrix} A_{I,I} & a_{n+1} & a_{n+2} \\ G_{2,I}^{out} & 0 & g_{n+2}^{out} \\ G_{1,I}^{out} & 0 & 0 \end{pmatrix} \begin{pmatrix} x_I \\ x_{n+1} \\ x_{n+2} \end{pmatrix} + \begin{pmatrix} f_1 \\ w_{f2} \\ w_{f1} \end{pmatrix} = \begin{pmatrix} x_I \\ x_{n+1} \\ x_{n+2} \end{pmatrix} \tag{5.62}$$

Note that the first matrix on the left-hand side is a square one with landfill and incineration occurring as both column and row elements. Solving for x yields:

$$\begin{pmatrix} x_I \\ x_{n+1} \\ x_{n+2} \end{pmatrix} = \begin{pmatrix} I - A_{I,I} & -a_{n+1} & -a_{n+2} \\ -G_{2,I}^{out} & 1 & -g_{n+2}^{out} \\ -G_{1,I}^{out} & 0 & 1 \end{pmatrix}^{-1} \begin{pmatrix} f_1 \\ w_{f2} \\ w_{f1} \end{pmatrix} \tag{5.63}$$

This corresponds to a special case of the Leontief EIO model (5.57), where two waste treatment processes are applied to three waste types.

5.3.2.3 Generalization to the Case of m Waste Types and k Treatment Processes: The Use of Allocation Matrix

The solutions (5.61) and (5.63) have been obtained above from the coefficients matrices, the original forms of which were not square (see Table 5.14), after having transformed them to square matrices by establishing the allocation of different waste types to different waste treatment processes. In the case of (5.61), all wastes were allocated to landfill, which is the only available treatment processes, whereas in the case of (5.63) wastes were allocated to incineration and landfill according to whether they are combustible.

Let there be m waste types and k treatment processes. The allocation of wastes to treatment processes can be formalized by using an allocation matrix $S = (s_{ij})$

of order $k \times m$, with s_{ij} referring to the share of waste j that is allocated to treatment process i, when the waste is subjected to waste treatment. Because waste for treatment has to be subjected to at least one treatment process, it follows:

$$\sum_{i=1}^{k} s_{ij} = 1, \, j = 1, \cdots m \tag{5.64}$$

or

$$\iota_k^{\top} S = \iota_m^{\top} \tag{5.65}$$

Henceforth, the subscript k attached to ι will be suppressed except for when its use appears indispensable.

Equation (5.61) corresponds to the case of S given by

$$S = \begin{pmatrix} 1 & 1 \end{pmatrix} = \iota^{\top} \tag{5.66}$$

which is obvious by the occurrence of $\iota^{\top}$ in (5.61). Alternatively, consider the balancing equations for waste that corresponds to Table 5.15:

$$G_{\mathrm{I}}^{\mathrm{out}} x_{\mathrm{I}} + w_f = \begin{pmatrix} w_1 \\ w_2 \end{pmatrix} \tag{5.67}$$

where w_i refers to the amount of waste i for treatment with $i = 1, 2$. Multiplication of both sides from the left by S given by (5.66) gives

$$\iota^{\top} G_{\mathrm{I}}^{\mathrm{out}} x_{\mathrm{I}} + \iota^{\top} w_f = \sum_{i=1}^{2} w_i = x_{n+1} \tag{5.68}$$

which is the second equation of (5.60), where the second equality follows from the fact that all wastes are landfilled.

Equation (5.63) corresponds to the case where S is given by

$$S = \begin{pmatrix} 0 & 1 & 1 \\ 1 & 0 & 0 \end{pmatrix} \tag{5.69}$$

Analogous to the case of (5.61), consider the balancing equation for the waste flow in Table 5.17:

$$\begin{pmatrix} G_{1,\mathrm{I}}^{\mathrm{out}} & 0 & 0 \\ G_{2,\mathrm{I}}^{\mathrm{out}} & 0 & 0 \\ 0 & 0 & g_{n+2}^{\mathrm{out}} \end{pmatrix} \begin{pmatrix} x_{\mathrm{I}} \\ x_{n+1} \\ x_{n+2} \end{pmatrix} + \begin{pmatrix} w_{f1} \\ w_{f2} \\ 0 \end{pmatrix} = \begin{pmatrix} w_1 \\ w_2 \\ w_3 \end{pmatrix} \tag{5.70}$$

where $G_{1,\mathrm{I}}^{\mathrm{out}}$ and $G_{2,\mathrm{I}}^{\mathrm{out}}$ refer to the first and second rows of $G_{\mathrm{I}}^{\mathrm{out}}$. Multiplication of S given by (5.69) from the left gives

$$\begin{pmatrix} G_{1,\mathrm{I}}^{\mathrm{out}} & 0 & g_{n+2}^{\mathrm{out}} \\ G_{2,\mathrm{I}}^{\mathrm{out}} & 0 & 0 \end{pmatrix} \begin{pmatrix} x_{\mathrm{I}} \\ x_{n+1} \\ x_{n+2} \end{pmatrix} + \begin{pmatrix} w_{f2} \\ w_{f1} \end{pmatrix} = \begin{pmatrix} w_2 + w_3 \\ w_1 \end{pmatrix} = \begin{pmatrix} x_{n+1} \\ x_{n+2} \end{pmatrix} \tag{5.71}$$

which is the balancing equation for waste treatment in (5.62).

It is now straightforward to generalize the above results. Write the balancing equation for m waste types with k treatment processes as

$$G_I^{out} x_I + G_{II}^{out} x_{II} + w_f = w \qquad (5.72)$$

Multiplication of S of order $k \times m$ from the left gives

$$S G_I^{out} x_I + S G_{II}^{out} x_{II} + S w_f = S w = x_{II} \qquad (5.73)$$

The system of balancing equations for goods and waste as a whole is then given by

$$\begin{pmatrix} A_{I,I} & A_{I,II} \\ S G_I^{out} & S G_{II}^{out} \end{pmatrix} \begin{pmatrix} x_I \\ x_{II} \end{pmatrix} + \begin{pmatrix} f_I \\ S w_f \end{pmatrix} = \begin{pmatrix} x_I \\ x_{II} \end{pmatrix} \qquad (5.74)$$

with the solution for x given by

$$\begin{pmatrix} x_I \\ x_{II} \end{pmatrix} = \begin{pmatrix} I - A_{I,I} & -A_{I,II} \\ -S G_I^{out} & I - S G_{II}^{out} \end{pmatrix}^{-1} \begin{pmatrix} f_I \\ S w_f \end{pmatrix} \qquad (5.75)$$

where

$$f_{II} = S w_f \qquad (5.76)$$

So far as the generation of waste and its treatment are concerned, (5.75) represents a generalization of (5.57) to the case where there is no one-to-one correspondence between wastes and treatment processes. Closing the model for waste flows for this general case still requires incorporation of waste recycling, to which we now turn.

5.3.2.4 Recycling

The above model is now extended to incorporate recycling (and reuse), which refers to the use of by-product as input in production and household consumption. In Section 3.2, by-product was classified into two types, by-product (I) and by-product (II), depending on whether there exists a production sector where the by-product is obtained as the primary product (Definition 3.1). It is thus useful to distinguish the case according to whether it deals with the use of by-product (I) or by-product (II).

The Recycling of By-Product (I)

A typical example of by-product (I) which is associated with waste treatment as considered in the previous section will be electricity generated by the heat produced in the incineration process (the heat is used to generate steam which is used to drive a turbine to produce electricity). For instance, Denmark has been a leading country in the use of the heat from incineration for more than a century, in localized combined heat and power facilities supporting district heating schemes [26, 49]. In

2005, waste incineration produced 4% of the electricity consumption and 13.7% of the total domestic heat consumption in Denmark [5].

A straightforward way to incorporate electricity from the heat of a waste incineration process is to resort to the Leontief model with by-product as described in Section 3.2.2, where electricity as by-product occurs as a negative input into the waste incineration process. Denoting by n the electric power generation sector, where electricity is obtained as the primary product, the unit process of waste incineration in Table 5.17 with electricity as by-product can be given by

$$\begin{pmatrix} a_{1,n+2} \\ a_{2,n+2} \\ \cdots \\ -a_{n,n+2} \\ 0 \\ 0 \\ g_{n+2}^{out} \end{pmatrix} \tag{5.77}$$

where $a_{n,n+2}$ refers to the surplus of electricity obtained from the incineration of a unit of waste of a given chemical composition (the amount of heat generated from the incineration of waste depends on its chemical composition, see Section 6.2 below for further details) per unit of time.

In general, the recycling of by-product (I) can be accounted for in (5.74) as a negative input from the sector, where the product is primarily produced, into the sector where it is obtained as by-product.

A word of caution is due, however, because, as has already been pointed out in Section 3.2.4, a mechanical application of this procedure may, under circumstances, result in a negative production, which is impossible in reality. For illustration, let product B be the by-product obtained in the production of product A. In the case where the final demand for product A is large relative to the final demand for product B, application of the above procedure could result in a negative production of product B. In this case, the relevant by-product should not be subjected to the above procedure, but should be subjected to the one designed for by-product (II), to which we now turn.

The Recycling of By-Product (II)

By-product (II) is characterized by the fact that it is not obtained as a primary product in any producing sector of the system under consideration. Ash (or molten slag, which is obtained by melting ash) generated by the incineration process can be mentioned as a typical example of by-product (II) which is associated with waste treatment. Ash or molten slag can be used (recycled) as a material for cement production within a limited range as a substitute for clay or sand.

Let sector $n - 1$ be the cement producing sector, where ash is used as an input. Its unit process is then given by

$$\begin{pmatrix} a_{1,n-1} \\ a_{2,n-1} \\ \cdots \\ a_{n,n-1} \\ 0 \\ 0 \\ 0 \\ 0 \\ 0 \\ g_{3,n-1}^{in} \end{pmatrix} \tag{5.78}$$

where $g_{3,n-1}^{in}$ refers to the input of ash per unit of cement. Subtracting the amount of waste recycled from the amount generated gives the net amount of waste generation. Assuming that the cement process is the only process where ash from the incineration process is used, and that the incineration process is the only process where ash is generated as by-product, the balancing equation for ash is given by

$$\underbrace{-g_{3,n-1}^{in}x_{n-1}}_{\text{ash used}} + \underbrace{g_{3,n+2}^{out}x_{n+2}}_{\text{ash generated}} = \underbrace{w_3}_{\text{ash for treatment}} \tag{5.79}$$

Recall that when not recycled ash is disposed to a landfill.

The above equation implies that if there is an excess supply of ash over the amount that can be absorbed by the cement process, the excess is treated as waste, and landfilled. This is in sharp contrast to the method for by-product (I) mentioned above, where it is assumed that the supply of by-product is always met by the equal amount of demand for it. Because of the explicit consideration of the adjustment mechanism of waste treatment sectors in dealing with an excess supply of by-product, the occurrence of a negative production as mentioned above can be avoided. In the above example, an excess supply of ash would not lead to a negative production of sand or gravel, but to the disposal of ash in a landfill.

A numerical example will be useful to demonstrate the difference between the present method and the one for by-product (I). In Section 3.2.4 an example was shown where the occurrence of negative input results in negative production (see (3.67)). Using the same numerical example as in (3.67), consider the following matrix where the standard Leontief matrix $A_{I,I}$ of order 3×3 is augmented from right by the 3×1 matrix of the input coefficients of waste treatment process, $A_{I,II}$, and from below by the row of net waste generation coefficients $G = (G_I, G_{II})$ with $G_I = (g_1, g_2, g_3)$ and $G_{II} = g_4$:

$$\begin{pmatrix} A_{I,I} & A_{I,II} \\ G_I & G_{II} \end{pmatrix} = \begin{pmatrix} a_{11} & a_{12} & a_{13} & a_{14} \\ a_{21} & a_{22} & a_{23} & a_{24} \\ a_{31} & a_{32} & a_{33} & a_{34} \\ g_1 & g_2 & g_3 & g_4 \end{pmatrix} = \begin{pmatrix} 0 & 0 & 0 & 0.001 \\ 0.4 & 0 & 0.3 & 0.001 \\ 0.1 & 0.1 & 0 & 0.001 \\ 0 & 0.3 & -0.15 & 0 \end{pmatrix} \tag{5.80}$$

Here, $g_2 = 0.3$ refers to the generation of by-product (II) per unit of Product 2, while $g_3 = -0.15$ refers to that 0.15 units of the by-product are used in the production of a unit of Product 3. Note that this corresponds to the case where product 1 is completely substituted by the by-product (see (3.61)). Let the level of final demand be given by

$$\begin{pmatrix} f_I \\ w_f \end{pmatrix} = \begin{pmatrix} 1 \\ 10 \\ 1 \\ 0 \end{pmatrix} \tag{5.81}$$

The system of balancing equations becomes

$$\begin{pmatrix} A_{I,I} & A_{I,II} \\ G_I & G_{II} \end{pmatrix} \begin{pmatrix} x_I \\ x_{II} \end{pmatrix} + \begin{pmatrix} f_I \\ w_f \end{pmatrix} = \begin{pmatrix} x_I \\ x_{II} \end{pmatrix} \tag{5.82}$$

with the equation for the by-product given by

$$G_I x_I + G_{II} x_{II} + w_f = 0.3 x_2 - 0.15 x_3 = w_1 = x_4 \tag{5.83}$$

where the equality follows from the fact that the excess of the by-product over the amount of demand for it is exclusively treated by a single waste treatment process, the activity level of which is x_4. Solving the system for x yields

$$\begin{pmatrix} x_1 \\ x_2 \\ x_3 \\ x_4 \end{pmatrix} = \begin{pmatrix} I - A_{I,I} & -A_{I,II} \\ -G_I & 1 - G_{II} \end{pmatrix}^{-1} \begin{pmatrix} f_I \\ w_f \end{pmatrix}$$

$$= \begin{pmatrix} 1.000 & 0.000 & 0.000 & 0.001 \\ 0.443 & 1.031 & 0.309 & 0.002 \\ 0.144 & 0.103 & 1.031 & 0.001 \\ 0.111 & 0.294 & -0.062 & 1.000 \end{pmatrix} \begin{pmatrix} 1 \\ 10 \\ 1 \\ 0 \end{pmatrix} = \begin{pmatrix} 1.0 \\ 11.1 \\ 2.2 \\ 3.0 \end{pmatrix} \tag{5.84}$$

which indicates that of the 3.33 ($=0.3 \times 11.1$) units of by-product generated only 0.3 ($=0.15 \times 2.2$) units are recycled, with the remaining 3.0 units being treated as waste. For comparison, application of (3.67) to this case yields a negative value for x_1:

$$x_I = (I - \check{A})^{-1} f_I \tag{5.85}$$

$$= \begin{pmatrix} 0.899 & -0.264 & 0.055 \\ 0.398 & 0.913 & 0.334 \\ 0.129 & 0.064 & 1.039 \end{pmatrix} \begin{pmatrix} 1 \\ 10 \\ 1 \end{pmatrix} = \begin{pmatrix} -1.7 \\ 9.9 \\ 1.8 \end{pmatrix} \tag{5.86}$$

In contrast, in (5.84) the occurrence of negative production is excluded for any non-negative final demand for products because the upper 3×4 matrix in the first parenthesis on its right-hand side does not contain negative elements.

On the other hand, when the final demand for Product 3 is large relative to the
final demand for Product 2, the demand for the by-product of Sector 2 could exceed
its available amount, and could result in a negative value of x_4. This implies that
the available amount of by-product is short of the amount required to meet the final
demand. In other words, the final demand is not feasible for the technology matrix.

General Representation

With recycling of by-product incorporated for both by-product (I) and by-product
(II), the system of balancing equations (5.74) now becomes

$$\begin{pmatrix} A_{I,I} & A_{I,II} \\ SG_I & SG_{II} \end{pmatrix} \begin{pmatrix} x_I \\ x_{II} \end{pmatrix} + \begin{pmatrix} f_I \\ Sw_f \end{pmatrix} = \begin{pmatrix} x_I \\ x_{II} \end{pmatrix} \tag{5.87}$$

with the solution for x given by

$$\begin{pmatrix} x_I \\ x_{II} \end{pmatrix} = \begin{pmatrix} I - A_{I,I} & -A_{I,II} \\ -SG_I & I - SG_{II} \end{pmatrix}^{-1} \begin{pmatrix} f_I \\ Sw_f \end{pmatrix} \tag{5.88}$$

Equation (5.88) represents an extended version of the Leontief EIO quantity model
that is applicable to waste management issues where there is no one-to-one correspondence between waste types and treatment processes. Because of this distinguishing feature, the approach is called Waste Input-output (WIO) [37, 38].

While (5.88) gives the level of waste treatment activity for a given set of final
demand and technology, it does not give the composition of waste for treatment.
The latter is provided by:

$$w = \begin{pmatrix} G_I & G_{II} \end{pmatrix} \begin{pmatrix} I - A_{I,I} & -A_{I,II} \\ -SG_I & I - SG_{II} \end{pmatrix}^{-1} \begin{pmatrix} f_I \\ Sw_f \end{pmatrix} + w_f \tag{5.89}$$

$$= \begin{pmatrix} \Xi_I & \Xi_{II} \end{pmatrix} \begin{pmatrix} f_I \\ Sw_f \end{pmatrix} + w_f \tag{5.90}$$

where Ξ_I is an $m \times n$ matrix, with $(\Xi_I)_{ij}$ referring to the net generation of waste i that
is directly and indirectly generated to satisfy a unit of product j for final delivery,
while Ξ_{II} is an $m \times k$ matrix, with $(\Xi_{II})_{io}$ referring to the net generation of waste i
that is directly and indirectly generated for a unit of waste treatment activity o.

The use of allocation matrix implies that a given amount of waste of different
types can be subjected to different treatment processes depending on the characteristics of the allocation matrix. For instance, waste plastics can be incinerated, or
landfilled (see Section 6.1.2.4 below for further details on this point). Furthermore,
as indicated by (5.89), the amount of recycling of waste of a particular type depends
on a large number of factors involving both demand and technology conditions, and
hence is likely to change when some of them, for instance, the composition of final
demand, changes. These observations imply that the way waste is divided according

to its treatment methods and recycling as used in the German PIOT (Section 5.2.3.3) would change depending on the allocation matrix and the composition of final demand, and should not be regarded as a fixed one.

5.3.2.5 The Constancy of S

The effectiveness of WIO as represented by (5.88) hinges upon the validity of assuming the independence of S from the level of final demand, because otherwise the model cannot be applied to analyzing the effects on production of a change in the final demand. A necessary condition for this is the presence of treatment capacity that is large enough to meet any levels of final demand under consideration, or the non-actualization of the limits of treatment. Suppose, for instance, that incineration and landfill happened to be the only options available to treat municipal solid waste (MSW) with the capacity of the former more limited (say due to the limited number of waste incinerators in operation) than that of the latter, and that incineration is a preferred treatment option in this municipality. If there is an increase in the amount of MSW, which exceeds the capacity of the incinerators, it would have to be shifted to a landfill. In this case, the share of landfill in the treatment of MSW would begin to increase as soon as the amount of MSW reaches the treatment capacity of incinerators.

As mentioned in Section 4.4.3, the presence of a productive capacity that is large enough to meet any levels of final demand under consideration, or the non-actualization of the limits of production, is a standard assumption in IOA. Following this standard practice of IOA, in the rest of this chapter it is assumed that treatment capacity is available that is large enough to meet any levels of final demand under consideration. In the next chapter, however, a more general approach will be shown where given limits of treatment capacity are explicitly taken into account.

Economic factors, technological factors, and institutional factors can be mentioned as major determinants of the elements of allocation matrix S. To the extent that these factors remain unchanged, it is likely that the elements of S remain constant as well. Economic factors refer to the choice of treatment processes, to which given waste are to be subjected, based on economic criteria such as cost minimization. As such, this involves the issue of the choice of technology, an issue which is to be addressed in the next chapter in conjunction with the extension of WIO. Leaving the consideration of economic factors for the next chapter, this chapter will be concerned with technological and institutional factors.

Technological Factors

The absence of a one-to-one correspondence between waste types and treatment methods is a general feature of waste treatment. Still, it does not imply that any combination of them is feasible. In fact, technological factors impose fundamental constraints on the set of possible combinations. Only a limited range of treatment processes can be applicable to some types of waste.

Any solid waste can in principle be landfilled. Disposing of liquid waste to a landfill, however, represents a technological mismatch because it runs at the risk of contaminating the soil and ground water. One of the main aims of waste incineration is, besides its disinfection, to reduce the volume of waste. Incineration of inorganic waste such as construction debris or inorganic sludge will then make little sense because it will not contribute to the reduction of waste volume, but will certainly contribute to deteriorating the efficiency of the incinerator. On the other hand, large shredding machines are designed to be applied to end of life vehicles or bulky appliances. Applying them to typical items of MSW such as garbage, discarded paper, or waste plastics would only contribute to the decline in the efficiency of their operation and the quality of recovered materials.

Technological factors thus provide fundamental constraints upon the degree to which the elements of S can vary.

Institutional Factors

Institutional factors refer to regulations on the way specific wastes are treated. The directive of European Union (EU) on landfill, Council Directive 99/31/EC, for instance, states that the following wastes should not be accepted in a landfill [12]:

- Liquid waste
- Flammable waste
- Explosive or oxidizing waste
- Hospital and other clinical waste which is infectious
- Used tires, with certain exceptions
- Any other type of waste which does not meet the acceptance criteria laid down in Annex II

In terms of S, this directive can be represented by

$$(S)_{i,\text{landfill}} = 0, \qquad (5.91)$$

where i refers to each waste in the above list. If incineration happens to be the only alternative option for treatment, the following would also hold:

$$(S)_{i,\text{incineration}} = 1.0 \qquad (5.92)$$

The ban on landfilling of waste suitable for incineration in Denmark that took effect in 1997 [6] can be mentioned as a precedent for this directive.

Another well known example of EU directives on waste is the one on waste electrical and electronic equipment (WEEE), Directive 2002/96/EC, which states the minimum rates of recovery for individual categories of WEEE, and the minimum rates of reuse and/or recycling of their component, material, and substance [13]:

- Large household appliances and automatic dispensers: recovery at 80% with reuse and recycling at 75%
- IT and telecommunications equipment, and consumer equipment: recovery at 75% with reuse and recycling at 65%
- Small household appliances, lighting equipment, electrical and electronic tools, toys, leisure and sports equipment, monitoring and control instruments: recovery at 70% with reuse and recycling at 50%

If landfill happens to be the only alternative to recovery (for reuse and recycling) for end of life consumer equipment included in the above list, say i, we would have

$$(S)_{i, \text{reuse \& recycling}} = .8, \text{ and } (S)_{i, \text{landfill}} = .2 \qquad (5.93)$$

The Japanese law for recycling of specified kinds of home appliances that was enforced in year 2001 can be mentioned as a precedent for this directive. Under the law, television (TV) sets, air conditioners, refrigerators, and washing machines have to be recovered, and specific percentages of their mass have to be reused or recycled [36], which implies:

$$(S)_{i, \text{reuse \& recycling}} = 1.0, \text{ and } (S)_{i, \text{landfill}} = 0.0 \qquad (5.94)$$

where i refers to each of the four items of appliances.

The regulation on the treatment of fly ash generated by waste incinerators can be mentioned as another example in Japan which refers to a specific form of S. In Japan, fly ash is designated as specially controlled waste due to harmful heavy metals contained in it, and is legislated to be subjected to further treatment, such as melting and solidification by cement, before disposal to a landfill, which implies

$$(S)_{i, \text{melting}} \geq 0, (S)_{i, \text{solidification}} \geq 0, \text{ and } (S)_{i, \text{landfill}} = 0.0 \qquad (5.95)$$

where i refers to fly ash from waste incinerators.

To the extent that certain regulations determine the way in which specific waste are allocated to specific treatment methods, the relevant elements of S would remain constant so long as the regulations are in effect (provided enough amounts of treatment capacity are available to meet the needs for treating the relevant waste).

The Distribution Among Waste Treatment Options: An International Comparison

Above, some examples of regulations on waste treatment have been shown. We now turn to empirical examples of country-specific allocation patterns of waste to treatment methods, which indicate relative constancy of the allocation pattern.

In developed countries, at least, incineration and (sanitary) landfilling represent two major treatment methods of MSW. The distribution of MSW among these treatment methods widely differs among countries depending on factors such as climate, land area, and availability of open space. Landfilling has been the dominant method

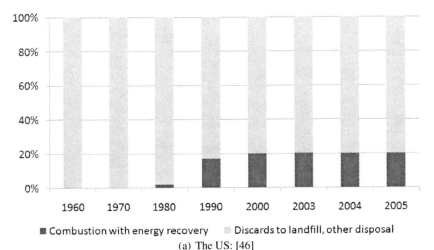

(a) The US: [46]

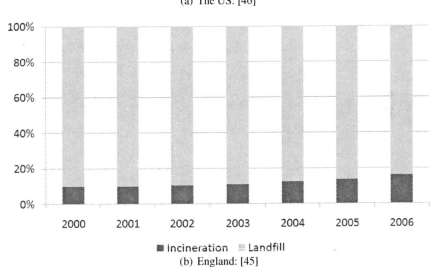

(b) England: [45]

Fig. 5.3 The Distribution Among Waste Treatment Options: Countries Where Landfill Is the Major Treatment Option.

in countries like Sweden, the UK (England), and the US, while incineration has been dominant in Denmark, Japan, and Switzerland [32].

Figure 5.3 shows the development over time of the distribution of incineration and landfill for the US and the UK (England), where landfilling is the major treatment method. On the other hand, Figure 5.4 shows the same for Denmark, Switzerland, and Japan, where incineration is the major treatment method. As these figures show, while the relative distribution among treatment options may differ among countries, it has been rather stable for each country.

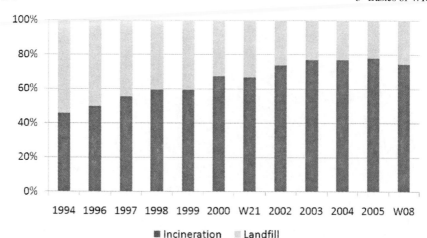

(a) Denmark: [6, 7]. W21: Targets for treatment in the Danish Government's Waste Management Plan 1998–2004, W08: Targets for the Danish government's Waste Strategy 2005–2008.

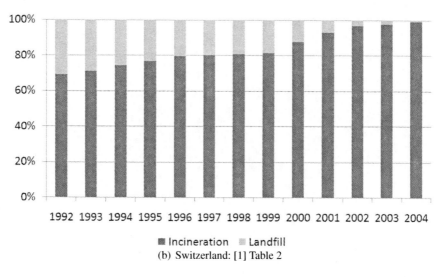

(b) Switzerland: [1] Table 2

Fig. 5.4 The Distribution Among Waste Treatment Options: Countries Where Incineration Is the Major Treatment Option.

In developing countries, the majority of MSW still ends up in uncontrolled landfills because of the absence of alternative treatment methods [34]. Changing the allocation pattern requires the introduction of alternative treatment methods, the construction of which would involve considerable costs and time.

These observations indicate that, unless there is a noticeable change in regulations for waste treatment, the elements of S would change rather slowly and at least would not significantly react to changes in current economic conditions, such as the

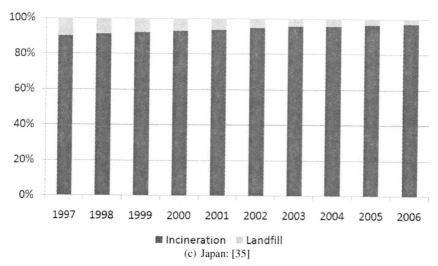

■ Incineration   ▨ Landfill
(c) Japan: [35]

Fig. 5.4 (continued).

level of final demand. It can be said that assuming the constancy of S is not unduly strong compared to the fundamental assumption in IOA of the constancy of input coefficients. If one is ready to accept the constancy of input coefficients as a useful working hypothesis, the same would apply to the use of a constant S as well. Furthermore, as we shall see later, the constancy, and hence the exogenous nature of S, makes it a convenient means of representing alternative policy scenarios with regard to the allocation of waste to alternative treatment methods.

References

1. Abfallstatistik (2004). *Zahlen und Entwicklungen der schweizerischen Abfallwirtschaft im Jahr 2004*. Bern: Bundesamt f¨ur Umwelt BAFU, Bern.
2. Bullard, C., & Herendeen, R. (1975). The energy cost of goods and services. *Energy Policy, 3*, 268–278.
3. Centre for Integrated Sustainability Analysis@ The University of Sydney. *Ecological systems.* http://www.isa.org.usyd.edu.au/research/ee.shtml. Cited 3 May 2008.
4. Daly, H. (1968). On economics as a life science. *Journal of Political Economy, 76*, 392–406.
5. Danish Energy Authority (2007). *Energy statistics 2005*. http://ens.dk/graphics/Publikationer/ Statistik_UK/Energy_statistics_2005/index.htm. Cited 22. July 2008.
6. Danish Environmental Protection Agency (2002). Waste Statistics 2000, Environmental Review No. 1 2002. Danish Environmental Protection Agency, Copenhagen 2005.
7. Danish Environmental Protection Agency (2007). Waste Statistics 2005, Environmental Review No. 6 2007. Danish Environmental Protection Agency, Copenhagen.
8. De Haan, M., & Keuning, S. J. (1996). Taking the environment into account: the NAMEA approach. *Review of Income Wealth, 42*, 131–148.
9. Diamond, J. (2005). *Collapse: How societies choose to fall or succeed*. New York: Viking.
10. Duchin, F. (1990). The conversion of biological materials and waste to useful products. *Structural Change and Economic Dynamics, 1*, 243–261.

11. Economic Input-Output Life Cycle Assessment, Green Design Institute, Carnegie Mellon University. http://www.eiolca.net/
12. European Commission, Environment, Waste. http://ec.europa.eu/environment/waste/landfill_index.htm. Cited 21 July 2008.
13. European Commission, Environment. http://ec.europa.eu/environment/waste/weee/index_en.htm. Cited 21 July 2008.
14. Federal Statistical Office, Germany, Endbericht zum Projekt A Physical Input-Output-Table for Germany 1995, Vertragsnummer 98/559/3040/B4/MM, 2001 Wiesbaden.
15. Gallego, B., & Lenzen, M. (2008). Estimating generalized regional input-output life cycle assessment case study. In M. Ruth & B. Davidsdottir (Eds.), *Dynamics of industrial ecosystems*. Boston, MIT Press.
16. Global Environment Centre Foundation Air pollution control technology in Japan. http://www.gec.jp/AIR/. Cited 28 July 2008.
17. Hannon, B. (1973). The structure of ecosystems. *Journal of Theoretical Biology, 41*, 535–546.
18. Hannon, B., & Joiris, C. (1989). A seasonal analysis of the southern North Sea ecosystem. *Ecology, 70*, 1916–1934.
19. Heijungs, R., & Suh, S. (2002). *The computational structure of life cycle assessment*. Dordrecht: Kluwer.
20. Hendrickson, C., Lave, L., & Matthews, S. (2006). *Environmental life cycle assessment of goods and services, an input-output approach*. Washington, DC: Resources for the Future.
21. Huang, G., Anderson, W., & Baetz, B. (1994). Environmental input-output analysis and its application to regional solid-waste management planning. *Journal of Environmental Management, 42*, 63–79.
22. Isard, W., Bassett, K., Choguill, C., Furtado, J., Izumita, R., Kissin, J., Romanoff, E., Seyfarth, R., & Tatlock, R. (1968). On the linkage of socio-economic and ecologic systems. *Papers in Regional Science, 21*, 79–99.
23. Isard, W. et al. (1972). *Ecological-economic analysis for regional development*. New York: Free Press.
24. Jin, D., Hoagland, P., & Dalton, T. (2003). Linking economic and ecological models for a marine ecosystem. *Ecological Economics, 46*, 367–385.
25. Keuning, S. J., van Dalen, J., & de Haan, M. (1999). The Netherlands' NAMEA; presentation, usage and future extensions. *Structural Change and Economic Dynamics, 10*, 15–37.
26. Kleis, H., Dalager, S. 100 Years of Waste Incineration in Denmark. http://www.wte.org/docs/100YearsofWasteIncinerationinDenmark.pdf. Cited 22 July 2008.
27. Lenzen, M. (2007). Structural path analysis of ecosystem networks. *Ecological Modelling, 200*, 334–342.
28. Leontief, W. (1970). Environmental repercussions and the economic structure: an input-output approach. *Review of Economics and Statistics, 52*, 262–271.
29. Leontief, W., & Ford, D. (1972). Air pollution and the economic structure: empirical results of input-output computations. In A. Bródy, & A. C. Carter (Eds.), *Conference on input-output techniques* (pp. 9–30). Geneva, Switzerland: North-Holland.
30. Leontief, W. (1973). National income, economic structure, and environmental externalities. In F. Moss (Ed.), *The measurement of economic and social performance, studies in income and wealth* (Vol. 38). New York: National Bureau of Economic Research.
31. Management and Coordination Agency, Government of Japan (2004). *2000 input-output tables*. Tokyo: The Federation of National Statistics Associations.
32. McDougall, F., White, P., Franke, M., & Hindel, P. (2001). *Integrated solid waste management: A life cycle inventory*. Oxford: Blackwell Science.
33. Meadows, D., Randers, J., & Behrebs, W. (1972). *The limits to growth*. New York: Universe Books.
34. Minh, N. H. et al. (2003). Open dumping site in Asian developing countries: a potential source of polychlorinated dibenzo-p-dioxins and polychlorinated dibenzofurans. *Environmental Science & Technology, 37(8)*, 1493–1502.
35. Ministry of Environment, The Government of Japan. http://www.env.go.jp/recycle/waste_tech/index.html

36. Ministry of International Trade and Industry, The Government of Japan. http://www.meti. go.jp/policy/kaden_recycle/en_cha/pdf/english.pdf. Cited 21 July 2008.
37. Nakamura, S. (1999). Input-output analysis of waste cycles. First International Symposium on Environmentally Conscious Design and Inverse Manufacturing, Proceedings. *IEEE Computer Society, Los Alamitos, 1999*, 475–480.
38. Nakamura, S., & Kondo, Y. (2002). Input-output analysis of waste management. *Journal of Industrial Ecology, 6(1)*, 39–63.
39. Nansai, K., Moriguchi, Y., & Tohno, S. (2002). *Embodied energy and emission intensity data for Japan using input-output tables (3EID)-inventory data for LCA-*. Tsukuba, Japan: National Institute for Environmental Studies.
40. National Institute of Environmental Studies, Tsukuba, Japan. http://www-cger.nies.go.jp/publication/D031/index.html
41. Nansai, K., Moriguchi, Y., & Tohno, S. (2003). Compilation and application of Japanese inventories for energy consumption and air pollutant emissions using input-output tables. *Environ. Sci. Technol., 37*, 2005–2015.
42. Peters, G. P. (2007). Efficient algorithms for life cycle assessment, input-output analysis, and Monte-Carlo analysis. *Int J LCA, 12(6)*, 373–380.
43. Proops, J., Faber, M., & Wagenhals, G. (1993). *Reducing CO_2 emissions: A comparative input-output study for Germany and the UK*. Berlin: Springer.
44. Stahmer, C., Kuhn, M., & Braun, N. (1998). Physical input-output tables for Germany, 1990. Eurostat Working Papers 2/1998/B/1, European Commission.
45. The UK Department for Environment, Food and Rural Affairs (2007). *Municipal waste statistics 2006/7*. http://www.defra.gov.uk/environment/statistics/wastats/bulletin07.htm. Cited Jan 2008.
46. United States Environmental Protection Agency (2006). *Municipal solid waste in the United States, 2005 facts and figures*. Washington, DC: United States Environmental Protection Agency.
47. U.S. Bureau of Economic Analysis, Industry Economic Accounts. http://www.bea.gov/industry/index.htm
48. Victor, P. (1972). *Pollution, economy and the environment*. London: Allen & Unwin.
49. http://en.wikipedia.org/wiki/Waste-to-energy
50. Wright, D. (1974). Goods and services: An input-output analysis. *Energy Policy, 2*, 307–315.

Chapter 6
WIO Analysis

Abstract This chapter is concerned with application of the WIO model to real data, the basic structure of which was shown in Chapter 5, and the derivation of the WIO cost-price model as well. First, as a real example of WIO table, a WIO table for a small city in northern Japan is shown, which is characterized by ambitious management programs of municipal solid waste (MSW) with a rate of recycling around 70%. Secondly, large scale WIO tables for Japan with 50 to about 400 endogenous sectors with dozens of waste types are introduced, and the results of some applications based on the WIO quantity model are shown. Also dealt with is the issue of adjusting for the imports of waste in WIO, which were not touched upon in Chapter 5. A distinguishing feature of a waste treatment process consists in its dependence on the chemical properties of the incoming waste. The third topic of this chapter deals with the consideration of this "dynamic nature of a waste treatment process" in WIO by use of engineering models. Fourth, the cost and price counterpart of the WIO quantity model, the WIO price model, is derived, and numerical examples are given for its illustration. The contents of this chapter make use of Nakamura [8], Nakamura and Kondo [10, 11].

6.1 WIO Tables and Analysis: Empirical Examples

This section shows real examples of WIO tables and some results obtained by applying the WIO quantity model to them. To help readers become familiar with the basic features of a WIO table, a simple WIO table for a small city is shown first for illustration, before turning to large scale Japanese WIO tables with several hundred endogenous sectors and dozens of waste types.

6.1.1 WIO for a City in Hokkaido, Japan

The first WIO table that was ever compiled using real data was that for City F in Hokkaido, Japan, a city with a population of around 20,000, the major industries of which are agriculture and tourism [8]. Besides its being an excellent place for skiing, the city has also become famous for its ambitious MSW management program with the rate of recycling of around 70% for the year 2005 [4], while the national average was around 20% (in the year 1997, it was at 56%, when the national average was only 8%). Instrumental in reaching this impressive rate of recycling were the separate collection of MSW, under which garbage and combustible waste were separately collected for composting and RDF production, and the recycling of the produced compost and RDF in local agriculture and service sectors (a public hotel and schools).

Owing to the rather simple structure of the city's economy combined with the extensive waste recycling policy, the example of City F will provide readers with a good introduction to the empirical application of WIO.

6.1.1.1 The Flow of Waste in the WIO Format

Table 6.1 shows a simplified flow of goods and waste in City F in the WIO format. This is a mixed unit table with the waste flows represented in a physical unit, while the flow of goods and services are in monetary terms. The flow of goods and services are rough estimates for the purpose of illustration, and should not be taken seriously. In particular, no inputs of goods and services occur in recycling and waste treatment sectors. Furthermore, the utilities and manufacturing sectors are omitted, because their products are imported from outside the city. The flows of waste were simplified by excluding discarded containers (cans, and bottles), metal items, bulky waste, as well as medical waste. For each waste item in Table 6.1, its generation is indicated by a positive value, and use (recycling and reuse) by a negative value. Accordingly, for each column of the panel "Waste input/output", the sum gives the net amount of waste that was directly generated from the corresponding sector/process.

The negative values for composting and RDF ($-2,808$ and -989) indicate that these sectors/processes contributed to the reduction of waste for treatment. On the other hand, the incineration sector generated 457×10^3 kg of ash.

Similarly, for each waste item in the panel, the row sum gives the net amount of its generation. For garbage and waste for RDF, the net amount was zero, because all of them were collected for recycling, and none was discarded for treatment. This does not imply, however, that they were recycled at the rate of 100%. In fact, the column of the composting sector indicates that of $3,269 \times 10^3$ kg of garbage collected for producing compost, 461×10^3 kg were not suitable for composting, and had to be discarded as general waste: the yield of composting was $0.85 = 2,808/3,269$. Similarly, of $1,156 \times 10^3$ kg of waste for RDF, 167×10^3 kg happened not to be suited for RDF, and had to be discarded as general waste. (In reality, the original amount of waste for RDF contained more than 300×10^3 kg of noncombustible waste, the

Table 6.1 A WIO Representation of the Flow of Goods Waste in City F.

	Industry		Recycling		Waste disposal		Final demand		Total
	Agr	Srv	Cmp	RDF	Inc	Lnd	Csm	Exp	
Agriculture*	50	10					15	220	295
Services*	40	300					520	950	1,810
Compost	2,808								2,808
RDF		989							989
				Waste input/output					
Garbage		1,148	−3269				2,121		0
Agricultural plastics[a]	757								757
Waste for RDF[b]		211		−1,156			945		0
General waste[c]		592	461	167			1,888		3,108
Ash		40[d]			457				457
Total	757	1,951	−2,808	−989	457		0	4,954	4,322

Units: The numbers in the rows denoted by * are in 100 million Japanese yen; all the others are in 1,000 kg. Cmp: composting, Inc: incineration, Lnd: landfill, Csm: consumption, Exp: export, RDF: refuse derived fuel. Source: [8] based on data for the year 1995. The numbers with * are rough estimates partly based on the IO table for Hokkaido, 1995. For the sake of simplicity, the flows that are associated with manufacturing and other industries are omitted (the production value of manufacturing was about one tenth of the service sector). [a]Plastics used for a greenhouse (mostly polyvinylchloride (PVC)). [b]Combustible waste collected for RDF production. [c]Combustible waste not suited for RDF production. [d]Ash generated by the use of RDF (estimated by assuming the ash content of 4%).

majority of which was sold as metal scrap: the yield of RDF was thus well below .7. This point was neglected for the purpose of simplifying the illustration.)

By its nature, the use of compost in the agriculture sector does not generate any waste (assuming the compost did not include any residue such as plastic, which does not decompose in soil). The use of RDF, however, generates ash, because its materials, paper, plastic, and textile products, do contain noncombustible portions, and leave ash behind when incinerated (see Table 6.21). In Table 6.1, 1 kg of RDF was assumed to leave 0.04 kg of ash when incinerated (ash content of 4%), which resulted in the generation of 40×10^3 kg of ash in the use phase.

Using the matrix notations in Table 5.13 (but without the assumption of a one-to-one correspondence between waste and treatment, and breaking the final demand into several categories), Table 6.1 can be represented as

$$
\begin{pmatrix}
\underbrace{X_{\mathrm{I,I}}}_{n \times n} & \underbrace{X_{\mathrm{I,II}}}_{n \times k} & \underbrace{F_{\mathrm{I}}}_{n \times o} \\
\underbrace{W_{\mathrm{I}}(= W_{\mathrm{I}}^{\mathrm{out}} - W_{\mathrm{I}}^{\mathrm{in}})}_{m \times n} & \underbrace{W_{\mathrm{II}}(= W_{\mathrm{II}}^{\mathrm{out}} - W_{\mathrm{II}}^{\mathrm{in}})}_{m \times k} & \underbrace{W_{\mathrm{f}}(= W_{f}^{\mathrm{out}} - W_{f}^{\mathrm{in}})}_{m \times o}
\end{pmatrix}
\tag{6.1}
$$

where I refers to agriculture, services, composting, and RDF, II to incineration and landfill, $n (= 2)$ to the number of producing sectors, $k (=2)$ to the number of waste treatment sectors, $o(=2)$ to the number of final demand categories, and $m(=5)$ to the number of waste types.

6.1.1.2 Allocation Matrices

Of the waste items shown in Table 6.1, agricultural plastics and ash are designated as waste for landfill, while the rest were designated as that for incineration. Table 6.2 gives the allocation matrix S for the waste items that are treated in City F.

Multiplication of S from the left converts the flow of waste in Table 6.1 to the flow of waste treatment activities:

$$SW_{\mathrm{I}} = \begin{pmatrix} 1 & 0 & 1 & 1 & 0 \\ 0 & 1 & 0 & 0 & 1 \end{pmatrix} \begin{pmatrix} 0 & 1148 & -3269 & 0 \\ 757 & 0 & 0 & 0 \\ 0 & 211 & 0 & -1156 \\ 0 & 592 & 461 & 167 \\ 0 & 40 & 0 & 0 \end{pmatrix}$$

$$= \begin{pmatrix} 0 & 1951 & -2808 & -989 \\ 757 & 40 & 0 & 0 \end{pmatrix} \tag{6.2}$$

$$SW_{\mathrm{II}} = \begin{pmatrix} 1 & 0 & 1 & 1 & 0 \\ 0 & 1 & 0 & 0 & 1 \end{pmatrix} \begin{pmatrix} 0 & 0 \\ 0 & 0 \\ 0 & 0 \\ 0 & 0 \\ 457 & 0 \end{pmatrix} = \begin{pmatrix} 0 & 0 \\ 457 & 0 \end{pmatrix} \tag{6.3}$$

$$SW_f = \begin{pmatrix} 1 & 0 & 1 & 1 & 0 \\ 0 & 1 & 0 & 0 & 1 \end{pmatrix} \begin{pmatrix} 2121 & 0 \\ 0 & 0 \\ 945 & 0 \\ 1888 & 0 \\ 0 & 0 \end{pmatrix} = \begin{pmatrix} 4954 & 0 \\ 0 & 0 \end{pmatrix} \tag{6.4}$$

Replacing the waste flows in Table 6.1 by the flows of waste treatment activities thus obtained yields an IO table with a square matrix of intermediate flows as the standard IO table (Table 6.3), the matrix representation of which is given by

$$\begin{pmatrix} X_{\mathrm{I,I}} & X_{\mathrm{I,II}} \\ SW_{\mathrm{I}} & SW_{\mathrm{II}} \end{pmatrix} = \begin{pmatrix} A_{\mathrm{I,I}}\hat{x}_{\mathrm{I}} & A_{\mathrm{I,II}}\hat{x}_{\mathrm{II}} \\ SG_{\mathrm{I}}\hat{x}_{\mathrm{I}} & SG_{\mathrm{II}}\hat{x}_{\mathrm{II}} \end{pmatrix} \tag{6.5}$$

Table 6.2 The Allocation Matrix S for the Waste Flow in Table 6.1.

	Waste types				
Treatment	Garbage	Agricultural plastics	RDF waste	General waste	Ash
Incineration	1	0	1	1	0
Landfill	0	1	0	0	1

Table 6.3 The Squared WIO Table of City F.

	Production				Treatment		Final demand		Total[a]
	Agr	Srv	Cmp	RDF	Inc	Lnd	Cns	Exp	
Agriculture*	50	10	0	0	0	0	15	220	295
Services*	40	300	0	0	0	0	520	950	1,810
Compost	2,808	0	0	0	0	0	0	0	2,808
RDF	0	989	0	0	0	0	0	0	989
Incineration	0	1,951	−2,808	−989	0	0	4,954	0	3,108
Landfill	757	40	0	0	457	0	0	0	1,254

Units: The numbers in the rows with * are in 100 million Japanese yen; all the others are in 10^3 kg. See Table 6.1 for the notations. [a]The level of production for producing sectors, and that of treatment for waste treatment sectors. Source: See Table 6.1.

Note that incineration and landfill now occur as both column and row elements. It is indicated that the production of compost and RDF contributed to the reduction of incineration by the amount of $3,800 \times 10^3$ kg: without these activities, the amount of incineration would have been twice as large as the actual one.

6.1.1.3 Some WIO Analysis: Attributing Waste to Final Demand Categories

From (6.5), division of the column elements of Table 6.3 by the corresponding activity level (production x_I and treatment x_{II}) gives the matrix of input coefficients (Table 6.4)

$$\begin{pmatrix} X_{I,I} & X_{I,II} \\ SW_I & SW_{II} \end{pmatrix} \begin{pmatrix} \hat{x}_I & 0 \\ 0 & \hat{x}_{II} \end{pmatrix}^{-1} = \begin{pmatrix} A_{I,I} & A_{I,II} \\ SG_I & SG_{II} \end{pmatrix} \tag{6.6}$$

which corresponds to the expression inside the first parenthesis on the left of (5.87). It indicates, among others, that the production of agriculture by the amount of 100 million Japanese yen requires 9.5×10^3 kg of compost, and generates 2.6×10^3 kg of waste for landfill (agricultural waste plastics), and that the incineration of 1 kg of waste generates 0.15 kg of ash, which indicates that (provided the combustible components are perfectly incinerated) the average ash content of waste entering the incinerator is about 15%.

Table 6.5 then gives the Leontief inverse matrix that corresponds to the inverse matrix on the right-hand side of (5.88). Several interesting results emerge. First, while the agriculture sector does not use any input of incineration directly (Table 6.4), its activity reduces incineration indirectly by the amount of 11.4×10^3 kg per unit (100 million Japanese yen). This reflects the fact that the agriculture sector is the single user of compost, and that compost is made of garbage, which would have been subjected to incineration were there no production of compost.

In contrast, the service sector does not contribute to any noticeable reduction in the amount of waste treatment. In particular, a unit of the final delivery of the service sector contributed to the generation of 130 kg of waste for landfill, whereas its direct effect in Table 6.4 was only 2 kg. Thirdly, the production of 1 kg of compost

Table 6.4 The Squared Matrix of WIO Input Coefficients: City F.

	Production				Treatment	
	Agriculture[a]	Services[a]	Compost	RDF	Incineration	Landfill
Agriculture[a]	0.1695	0.0055	0.0000	0.0000	0.0000	0.0000
Services[a]	0.1356	0.1657	0.0000	0.0000	0.0000	0.0000
Compost	9.5186	0.0000	0.0000	0.0000	0.0000	0.0000
RDF	0.0000	0.5464	0.0000	0.0000	0.0000	0.0000
Incineration	0.0000	1.0779	−1.0000	−1.0000	0.0000	0.0000
Landfill	2.5661	0.0022	0.0000	0.0000	0.1470	0.0000

Units: [a] 100 million Japanese yen, 10^3 kg for all others.

Table 6.5 The Matrix of Leontief WIO Inverse Coefficients: City F.

	Production				Treatment	
	Agriculture[a]	Services[a]	Compost	RDF	Incineration	Landfill
Agriculture[a]	1.2054	0.0080	0.0000	0.0000	0.0000	0.0000
Services[a]	0.1959	1.2000	0.0000	0.0000	0.0000	0.0000
Compost	11.4736	0.0760	1.0000	0.0000	0.0000	0.0000
RDF	0.1070	0.6557	0.0000	1.0000	0.0000	0.0000
Incineration	−11.3695	0.5618	−1.0000	−1.0000	1.0000	0.0000
Landfill	1.4257	0.1293	−0.1470	−0.1470	0.1470	1.0000

Units: [a] 100 million Japanese yen, 10^3 kg for all others.

reduced the amount of waste going to a landfill by 0.14 kg, which corresponds to the ash content of 1 kg of garbage: if garbage were incinerated instead of being composted, this amount of ash would have been generated, and landfilled. Note that the same value is obtained for the production of RDF as well, because of the use of the same ash content for both garbage and waste for RDF in this example. Note also that the generation of ash in the use phase of RDF is counted as output from its user, the service sector.

Within the framework of standard IOA, any production is ultimately attributed to the final demand for goods and services. Similarly, within the framework of WIO, any production and waste generation can be attributed to the final demand for goods and services. This "attribution" can be facilitated by:

$$\begin{pmatrix} I - A_{I,I} & -A_{I,II} \\ -SG_I & I - SG_{II} \end{pmatrix}^{-1} \begin{pmatrix} f_I^i \\ Sw_f^i \end{pmatrix} \tag{6.7}$$

with f_I^i and w_f^i referring to the column i of F_I and W_f with $i = C$ for consumption, and $i = E$ for export. Table 6.6 gives the results. It is noteworthy that the economy of City F is mostly driven by export, including the production of recycled products. With regard to the generation of waste and its treatment, however, the effects of

Table 6.6 Attributing Production and Treatment to Final Demand Category: City F.

| | Final demand categories | | |
Sector	Consumption	Export	Total
Agriculture	22	273	295
Services	627	1,183	1,810
Compost	212	2,596	2,808
RDF	343	646	989
Incineration	5,076	−1,968	3,108
Landfill	817	436	1,254

Units: 10^3 kg except for agriculture and services.

local consumption dominated those of export. In particular, household consumption resulted in the incineration of waste of about $5,000 \times 10^3$ kg, whereas the export reduced it by almost $2,000 \times 10^3$ kg. Furthermore, while the amount of landfill invoked by the export was positive, that amount was about 53% of that invoked by household consumption.

For a better understanding of the above findings from Tables 6.5 and 6.6, it will be useful to decompose the level of waste treatment into its components, that is, into the amounts of individual waste items that are treated in each treatment process. Recall from (5.90) that the net amount of generation of waste i that is directly and indirectly generated to satisfy a unit of product j for final delivery or waste treatment is given by $(\Xi)_{ij}$ with

$$\Xi = \begin{pmatrix} G_I & G_{II} \end{pmatrix} \begin{pmatrix} I - A_{I,I} & -A_{I,II} \\ -SG_I & I - SG_{II} \end{pmatrix}^{-1} \tag{6.8}$$

The results are shown in Table 6.7. Several findings can be pointed out. First, the large effect of the agriculture sector on reducing the amount of incineration is attributed to its reduction of garbage for incineration by its use of compost made of garbage: the production of the agriculture sector in the amount of one million Japanese yen absorbs 13 kg of garbage. Secondly, the avoidance of incineration reduces the generation of ash as well, which explains the above finding of the reduced requirement for landfill of the agriculture sector when the indirect effects are considered. Third, the use of RDF in the service sector does not make it 'waste neutral' because of its generation of general waste in the production phase of RDF (the occurrence of waste items that are flammable but not suited for RDF production), and the generation of ash in the use phase.

Finally, based on these results, Table 6.8 gives the results of attributing net waste generation by waste type to final demand category by use of (5.89). It is found that it is export that mostly contributes to the reduction of waste for treatment. It is concluded that for the economy of City F, export played a vital role in terms of both economy and waste recycling.

Table 6.7 Net Generation of Waste Per Unit of Product/Treatment for Final Delivery: Ξ.

	Agriculture	Services	Compost	RDF	Incineration	Landfill
Waste for incineration						
Garbage	−13.23	0.67	−1.16	0.00	0.00	0.00
RDF waste	−0.10	−0.63	0.00	−1.17	0.00	0.00
General waste	1.97	0.52	0.16	0.17	0.00	0.00
Total	−11.37	0.56	−1.00	−1.00	0.00	0.00
Waste for landfill						
Agricultural plastics	3.09	0.02	0.00	0.00	0.00	0.00
Ash	−1.67	0.11	−0.15	−0.15	0.15	0.00
Total	1.43	0.13	−0.15	−0.15	0.15	0.00

Units: 10^3 kg except for Agriculture and Services, the products of which are in units of 100 million Japanese yen.

Table 6.8 Attributing Waste Generation to Final Demand Categories.

Waste for incineration					
	Consumption			Export	Total
	Indirect	Direct	Total		
Garbage	151	2,121	2,272	−2,272	0
RDF waste	−327	945	618	−618	0
General waste	298	1,888	2,186	922	3,108
Total	122	4,954	5,076	−1,968	3,108

Waste for landfill					
	Consumption			Export	Total
	Indirect	Direct	Total		
Agricultural plastics	57	0	57	700	757
Ash	760	0	760	−263	497
Landfill	817	0	817	436	1,254

Units: 10^3 kg.

6.1.2 WIO Tables for Japan

We now turn to a WIO table that can be used for practical purposes. WIO tables for Japan with dozens to hundreds of sectors have been compiled by the authors, which are available on the Internet [12]. The main features of the Japanese WIO tables are explained. Before proceeding to analysis, the issue about the treatment of imports of waste is addressed, an issue that was previously not dealt with. The part on analysis includes topics such as final demand origins of waste, and the waste footprints of products.

6.1.2.1 Overview of Available National Tables

Up to now, national WIO tables have been compiled for Japan for the years 1990 [9], 1995 [2,10], and 2000 [3], based on, among others, Japanese IO tables published by

Table 6.9 Waste Items in WIO 1990.

1	Kitchen waste	10	Waste rubber	19	Waste acid
2	Waste paper	11	Animal and plant residue	20	Waste alkali
3	Waste textiles	12	Dust	21	Construction debris
4	Waste plastics	13	Ash	22	Animal waste
5	Iron scrap	14	Slag	23	Carcass
6	Nonferrous metal scrap	15	Waste wood	24	Bulky waste
7	Glass bottles	16	Organic sludge	25	End of life vehicles
8	Glass waste	17	Inorganic sludge	26	Molten slag
9	Pottery waste	18	Waste oil	27	Shredding residue

the Japanese Government, data on waste flows published by the Ministry of the Environment, engineering information on waste treatment, and industry reports from various sources (see the references for details of their compilation). The WIO tables for 1995 and 2000 are available on the Internet [12].

The 1990 WIO table consisted of 52 producing sectors, 3 basic waste treatment processes (shredding, incineration, and landfill), and 27 waste types (Table 6.9). Shredding was further distinguished by its feed, say furniture, appliances, and vehicles, with resulting shredding residue also distinguished accordingly. Incineration was distinguished into several types, too, depending on the size of incinerator, method of energy recovery, and treatment of incineration residue.

As mentioned above (Section 5.2.3.1), the original Japanese IO tables contain two waste treatment sectors distinguished by institutional factors (public and private). In the WIO tables, the rows and columns corresponding to these sectors were entirely removed for reasons mentioned in Section 5.2.3.1, and have been replaced by columns referring to individual waste treatment processes, the unit processes of which were obtained primarily based on engineering information (see Section 6.2 below for this point). The corresponding row elements referring to the input of waste treatment services into producing and treatment sectors can then be obtained by applying the allocation matrix S to the matrices W_I and W_{II}, as in (6.5).

In the 1995 WIO table, the sectoral classification was disaggregated into 78 producing sectors, treatment methods were extended to five by adding composting and gasification, and waste classification was also extended with home appliances divided into small electronics, TV sets, refrigerators, washing machines, and air conditioners, and nonferrous metal scrap distinguished for copper and aluminum.

In the WIO table for the year 2000, the sectoral classification was matched to the basic classification in the Japanese IO table of 396 producing sectors, and the classification of waste was also substantially extended by breaking down waste according to its source, MSW from households, MSW from commercial sources, and industrial waste (Table 6.10). Note that shredding residue is distinguished into seven types according to the type of waste entering the shredding process.

All the waste items in the WIO tables are counted as by-product (II) (see Definition 3.2 in Section 3.2.1), and occur in the matrix W. Electricity from waste incineration produced by the heat of waste, on the other hand, is counted as by-product (I), and occurs in the matrix X as a negative input into the corresponding incineration sector.

Table 6.10 Waste Classification in WIO 2000.

MSW (home)	MSW (commercial)	Industrial waste
1 Kitchen waste	1 Kitchen waste	1 Ash
2 Newspapers	2 Newspapers and magazines	2 Sludge
3 Magazines		3 Waste oil
4 Corrugated cardboard	3 Corrugated cardboard	4 Waste acid
5 Paper package for beverages	4 Waste paper (high quality)	5 Waste alkali
6 Paper boxes, paper bags, and wrapping paper	5 Waste paper (others)	6 Waste plastics
7 Other paper (letter and diaper, etc.)		7 Waste paper
8 Waste textiles	6 Waste textiles	8 Waste wood
9 PET bottles	7 PET bottles	9 Waste textiles
10 Bottles other than PET	8 Styrene foams	10 Animal and plant residue
11 Plastic cups and trays	9 Other plastics	11 Waste rubber
12 Plastic bags, seats, buffer materials, wrappings		12 Iron scrap
13 Other plastics		13 Copper scrap
14 Steel cans	10 Steel cans	14 Aluminum scrap
15 Aluminum cans	11 Aluminum cans	15 Lead and zinc scrap
16 Iron other than cans	12 Other metals	16 Other nonferrous s metal scrap
17 Nonferrous metals other than can		17 Waste glass and pottery
18 Returnable bottles	13 Returnable bottles	18 Slag
19 One-way bottles	14 One-way bottles	19 Debris
20 Other glass waste	15 Other glass waste	20 Animal waste
21 Waste pottery		21 Carcass
22 Waste rubber and leather products		22 Dust
23 Green waste	16 Green waste	23 Molten slag
24 Waste textiles	17 Others	
25 Waste wood products		
26 Discarded bikes and stoves		24 Shredding residue (bikes and stoves)
27 Discarded small electric appliances		25 Shredding residue (small electric appliances)
28 Discarded TV sets		26 Shredding residue (TV sets)
29 Discarded refrigerators		27 Shredding residue (refrigerators)
30 Discarded washing machines		28 Shredding residue (washing machines)
31 Discarded air conditioners		29 Shredding residue (air conditioners)
32 Discarded automobiles		30 Shredding residue (automobiles)

The text inside the parenthesis after Shredding residue refers to the origin for the residue.

6.1.2.2 WIO Table for Japan for the Year 2000

Tables 6.12 and 6.11 show an aggregated version of the WIO table for Japan which consists of 13 producing sectors, 3 waste treatment methods (shredding, incineration, and landfilling), and 13 waste types. The equality between the number of types of producing sectors and of waste is a mere coincidence and has no deeper meaning. Final demand is divided into consumption (CONS), fixed investment (INVS), and export (EXPT). The entries in Table 6.12 refer to the flow of goods & services, and are measured in a monetary unit (billion yen), while those in Table 6.11 refer to the flow of waste, and are measured in a physical unit (10^6 kg). The emission of GWP (global warming potential, in 10^3 kg-C) is based on Matsuto [6] for waste treatment sectors, and [13] for all the other sectors. The negative entry of utilities in the column of incineration sector represent the supply of electricity and/or heat (by-product (I)) to be used in other sectors.

In terms of the way imports are accounted for, the national WIO tables are of a competitive import type regarding waste flows, due to the lack of data on waste flows by origin. Accordingly, the values in Tables 6.12 and 6.11 include imports for both goods & services and waste. The column termed IMPT refers to the imports of goods & services and waste.

Note that in Table 6.11 the imports of waste occur in the upper panel referring to the generation of waste, because they refer to the supply of waste from foreign countries to Japan. On the other hand, the exports of waste occur in the lower panel referring to the use of waste, because it refers to the use of waste generated in Japan by foreign countries.

Of the 13 types of waste, "ash" (ash, dust, and slag), "sld" sludge , "oil" (waste oil, acid, and alkali), "cons" (construction debris), and "dst" (shredder residue) are solely generated by industry sectors, whereas the consumption sector is the sole generator of "blk" (bulky waste) including discarded appliances, and "vhc" (discarded vehicles). The remaining six types of waste are generated from both industry and final demand sectors. Recall that an entry in the flow table of waste (the lower half of the table) refers to the net generation of waste (see (6.1)). The row labeled "metl", for instance, indicates that fixed investment generated 23×10^9 kg of metal scrap, while the basic metals industry (METL) used 41.2×10^9 kg thereof as materials. The negative entry of "gls" (waste glass) into "FOOD" (food, beverages, feed, and textiles) by the amount of $.6 \times 10^9$ kg refers to the retrieval of used glass bottles for beverages, the largest generator of which is the final demand.

The net generation of waste from each waste treatment sector represents the outputs of its conversion processes. Incineration (INCN) converts its waste feedstock into 4.4×10^9 kg of ash. Shredding (SHRD) converts discarded durables (automobiles, appliances, and other bulky wastes) into several waste materials such as 4.9×10^9 kg of metal scrap (metl), $.9 \times 10^9$ kg of wood waste (wod), $.3 \times 10^9$ kg of waste plastics, and 2.1×10^9 kg of shredder residue (rsd). Because landfilling is regarded as the final form of waste disposal, it involves no waste conversion process and hence no entry of waste occurs in its column (issues of leachate are not considered in this book). The inputs from production sectors into waste treatment refer

Table 6.11 WIO Table for Japan, 2000: The Flow of Goods & Services.

	AGRC	MING	FOOD	WOOD	CHEM	CMNT	METL	METP	MCHN	CNST	UTIL	SRVC	TRNS	INCN	LNDF	SHRD	CONS	INVS	EXPT	IMPT	Total
AGRC	1,558	0	7,617	579	125	0	0	0	105	153	0	1,343	2	0	0	0	3,966	967	72	−2,119	14,370
MING	0	3	0	20	5,602	594	837	8	22	676	2,014	6	0	0	0	0	0	−11	11	−8,669	1,114
FOOD	1,138	8	7,319	126	258	27	1	41	315	200	8	7,641	113	0	0	0	35,620	481	778	−8,066	46,008
WOOD	216	7	1,187	6,943	716	212	6	216	1,329	3,856	139	9,869	711	0	0	0	2,505	536	344	−1,958	26,834
CHEM	966	35	1,790	1,452	15,270	349	197	630	6,083	2,214	931	9,467	4,825	10	13	2	9,133	27	4,842	−5,946	52,290
CMNT	17	0	172	95	242	733	119	125	1,242	4,775	18	403	3	17	0	0	367	−60	585	−397	8,457
METL	0	0	0	2	81	29	1,623	4,924	626	1	0	2	0	0	0	0	61	−8	220	−1,551	6,011
METP	18	28	806	349	450	150	4	7,336	9,517	8,818	33	842	91	0	0	0	472	561	2,705	−1,123	31,057
MCHN	93	13	149	84	69	44	0	140	46,617	1,679	32	7,909	601	1,155	204	36	17,174	37,764	37,004	−15,072	135,694
CNST	81	9	104	114	272	125	48	266	365	199	1,234	5,490	648	0	140	0	0	68,331	0	0	77,427
UTIL	91	40	641	671	1,660	338	396	847	1,589	441	1,398	6,874	1,027	−5	4	4	7,454	0	31	−2	23,497
SRVC	1,475	184	6,414	3,773	7,837	1,413	577	3,696	22,285	12,839	3,971	74,922	16,869	16	124	3	284,393	21,149	6,473	−6,085	462,329
TRNS	631	386	1,535	1,139	1,525	640	251	965	2,911	4,936	609	20,080	8,522	0	0	0	23,249	771	4,313	−3,011	69,452
CO2	16,885	716	18,396	20,958	91,670	68,434	143,706	33,838	18,281	14,342	382,146	82,574	211,022	56,467	35,863	90	173,319				1,368,707

Units: Goods & services in billion Japanese yen, CO_2 (GWP) in 10^6 kg-C. AGRC: agriculture, fishery, forestry, MING: ore, coal, petroleum, FOOD: food, beverage, feed, textile, WOOD: wood & products, paper & products, CHEM: chemicals & products, plastics, rubber products CMNT: cement, glass, pottery, earthen products, METL: iron & steel nonferrous metals, METP: metal products, MCHN: machinery and equipment, CNST: construction, & civil engineering, UTIL: utilities, SRVC: services, TRNS: transport and communications

Table 6.12 WIO Table for Japan, 2000: The Flow of Waste.

Waste	AGRC	MING	FOOD	WOOD	CHEM	CMNT	METL	METP	MCHN	CNST	UTIL	SRVC	TRNS	INCN	LNDF	SHRD	CONS	INVS	EXPT	IMPT	Total
The generation of waste W^{out}																					
grb	533	1	792	21	17	6	1	17	68	131	7	4,870	82	0	0	0	9,328	0	0	0	15,874
pap	1,084	8	540	5,791	249	86	18	300	989	1,023	44	8,434	447	0	0	119	16,755	0	0	278	36,165
pls	384	3	516	603	1,237	125	6	605	1,002	1,034	14	2,718	189	0	0	290	4,317	0	0	15	13,057
metl	100	2	91	40	43	12	70	16,763	8,688	1,224	11	977	41	0	0	4,865	1,830	23,064	0	661	58,483
gls	158	0	86	111	101	3	173	147	261	2,042	17	1,528	45	0	0	0	2,741	0	0	0	9,523
wod	90,686	0	4,015	2,335	54	2,109	1	12	37	3,208	1	316	6	0	0	971	855	0	0	0	102,500
ash	0	1,005	89	527	691	327	41,307	2,943	1,129	72	7,245	14	1	4,417	0	0	0	0	0	0	66,743
slg	0	2,672	1,874	4,089	2,342	1,465	363	1,398	739	2,295	14,136	102	81	0	0	0	0	0	0	6,977	31,556
oil	3	5	834	141	2,489	261	210	1,168	1,692	35	18	823	86	0	0	0	0	0	0	0	7,765
cons	0	0	0	0	0	0	0	0	0	58,917	0	0	0	0	0	0	0	0	0	0	58,917
blk	0	0	0	0	0	0	0	0	0	0	0	0	0	0	0	0	3,274	0	0	0	3,274
vhc	0	0	0	0	0	0	0	0	0	0	0	0	0	0	0	0	5,540	0	0	0	5,540
dst	0	0	0	0	0	0	0	0	0	0	0	0	0	0	0	2,180	0	0	0	0	2,180
The use of waste W^{in}																					
grb	0	0	0	0	0	0	0	0	0	0	0	0	0	0	0	0	0	0	0	0	0
pap	0	0	239	17,400	0	0	0	0	0	0	0	0	0	0	0	0	0	120	372	0	18,131
pls	0	0	0	0	1,550	0	235	0	0	0	34	0	0	0	0	0	0	0	300	0	2,119
metl	0	0	0	0	27	0	41,172	6,655	0	0	0	0	0	0	0	0	0	0	3,054	0	50,908
gls	0	0	651	0	0	3,764	0	0	0	0	0	0	0	0	0	0	0	0	0	0	4,415
wod	0	0	87,149	2,056	0	0	0	0	0	0	0	0	0	0	0	0	0	0	0	0	89,204
ash	0	0	0	0	474	26,682	3,957	0	0	19,126	0	0	0	0	0	0	0	0	422	0	50,659
sld	0	0	6,166	0	0	8,054	0	0	0	0	0	0	0	0	0	0	0	0	0	0	14,220
oil	0	0	0	0	1,959	0	0	0	0	0	0	0	0	0	0	0	0	0	0	0	1,959
blk	0	0	0	0	0	0	0	0	0	48,212	0	0	0	0	0	0	0	0	0	0	48,212
vhc	0	0	0	0	0	0	0	0	0	0	0	0	0	0	0	0	0	0	0	0	0
rsd	0	0	0	0	0	0	0	0	0	0	0	0	0	0	0	0	0	0	0	0	0

Units: 10^6 kg. grb: kitchen waste, pap: waste paper, pls: waste plastics, metl: metal scrap, gls: waste glass, wod: waste wood, ash: ash, dust, slag, slg: molten slag sld: sludge oil: waste oil, cons: construction waste, blk: bulky waste, vhc: end of life vehicles, rsd: shredder residue. See Table 6.11 for the notations for the sectors at the top of the table.

to those that are needed for running the treatment processes, such as chemicals, electricity, fuels, and machine parts. The negative input of electricity (UTIL) into incineration (INCN) indicates that the amount of electricity generated from waste heat exceeded the demand for electricity for that sector.

6.1.2.3 Accounting for Competitive Imports

As mentioned above, the national WIO tables for Japan are of a competitive import type, the details of which have already been dealt with in Section 3.1.1.1. The occurrence of physical waste flows in WIO tables, however, introduces a new feature, which cannot be dealt with in a straightforward fashion, and calls for special treatment, to which we now turn.

In discussing the import and/or export of waste, it is important to notice that the transboundary movements of hazardous waste are banned by the Basel convention [1]. Accordingly, the import and/or export of waste to be discussed below should be understood to deal with by-product of type II, and exclude hazardous waste.

Adding the imports of waste, w^m, the balancing equations for the supply and use of waste can be represented, following the notations in (6.1), as

$$\text{Supply:} \quad w^{\text{out}} = (W_{\text{I}}^{\text{out}} + W_{\text{II}}^{\text{out}})\iota + w^m + W_f \iota \tag{6.9}$$

$$\text{Use:} \quad w^{\text{in}} = (W_{\text{I}}^{\text{in}} + W_{\text{II}}^{\text{in}})\iota + w_f^e \tag{6.10}$$

$$= \underbrace{(W_{\text{I}}^{m\,\text{in}} + W_{\text{II}}^{d\,\text{in}})\iota}_{\text{domestic origin}} + \underbrace{(W_{\text{I}}^{m\,\text{in}} + W_{\text{II}}^{m\,\text{in}})\iota}_{\text{foreign origin}} + \underbrace{w_f^e}_{\text{export}}, \tag{6.11}$$

where the use of waste in endogenous sectors, W_{I}^{in} and $W_{\text{II}}^{\text{in}}$, is distinguished by its origin into two components, that is, domestic and import. For the sake of simplicity, it is assumed in (6.11) that waste is not used in the final demand sectors (which approximately holds in Table 6.11).

Assuming also that the imports of waste are not exported (which sounds reasonable), and noting that $W_{\text{I}}^{\text{out}}$ and $W_{\text{II}}^{\text{out}}$ do not include imports, it follows that

$$w^m = (W_{\text{I}}^{m\,\text{in}} + W_{\text{II}}^{m\,\text{in}})\iota \tag{6.12}$$

Inserting this into (6.11), and rewriting gives

$$\underbrace{w^{\text{in}} - w_f^e}_{\text{Domestic use of waste}} = \underbrace{(W_{\text{I}}^{m\,\text{in}} + W_{\text{II}}^{d\,\text{in}})\iota}_{\text{Domestic origin}} + \underbrace{w^m}_{\text{Foreign origin}} \tag{6.13}$$

Following (3.6), define the share of imports in the domestic use of waste:

$$\widehat{\mu^w} = \widehat{w^m}(\widehat{w^{\text{in}} - w_f^e})^{-1} \tag{6.14}$$

Table 6.13 The Share of Imports in the Domestic Use of Waste, μ^w.

grb:	kitchen waste	0.000
pap:	waste paper	0.016
pls:	waste plastics	0.008
metl:	metal scrap	0.014
gls:	waste glass	0.000
wod:	waste wood	0.000
ash:	ash, dust, slag, molten slag	0.139
sld:	sludge	0.000
oil:	waste oil, acid, & alkali	0.000
cons:	construction waste	0.000
blk:	bulky waste	0.000
vhc:	end of life vehicles	0.000
rsd:	shredder residue	0.000

Table 6.13 shows the values of μ^w obtained from Table 6.12. It is noteworthy that 14% of the supply of ash, slag, and molten slag in Japan came from imports (to be exact, it dealt with slag).

Assuming the constancy of μ^w, the imports of waste will be given by

$$w^m = \widehat{\mu^w}(w^{in} - w_f^e) = \widehat{\mu^w}(W_I^{in} + W_{II}^{in})\iota \qquad (6.15)$$

where the second equality follows from (6.10). Substitution of this into (6.9) gives:

$$w^{out} = \left(W_I^{out} + W_{II}^{out} + \widehat{\mu^w}(W_I^{in} + W_{II}^{in})\right)\iota + w_f \qquad (6.16)$$

where $w_f = W_f\iota$. Finally, subtracting (6.10) from this equation, one obtains the following expression for the net generation of waste adjusted for competitive imports of waste:

$$w = \left(W_I^{out} - (I - \widehat{\mu^w})W_I^{in}\right)\iota + \left(W_{II}^{out} - (I - \widehat{\mu^w})W_{II}^{in}\right)\iota + w_f - w_f^e \qquad (6.17)$$

$$= \left(W_I^d + W_{II}^d\right)\iota + w_f^d \qquad (6.18)$$

where W_I^d and W_{II}^d stand for the terms in the first and second parenthesis in (6.17), and w_f^d stands for $w_f - w_f^E$. Using waste generation coefficients matrices, these equations can be represented as

$$w = \left(G_I^{out} - (I - \widehat{\mu^w})G_I^{in}\right)x_I + \left(G_{II}^{out} - (I - \widehat{\mu^w})G_{II}^{in}\right)x_{II} + w_f^d \qquad (6.19)$$

$$= G_I^d x_I + G_{II}^d x_{II} + w_f^d \qquad (6.20)$$

Multiplication of a given allocation matrix S from the left of both sides of (6.18) (or (6.20)) then converts w to x_{II}:

$$x_{II} = Sw = S(W_I^d + W_{II}^d)\iota + Sw_f^d, \quad \text{or} \qquad (6.21)$$

$$= SG_I^d x_I + SG_{II}^d x_{II} + Sw_f^d \qquad (6.22)$$

Consequently, the counterpart of (5.87) with competitive imports of waste is given by:

$$\begin{pmatrix} x_{\mathrm{I}} \\ x_{\mathrm{II}} \end{pmatrix} = \begin{pmatrix} A_{\mathrm{I}}^d & A_{\mathrm{II}}^d \\ SG_{\mathrm{I}}^d & SG_{\mathrm{II}}^d \end{pmatrix} \begin{pmatrix} x_{\mathrm{I}} \\ x_{\mathrm{II}} \end{pmatrix} + \begin{pmatrix} f^d + x^e \\ Sw_f^d \end{pmatrix}, \tag{6.23}$$

where $A_i^d = (I - \widehat{\mu})A_{\mathrm{I},i}$ for $i = \mathrm{I}, \mathrm{II}$, $f^d = (I - \widehat{\mu})f_{\mathrm{I}}$, and x^e refers to the export of goods and services, with the solution given by

$$\begin{pmatrix} x_{\mathrm{I}} \\ x_{\mathrm{II}} \end{pmatrix} = \begin{pmatrix} I - A_{\mathrm{I}}^d & -A_{\mathrm{II}}^d \\ -SG_{\mathrm{I}}^d & I - SG_{\mathrm{II}}^d \end{pmatrix}^{-1} \begin{pmatrix} f^d + x^e \\ Sw_f^d \end{pmatrix} \tag{6.24}$$

6.1.2.4 The Allocation Matrix

Two Examples with Different Sorting Patterns

Table 6.14 gives two examples of the allocation matrix S, S_0 and S_1. Because the present case involves 13 types of waste and 3 types of waste treatment, these matrices are of order 3×13. The matrix S_1 refers to a hypothetical case where combustible and noncombustible waste items are completely sorted from each other and no mixing up takes place in their treatment. On the other hand, S_0 gives a more realistic picture with regard to the first six waste items and bulky waste, which is based on estimates of the national average from the 1990s (see [10] for details).

In reality, the separation of waste between combustible and noncombustible items cannot be perfect. Consequently, it is unavoidable that some portions of combustible waste ends up in a landfill, while some portions of noncombustible waste find their

Table 6.14 Examples of the Allocation Matrix.

Waste\treatment		Treatment					
		S_0: imperfect sorting			S_1: perfect sorting		
	Waste item	Incineration	Landfill	Shredding	Incineration	Landfill	Shredding
grb	kitchen waste	0.900	0.100	0.000	1.000	0.000	0.000
pa	waste paper	0.931	0.069	0.000	1.000	0.000	0.000
pls	waste plastics	0.589	0.411	0.000	1.000	0.000	0.000
metl	metal scrap	0.006	0.994	0.000	0.000	1.000	0.000
gls	waste glass	0.036	0.964	0.000	0.000	1.000	0.000
wood	waste wood	0.997	0.003	0.000	1.000	0.000	0.000
ash	ash, dust, slag	0.000	1.000	0.000	0.000	1.000	0.000
sld	sludge,	0.000	1.000	0.000	0.000	1.000	0.000
oil	waste oil	1.000	0.000	0.000	1.000	0.000	0.000
cons	construction waste	0.000	1.000	0.000	0.000	1.000	0.000
blk	bulky waste	0.067	0.051	.882	0.000	1.000	0.000
vhc	end of life vehicles	0.000	0.000	1.000	0.000	0.000	1.000
dst	shredder residue	0.000	1.000	0.000	0.000	0.000	1.000

Source: [10].

way into an incinerator. For instance, about 7% of waste paper ended up in a landfill, although it is combustible. On the other hand, about 4% of waste glass was subjected to incineration, although it has no combustible components.

Chemically, waste plastics are combustible (see Table 6.21). Notwithstanding this, about 40% of them were landfilled. This number is too large to be attributed to mistakes in separation that were committed nondeliberately. As a matter of fact, in quite a number of Japanese municipalities including the Tokyo metropolis, waste plastics were deliberately landfilled until quite recently, because of the fear of a potential increase in the emission of hazardous substances (such as dioxins) and of the possible damage to incinerators due to the high temperature that results from incinerating large amounts of plastics. (Note that with technological improvements in incinerators and the closing down of small-scale incinerators with an insufficient control capability of flue gases, the situation has changed significantly, and the share of plastics incinerated is increasing.)

The Flow of Waste Treatment Activities

Table 6.15 gives the flow of waste treatment activities obtained by applying the allocation matrices S_0 (the upper panel of the table) and S_1 to the flow of waste in Table 6.12 after having adjusted for the imports of waste as in (6.18):

	Production Waste treatment Final demand
Waste treatment	SW_{I}^d $\qquad$ SW_{II}^d $\qquad$ SW_f^d $\quad$ $-w_f^E$

where W_f^d refers to W_f except export, w_f^E. It follows that under S_0 the amounts of waste for incineration and for landfill were 57.6×10^9 and 65.7×10^9 kg, whereas under S_1 they were 63.9×10^9 and 59×10^9 kg.

The Composition of Waste Allocated to Each Treatment Process

In waste treatment, a change in the composition of waste subjected to a particular process, say incineration, can exert nonnegligible effects on the input-output relationships of the process. As we shall see below in Section 6.2, this is in particular the case for incineration. For a given allocation matrix, the composition of waste subjected to each of the three processes is given by the following $k \times m$ matrix

$$S\widehat{w} \tag{6.25}$$

the results of which are shown in Table 6.16. The transition of the allocation matrix from S_0 to S_1 increases the proportion of plastics in the total waste for incineration from 13% to 17%, while the proportions of all the remaining waste items decrease or remain unchanged. As will be discussed in Section 6.2 below, this change in the composition of waste for incineration leads to a nonnegligible change in the unit process of incineration.

Table 6.15 The Flow of Waste Converted to the Flow of Treatment by the Allocation Matrix S.

	AGRC	MING	FOOD	WOOD	CHEM	CMNT	METL	METP	MCHN	CNST	UTIL	SRVC	TRNS	INCN	LNDF	SHRD	CONS	INVS	EXPT	Total
								Conversion by S_0												
Incineration	85,160	16	−74,326	−9,524	626	257	−2,467	2,545	3,957	5,048	54	15,312	711	0	0	1,509	28,111	1,345	−740	57,594
Landfill	7,789	3,683	−11,038	3,999	2,665	−30,658	372	14,245	10,646	253	21,405	4,473	267	4,417	0	6,917	8,102	21,598	−3,409	65,725
Shredding	0	0	0	0	0	0	0	0	0	0	0	0	0	0	0	0	8428	0	0	8.428
								Conversion by S_1												
Incineration	92,690	17	−80,687	−10,291	548	481	4	2,102	3,786	5,432	49	17,163	809	0	0	1,380	31,255	−120	−672	63,945
Landfill	259	3,681	−4,677	4,767	2,743	−30,881	−2,098	14,687	10,816	−131	21,409	2,622	169	4,417	0	7,045	4,571	23,064	−3,476	58,987
Shredding	0	0	0	0	0	0	0	0	0	0	0	0	0	0	0	0	8,814	0	0	8.814

Units: 10^6 kg. See Table 6.11 for the notations.

Table 6.16 Allocation of Waste to Treatment Under Alternative Allocation Matrices.

	grb	pap	pls	metl	gls	wod	ash	sld	oil	cons	blk	vhc	rsd	Total
							Allocated by S_0							
Incineration	14,287	16,497	7,536	478	546	12,227	0	0	5,806	0	218	0	0	57,594
Landfill	1,587	1,535	3,402	7,097	4,561	1,069	16,084	17,336	0	10,706	168	0	2,180	65,725
Shredding	0	0	0	0	0	0	0	0	0	0	2,888	5,540	0	8,428
							Shares							
Incineration	0.25	0.29	0.13	0.01	0.01	0.21	0.00	0.00	0.10	0.00	0.00	0.00	0.00	1.00
Landfill	0.02	0.02	0.05	0.11	0.07	0.02	0.24	0.26	0.00	0.16	0.00	0.00	0.03	1.00
Shredding	0.00	0.00	0.00	0.00	0.00	0.00	0.00	0.00	0.00	0.00	0.34	0.66	0.00	1.00
							Allocated by S_1							
Incineration	15,874	18,032	10,938	0	0	13,295	0	0	5,806	0	0	0	0	63,945
Landfill	0	0	0	7,575	5,107	0	16,084	17,336	0	10,706	0	0	2,180	58,987
Shredding	0	0	0	0	0	0	0	0	0	0	3,274	5,540	0	8,814
							Shares							
Incineration	0.25	0.28	0.17	0.00	0.00	0.21	0.00	0.00	0.09	0.00	0.00	0.00	0.00	1.00
Landfill	0.00	0.00	0.00	0.13	0.09	0.00	0.27	0.29	0.00	0.18	0.00	0.00	0.04	1.00

Units: 10^6 kg, INC: incineration, LND: landfill, SHR: shredding. The numbers referring to shredding under S_1 are omitted because they are the same as under S_1. See Table 6.12 for the notations for waste.

Because of this, the rest of this subsection will deal with the case where the allocation matrix is fixed at S_0. The case of the shift of the allocation matrix to S_1 will be dealt with in Section 6.2.3.

6.1.2.5 WIO Analysis

Stacking of the flow matrix of goods and services (Table 6.11) from below by the flow matrix of waste treatment (Table 6.15) gives the WIO table with a square intermediate flow matrix of order 16×16. Dividing its column elements by the corresponding level of activity then yields the coefficients matrices occurring in the first parenthesis of the right of (6.23) (Table 6.17).

The "subtotal" at the bottom of Table 6.17 gives for each column the sum over the three waste treatment processes, and hence the amount of waste for treatment that was generated directly to produce a unit of its product or activity. With 6.5×10^3 kg per million yen, the agricultural sector (AGRC) generated the largest amount of waste per unit of its product, of which 5.9×10^3 kg was incinerated. Next to the agricultural sector was the mining sector (MING) with 3.3×10^3 kg per million yen, all of which went to landfills. On the other hand, the cement-producing sector (CMNT) absorbed 3.6×10^3 kg of waste per million yen of its product, which otherwise would have gone to landfills. The cement sector was followed by the food-producing sector (FOOD), which absorbed 1.9×10^3 kg of waste per million yen of its product, the most part of which would otherwise have been incinerated.

The Waste Footprints of a Product: Tracing the Origin of Waste to Final Products

Table 6.18 gives the Leontief inverse matrix. For each column, the sum over the three waste treatment processes gives the amount of waste for treatment that was generated directly and indirectly to deliver a unit of its product to the final demand. Comparison with the corresponding values for Table 6.17 that refer to direct requirements only indicates that, except for cement (CMNT), metal (METL), and construction (CNST), consideration of the indirect effects tended to reinforce the direct effects. This was in particular the case for the transport sector (TRNS) and the service sector (SRVC), where the directly generated amount was augmented by 500% (from 14 to 72 kg per million yen), and by 200% (from 43 to 87 kg per million yen), respectively.

The result for the construction sector (CONST) was the exact opposite to this. In terms of the direct effects, this sector was a net generator of waste with none of its input coefficients taking a negative value for waste treatment. When the indirect effects were taken into account, however, this sector turned out to be a "green one", which contributed to the reduction of waste for treatment by 30 kg per million yen of its product. The main reason behind this is the fact that this sector is the largest single user of cement, the production of which absorbs the largest amount of waste per million yen of its product.

Table 6.17 The Matrix of Squared Input/Generation Coefficients Under S_0.

	AGRC	MING	FOOD	WOOD	CHEM	CMNT	METL	METP	MCHN	CNST	UTIL	SRVC	TRNS	INC	LND	SHR
AGRC	0.0945	0.0004	0.1442	0.0188	0.0021	0.0000	0.0000	0.0000	0.0007	0.0017	0.0000	0.0025	0.0000	0.0000	0.0000	0.0000
MING	0.0000	0.0003	0.0000	0.0001	0.0121	0.0079	0.0157	0.0000	0.0000	0.0010	0.0097	0.0000	0.0000	0.0000	0.0000	0.0000
FOOD	0.0672	0.0058	0.1350	0.0040	0.0042	0.0027	0.0002	0.0011	0.0020	0.0022	0.0003	0.0140	0.0014	0.0000	0.0000	0.0000
WOOD	0.0140	0.0057	0.0240	0.2409	0.0127	0.0234	0.0010	0.0065	0.0091	0.0464	0.0055	0.0199	0.0095	0.0000	0.0000	0.0000
CHEM	0.0598	0.0276	0.0346	0.0481	0.2595	0.0367	0.0291	0.0180	0.0398	0.0254	0.0352	0.0182	0.0617	0.0002	0.0002	0.0002
CMNT	0.0011	0.0002	0.0036	0.0034	0.0044	0.0825	0.0188	0.0038	0.0087	0.0587	0.0007	0.0008	0.0000	0.0003	0.0000	0.0000
METL	0.0000	0.0000	0.0000	0.0001	0.0012	0.0027	0.2129	0.1251	0.0036	0.0000	0.0000	0.0000	0.0000	0.0000	0.0000	0.0000
METP	0.0012	0.0241	0.0168	0.0125	0.0083	0.0170	0.0007	0.2272	0.0675	0.1096	0.0013	0.0018	0.0013	0.0174	0.0027	0.0038
MCHN	0.0056	0.0098	0.0028	0.0027	0.0011	0.0045	0.0000	0.0039	0.2980	0.0188	0.0012	0.0148	0.0075	0.0000	0.0000	0.0000
CNST	0.0056	0.0081	0.0023	0.0042	0.0052	0.0148	0.0079	0.0086	0.0027	0.0026	0.0525	0.0119	0.0093	0.0000	0.0021	0.0000
UTIL	0.0063	0.0357	0.0139	0.0250	0.0318	0.0399	0.0658	0.0117	0.0057	0.0057	0.0595	0.0149	0.0148	-0.0001	0.0001	0.0004
SRVC	0.1013	0.1627	0.1376	0.1388	0.1479	0.1648	0.0948	0.1174	0.1621	0.1636	0.1668	0.1599	0.2397	0.0003	0.0019	0.0004
TRNS	0.0420	0.3316	0.0319	0.0406	0.0279	0.0723	0.0399	0.0297	0.0205	0.0609	0.0248	0.0415	0.1173	0.0000	0.0000	0.0000
INC	5.9262	0.0144	-1.6155	-0.3549	0.0120	0.0304	-0.4104	0.0819	0.0292	0.0652	0.0023	0.0331	0.0102	0.0000	0.0000	0.1790
LND	0.5420	3.3070	-0.2399	0.1490	0.0510	-3.6251	0.0620	0.4587	0.0785	0.0033	0.9110	0.0097	0.0038	0.0767	0.0000	0.8207
SHR	0.0000	0.0000	0.0000	0.0000	0.0000	0.0000	0.0000	0.0000	0.0000	0.0000	0.0000	0.0000	0.0000	0.0000	0.0000	0.0000
Subtotal[a]	6.4683	3.3214	-1.8554	-0.2059	0.0629	-3.5947	-0.3484	0.5406	0.1076	0.0685	0.9132	0.0428	0.0141	0.0767	0.0767	0.9997

[a] the amount of total inputs from waste treatment processes (10^6 kg per billion Japanese yen. AGRC: agriculture, fishery, forestry, MING: ore, coal, petroleum, FOOD: food, beverage, feed, textile, WOOD: wood & products, paper & products, CHEM: chemicals & products, plastics, rubber products CMNT: cement, glass, pottery, earthen products, METL: iron & steel nonferrous metals, METP: metal products, MCHN: machinery and equipment, CNST: construction, & civil engineering, UTIL: utilities, SRVC: services, TRNS: transport and communications.

Table 6.18 The WIO Inverse Matrix Under S_0.

	AGRC	MING	FOOD	WOOD	CHEM	CMNT	METL	METP	MCHN	CNST	UTIL	SRVC	TRNS	INC	LND	SHR
AGRC	1.121	0.005	0.189	0.031	0.007	0.004	0.002	0.003	0.005	0.006	0.002	0.008	0.003	0.000	0.000	0.000
MING	0.002	1.002	0.002	0.002	0.017	0.010	0.022	0.005	0.002	0.003	0.011	0.001	0.002	0.000	0.000	0.000
FOOD	0.093	0.015	1.176	0.014	0.012	0.009	0.005	0.007	0.010	0.009	0.006	0.021	0.009	0.000	0.000	0.000
WOOD	0.037	0.028	0.051	1.331	0.034	0.046	0.013	0.023	0.032	0.077	0.021	0.037	0.028	0.001	0.000	0.001
CHEM	0.125	0.091	0.089	0.108	1.373	0.078	0.071	0.060	0.102	0.068	0.069	0.044	0.112	0.002	0.001	0.002
CMNT	0.008	0.004	0.006	0.007	0.008	1.092	0.028	0.012	0.017	0.067	0.006	0.003	0.003	0.001	0.000	0.000
METL	0.005	0.007	0.005	0.005	0.005	0.009	1.272	0.207	0.028	0.025	0.003	0.002	0.002	0.000	0.000	0.000
METP	0.025	0.042	0.030	0.027	0.020	0.030	0.007	1.301	0.131	0.151	0.015	0.010	0.009	0.002	0.001	0.001
MCHN	0.182	0.041	-0.007	0.007	0.012	0.002	-0.005	0.018	1.436	0.038	0.015	0.029	0.022	0.025	0.004	0.013
CNST	0.016	0.028	0.010	0.014	0.015	0.017	0.020	0.022	0.013	1.012	0.063	0.018	0.018	0.000	0.002	0.002
UTIL	0.023	0.057	0.030	0.047	0.055	0.058	0.099	0.061	0.036	0.027	1.073	0.025	0.030	0.001	0.000	0.001
SRVC	0.255	0.373	0.273	0.289	0.295	0.284	0.217	0.267	0.356	0.304	0.269	1.244	0.373	0.007	0.004	0.007
TRNS	0.083	0.405	0.073	0.085	0.071	0.116	0.086	0.075	0.065	0.105	0.056	0.066	1.160	0.001	0.001	0.001
INC	6.498	0.030	-0.784	-0.298	0.036	0.035	-0.511	0.030	0.056	0.077	0.015	0.043	0.023	1.001	0.001	0.180
LND	1.124	3.391	-0.202	0.235	0.170	-3.841	0.112	0.651	0.172	-0.107	1.013	0.043	0.049	0.079	1.000	0.836
SHR	0.000	0.000	0.000	0.000	0.000	0.000	0.000	0.000	0.000	0.000	0.000	0.000	0.000	0.000	0.000	1.000
Subtotal[a]	7.622	3.421	-0.986	-0.063	0.205	-3.807	-0.399	0.681	0.227	-0.030	1.028	0.087	0.072	0.079	0.000	1.016

[a] Total inputs from waste treatment processes.

For the waste treatment processes, own inputs are excluded.

AGRC: agriculture, fishery, forestry, MING: ore, coal, petroleum, FOOD: food, beverage, feed, textile, WOOD: wood & products, paper & products, CHEM: chemicals & products, plastics, rubber products CMNT: cement, glass, pottery, earthen products, METL: iron & steel nonferrous metals, METP: metal products, MCHN: machinery and equipment, CNST: construction, & civil engineering, UTIL: utilities, SRVC: services, TRNS: transport and communications.

As is indicated by (6.24) (or (5.90)), for any waste there exist some products or waste treatment activities the final demand for which invoked its generation. Differently stated, the generation of any waste can be traced back to certain products or waste treatment activities, the final demand for which gave the initial stimulus, the interindustry repercussion effects of which have resulted in its generation. From (6.24) the product origin of waste for treatment can be obtained by

$$
\begin{pmatrix} B_{\mathrm{II,I}} & B_{\mathrm{II,II}} \end{pmatrix} \begin{pmatrix} \widehat{f_1} & 0 \\ 0 & \widehat{Sw_f^d} \end{pmatrix} \tag{6.26}
$$

where $B_{\mathrm{II,I}}$ and $B_{\mathrm{II,II}}$ refer to the first and second terms at the bottom of the inverse matrix in (6.24).

Figure 6.1 shows the results. (Table 6.19 gives the values). The largest amount of waste for incineration is attributed to the final demand for the agricultural sector (AGRC), followed by the service sector (SRVC). About 80% of the waste for incineration can be attributed to the final demand for these two sectors. On the other hand, 95% of the reduction in the waste for incineration is invoked by the final demand for the food sector. Arithmetically, 87% of the waste for incineration generated by the agricultural sector can be balanced with the amount absorbed by the food sector.

Concerning the waste for landfill, 60% of its generation is attributed almost equally to the final demands for machinery (MCHN) and service (SRVC), followed by the final demand for utilities (UTIL) with a share of 17%. As for the absorption of the waste for landfill, the final demand for construction occupied the largest share

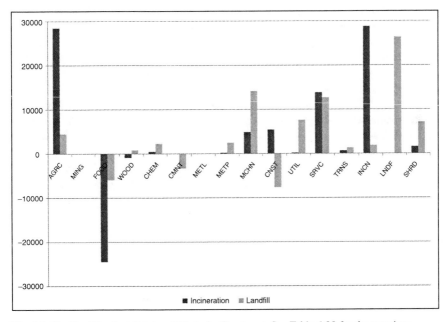

Fig. 6.1 The Final Product Origin of Waste for Treatment. See Table 6.20 for the notations.

Table 6.19 Generation of Waste Attributed to Final Demand Categories Under S_0.

	Consumption			Investment			Export		
	Indirect	Direct	Total	Indirect	Direct	Total	Indirect	Direct	Total
grb	5,273	9,328	14,601	907	0	907	366	0	366
pap	2,052	16,755	18,807	−441	−120	−561	158	−372	−214
pls	4,053	4,317	8,371	2,156	0	2,156	711	−300	411
metl	1,142	1,830	2,973	−8,823	23,064	14,241	−6,584	−3,054	−9,638
gls	965	2,741	3,706	1,338	0	1,338	63	0	63
wod	1,302	855	2,157	10,608	0	10,608	530	0	530
ash	12,713	0	12,713	−6,802	0	−6,802	10,594	−422	10,172
sld	11,491	0	11,491	3,537	0	3,537	2,308	0	2,308
oil	2,689	0	2,689	1,800	0	1,800	1,317	0	1,317
cons	907	0	907	9,683	0	9,683	116	0	116
blk	0	3,274	3,274	0	0	0	0	0	0
vhc	0	5,540	5,540	0	0	0	0	0	0
rsd	2,180	0	2,180	0	0	0	0	0	0

Units: 10^6 kg. See Table 6.14 for the notations.

(45% in the total amount of waste absorbed), followed by the final demands for food products (35%) and cement (20%). Notice that the contribution of the cement sector refers to that attributed to its direct use in the final demand. On the other hand, the indirect use of cement in construction & civil engineering, which constitutes the overwhelming majority of its use, is attributed to the construction sector.

The above results indicate two important points. The first point refers to the substantial contribution of the food and construction sectors to the reduction in the amount of waste for treatment through their use of waste generated in other sectors of the economy. Without the absorption by these two sectors, the waste for landfill would have been 21% larger, and without the food sector the waste for incineration would have been 42% larger. The second point refers to the fact that the service sector occupies a significant share in the generation of waste for treatment. In particular, the share in waste generation of the service sector is the second largest in both waste for incineration and waste for landfill. It follows that regarding the generation of waste the service sector is not as green as it might appear (similar results were obtained by Suh [16] for the emission of GWP in the US, and by Sánchez-Chóliz and Duarte [15] for major atmospheric and aquatic pollutants in Spain). In particular, an increase in the share of the expenditure for services in GDP (gross domestic product) would not lead to a "dematerialization" or "decoupling from material basis" of the economy in the form of a reduction in waste.

Besides the amount of waste subjected to each treatment, one may also be interested in its content. By use of (5.90), one can break down the waste for respective treatment into different waste types as

$$\Xi \left(\frac{\widehat{f^d + x^e}}{\widehat{Sw_f^d}} \right), \tag{6.27}$$

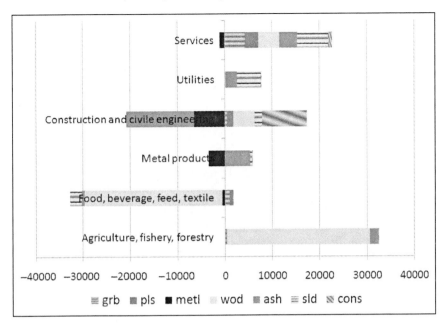

Fig. 6.2 The Waste Footprint of Selected Products. See Table 6.20 for the notations.

where

$$\Xi = \begin{pmatrix} G_{\mathrm{I}}^d & G_{\mathrm{II}}^d \end{pmatrix} \begin{pmatrix} I - A_{\mathrm{I}}^d & -A_{\mathrm{II}}^d \\ -SG_{\mathrm{I}}^d & I - SG_{\mathrm{II}}^d \end{pmatrix}^{-1} \tag{6.28}$$

Because this gives the overall flow of each waste associated with (or embodied in) the delivery of a final product, it can be called the "waste footprints of a product". Figure 6.2 shows the results for six producing sectors with significant contributions to the waste flow (see Table 6.20 for the full results).

It was conjectured above that about 90% of the waste for incineration that was generated by the agricultural sector can be balanced with the amount absorbed by the food sector. This balancing would be possible if it dealt with the same type of waste. Figure 6.2 indicates that this is in fact the case: it deals with organic waste. The construction sector is the largest generator of construction debris, but is at the same time the largest user of both "ash" and metal scrap. The metal products sector is the largest generator of "ash", followed by services and utilities (note that slag is contained in "ash"). Besides its significant share in the amount of waste generated as mentioned above, the service sector is also characterized by the occurrence of many types of waste, with none dominating the other items.

Tracing the Origin of Waste to Final Demand Categories

A slight modification of (6.27) enables one to associate the amount of individual waste to its final demand category of origin (household consumption, investment,

Table 6.20 The Net Generation of Waste Per Unit of Product/Treatment for Final Delivery: The Matrix Ξ Under S_0.

	AGRC	MING	FOOD	WOOD	CHEM	CMNT	METL	METP	MCHN	CNST	UTIL	SRVC	TRNS	INC	LND	SHR
grb	0.046	0.006	0.030	0.006	0.004	0.004	0.003	0.004	0.005	0.006	0.004	0.014	0.006	0.000	0.000	0.000
pap	0.077	0.007	0.007	-0.553	-0.001	-0.001	0.004	0.010	0.006	-0.009	0.000	0.009	0.004	0.000	0.000	0.014
pls	0.035	0.007	0.021	0.033	-0.004	0.019	-0.046	0.020	0.015	0.020	0.002	0.009	0.006	0.000	0.000	0.035
metl	-0.005	-0.028	-0.019	-0.018	-0.028	-0.045	-8.573	-0.966	-0.050	-0.097	-0.009	-0.004	-0.005	-0.001	0.000	0.577
gls	0.012	0.005	-0.012	0.006	0.003	-0.212	0.033	0.012	0.003	0.017	0.003	0.004	0.002	0.000	0.000	0.000
wod	6.908	0.005	-0.929	0.185	0.023	0.007	0.004	0.005	0.012	0.064	0.008	0.013	0.007	0.000	0.000	0.115
ash	0.520	0.959	-0.031	0.030	0.054	-2.840	7.952	1.418	0.169	-0.206	0.331	0.013	0.015	0.079	0.000	0.015
sld	0.018	2.447	-0.079	0.236	0.136	-0.779	0.173	0.117	0.038	0.023	0.677	0.022	0.031	0.000	0.000	0.001
oil	0.008	0.010	0.025	0.010	0.016	0.037	0.047	0.058	0.026	0.012	0.003	0.004	0.004	0.000	0.000	0.000
cons	0.002	0.004	0.001	0.002	0.002	0.002	0.003	0.003	0.002	0.140	0.009	0.002	0.002	0.000	0.000	0.000
blk	0.000	0.000	0.000	0.000	0.000	0.000	0.000	0.000	0.000	0.000	0.000	0.000	0.000	0.000	0.000	0.000
vhc	0.000	0.000	0.000	0.000	0.000	0.000	0.000	0.000	0.000	0.000	0.000	0.000	0.000	0.000	0.000	0.000
rsd	0.000	0.000	0.000	0.000	0.000	0.000	0.000	0.000	0.000	0.000	0.000	0.000	0.000	0.000	0.000	0.259
Total	7.622	3.421	-0.986	-0.063	0.205	-3.807	-0.399	0.681	0.227	-0.030	1.028	0.087	0.072	0.080	0.001	1.016

Units: 10^6 kg per one billion Japanese yen for producing sectors, and 10^6 kg per 10^6 kg of waste treated for waste treatment sectors.
AGRC: agriculture, fishery, forestry, MING: ore, coal, petroleum, FOOD: food, beverage, feed, textile, WOOD: wood & products, paper & products, CHEM: chemicals & products, plastics, rubber products CMNT: cement, glass, pottery, earthen products, METL: iron & steel nonferrous metals, METP: metal products, MCHN: machinery and equipment, CNST: construction, & civil engineering, UTIL: utilities, SRVC: services, TRNS: transport and communications.
grb: kitchen waste, pap: waste paper, pls: waste plastics, metl: metal scrap, gls: waste glass, wod: waste wood, ash: ash, dust, slag, molten slag sld: sludge, oil: waste oil, cons: construction waste, blk: bulky waste, vhc: end of life vehicles, dst: shredder residue.

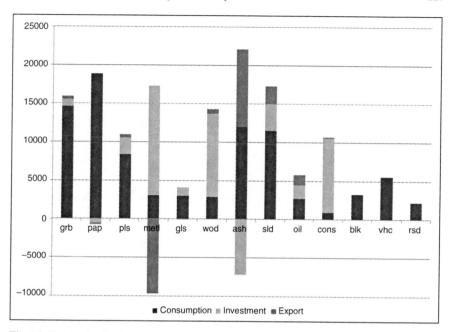

Fig. 6.3 Tracing the Origin of Waste to Final Demand Categories. See Table 6.20 for the notations.

and export):

$$\Xi \begin{pmatrix} F^{\mathrm{d}} + x^{e} \\ SW_{f}^{d} \end{pmatrix} \tag{6.29}$$

The results in Figure 6.3 indicate several findings. First, household consumption turned out to be the largest source of waste for treatment for 9 out of the 13 waste types, except for metal scrap, wood waste, waste oil, and construction waste. While investment expenditure is the major source of waste for treatment for metal scrap, wood waste, and construction waste, it is the major source of reducing ash. Finally, export contributes to a significant reduction in the generation of metal scrap, but on the other side is a major source of the generation of ash. These findings indicate that a fluctuation in the expenditure composition of GDP can have significant effects on the pattern of waste generation.

6.1.2.6 The Alternative Approach Based on Unit Processes

Above, the matrices of input/generation coefficients were obtained from the flow tables (Tables 6.11 and 6.12). One could also have proceeded in exactly the opposite way: starting from a given matrix of unit processes, the allocation matrix, and a vector of final demand, the vector of activity levels can be obtained as in (5.88), and the flow tables obtained as

$$\begin{pmatrix} A_{I,I}\widehat{x_I} & A_{I,II}\widehat{x_{II}} \\ SG_I\widehat{x_I} & SG_{II}\widehat{x_{II}} \end{pmatrix} \tag{6.30}$$

In particular, it should be noted that even for Tables 6.11 and 6.12 the columns of waste treatment processes were obtained in this way, where the system engineering model of waste treatment due to Matsuto [6] was used in obtaining the relevant unit processes.

6.2 The Dynamic Nature of Waste Treatment Processes

In Section 5.3.2.4, a simple example of the unit process of waste incineration was introduced where electricity and residue occur as by-product (see (5.77)). Figure 6.4 shows the flow diagram of a typical waste incineration process, where waste of a given chemical composition is converted to ash and gaseous emissions, and electricity is obtained by use of the waste heat. On the other hand, the input of labor is omitted, because in a modern incineration plant its level does not usually vary with the inputs shown in the figure.

Slaked lime ($Ca(OH)_2$) is used for removing HCl and SO_x contained in the flue gas, with the chemical reactions of the former given by

$$Ca(OH)_2 + 2HCl \rightarrow CaCl_2 + 2H_2O \tag{6.31}$$

For the removal process of SO_x, see (5.22).

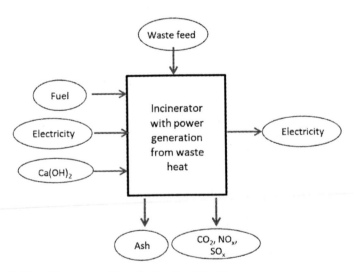

Fig. 6.4 The Flow Diagram of a Waste Incineration Process.

A distinguishing feature of waste treatment processes, in particular incineration, consists in that the combinations of their inputs and outputs depend on the chemical composition of waste to be treated. In fact, the levels of all the inputs and outputs occurring in Figure 6.4 depend on the chemical composition of the waste to be incinerated. This implies that the coefficients occurring in (5.77) may change when the chemical composition of the waste changes. This feature is in sharp contrast to the standard production process such as steel mills or petrochemical plants, where, for a given technology, the amounts of inputs are largely determined by the amount of its primary product.

In the language of LCA, this difference can be attributed to the difference in the functional unit as was pointed out by McDougall et al. [7]:

··· in a product LCI the functional unit is defined by the output (product) of the system, in an LCI of waste the functional unit is defined in terms of the system's input, i.e. the waste. (p.106).

In a production process the functional unit refers to the primary product, say, a unit of steel, or chemical product, the physical and chemical characteristics of which should be at a given level. In a waste treatment process, however, the functional unit refers to the waste, the physical and chemical characteristics of which may fluctuate due to the very nature of their being waste.

As an example, Table 6.21 shows the higher heating value (kJ/kg), H_H, and chemical composition (%) of MSW based on Japanese data. Using the Steuer equation, H_H is given as follows (Tanaka [17], p.89):

$$H_H = 339.4\left(C - (3/8)O\right) + 238.8 \cdot (3/8)O + 1435.1(H - O/16) + 94.3S \quad (6.32)$$

where the atomic symbols refer to the composition (%) of each element in the table.

Table 6.21 Heat Value and Chemical Composition of Municipal Solid Waste.

	Paper	Plastics	Garbage	Textiles	Green waste	Others
Combustibles[a]	89.61	94.1	87.29	97.26	98.01	76.86
C	42.1	72.11	45.12	51.79	49.11	40.05
H	6.56	11.1	6.12	6.6	6.35	5.11
N	0.35	0.55	3.15	3.67	0.78	2.18
S	0.03	0.04	0.09	0.22	0.01	0.07
Cl	0.19	3.37	0.3	0.49	0.14	0.27
O	40.38	6.93	32.51	34.49	41.62	29.18
Ash content[a]	10.39	5.9	12.71	2.74	1.99	23.14
Higher heat value[b]	17,260	37,380	18,740	21,420	19,900	16,030

[a]% in weight, [b]dry solid, kJ/kg, Source: [17], p.87.

6.2.1 A System Engineering Representation
of the Incineration Process

In the following, the quantitative aspects of the dependence of process inputs and outputs on the chemical composition of incoming waste will be illustrated based on the detailed system engineering studies on waste incineration processes conducted by Tanaka [17] and Matsuto [6]. The illustration will be concerned with the inputs and outputs that occur in Figure 6.4, with ash further classified into bottom ash and fly ash (dust). It should be noted that all the numbered equations in this paragraph are based on Chapter 3 of Matsuto [6].

First, the amount of $Ca(OH)_2$ required for reducing HCl emission, U_{Ca} (kg/year), depends on the level of HCl in the flue gas, X_{HCL} (kmol/year), which in turn depends on the amount of waste treated per year, Q, and the Cl content of the waste, which is determined by the chemical composition of the waste only:

$$U_{Ca} = X_{HCL} \times 101 \tag{6.33}$$

$$X_{HCL} = b_{HCL} Q Cl/35.5 \tag{6.34}$$

where b_{HCL} is a parameter specific to HCl treatment methods.

In Table 6.21, plastics are characterized by the highest Cl content. It should be noted, however, that this number is somewhat misleading because not all plastic resins contain Cl. In fact, except for PVC (polyvinyl chloride), the monomer of which (vinyl chloride) has the chemical formula $CH_2 = CHCl$, the major plastic resins such as PE (polyethylene), PP (polypropylene), and PS (polystyrene) are free of Cl: the monomers of PE, PP, and PS are respectively $CH_2 = CH_2$, $CH_2 = CHCH_3$, and $CH_2 = CHC_6H_5$.

The Cl content of waste is thus significantly affected by the percentage of plastic products based on PVC in the waste. On the other hand, the level of SO_x depends on the S content of the waste and fuel. The input of fuel, on the other side, depends on the heat value of the waste. No input of fuel is required for incineration when the lower heat value, H_L, is higher than 4,184 kJ/kg [6]. H_L is related to the higher value H_H as

$$H_L = H_H - 25(9H + W) \tag{6.35}$$

with W referring to the water content of the waste.

Closely related to the heat value of waste is the amount of electricity that can be obtained from the heat of incineration. Writing γ for the thermal efficiency of the boiler, δ for the power generation efficiency of the turbine, and neglecting steam use within the process for the sake of simplicity, the amount of power obtained from the heat of incineration, G (kWh/year), is given by

$$G = H_L Q \gamma/(h_s - h_n) \times h_s/717 \times \delta/860 \tag{6.36}$$

where h_s and h_n are parameters referring to the properties of the boiler.

The amount of residue, Q_w, depends on, besides the amount of waste, its ash (noncombustible) content, α (Table 6.21), and incinerator types (a stoker system or a fluidized bed system):

$$Q_w = \alpha Q / (1 - b_{inc}) \tag{6.37}$$

where b_{inc} is a parameter specific to incinerator types. The amounts of bottom ash, Q_{w1}, and fly ash, Q_{w2}, are then given by

$$Q_{w1} = (1 - r_{inc}) Q_w \tag{6.38}$$

$$Q_{w2} = r_{inc} Q_w + Q_{CaO} \tag{6.39}$$

where r_{inc} is a parameter specific to incinerator types, and Q_{CaO} stands for the sum of $Ca(OH)_2$ that remained without entering the reaction process (6.31) and that regenerated (by the opposite $\leftarrow$ reaction) after the reaction:

$$Q_{CaO} = \left((3 - r_{HCl}) \times 37 + r_{HCl} \times 55.5 \right) X_{HCL} \tag{6.40}$$

where r_{HCl} refers to the rate of HCl reduction, and X_{HCL} is given by (6.34).

Finally, the amount of electricity used in the process, G_p, is given by

$$G_p = (1 + \varepsilon) \zeta_1 Q + \zeta_2 Q_{w2} \tag{6.41}$$

where ζ_1 and ζ_2 refer to the use of electricity per unit of waste and per unit of fly ash (which are assumed constants), while ε is a parameter that is specific to the type of incinerator, the ways combustion gases are cooled, HCl is treated, and electricity is generated. The surplus of G given by (6.36) over G_p is the amount of electricity that can be sold outside the plant.

6.2.2 Implications for WIO

The above system of equations based on engineering data indicates that, given the type of incinerator and the technical parameters specific to the incinerator and its associated power generation system (the boiler and the turbine), all the inputs and outputs that occur in Figure 6.4 are determined by the chemical characteristics of the waste. This implies that a significant change in the chemical characteristics of the waste would result in a nonnegligible change in the amounts of inputs and outputs, that is, the input- and output coefficients in (5.77) would change with a change in the chemical characteristics of the waste.

According to Table 6.21, plastics are characterized by the highest heat value and a relatively low ash content. Consider now the case where the share of plastics in waste increases with the relative shares of other waste items kept constant, say by a change in waste policy which shifts the treatment of plastics from landfill to incineration. From (6.36) it then follows that there would be an increase in the generation of electricity due to an increase in H_L, while whether or not this results in a net increase in the amount of electricity for sale has to be determined under consideration of (6.41).

When it comes to the amount of bottom ash and fly ash, the effects are not as straightforward as the generation of electricity, because the possible increase in the Cl content can increase the amount of fly ash per waste via (6.39) and (6.40), while the relatively low ash content contributes to reducing the amount of residue per waste.

Accordingly, the generation coefficient of residue, g_{n+2}^{out} in (5.77), would change when there is a significant change in the proportion of plastics. When plastics include a nonnegligible amount of PVC, the effects would be a result of mutually opposing effects, the direction of which can only be evaluated quantitatively. If plastic products based on PVC are avoided to enter the incineration process, however, the net result would be a reduction in the amounts of both bottom and fly ash, and an increase in the amount of electricity for sale (it should also be noted that PVC has a lower heat value compared to other major resins such as PP).

In WIO, the allocation of waste to treatment processes is represented by the allocation matrix S. As we have already seen in Section 5.3.2.5, S will change when there is a change in waste policy and/or regulations on the way waste is treated. A change in S is thus likely to affect the input and output coefficients of a specific treatment process, say incineration because of its effects on the chemical composition of the waste entering the process. If a change in S is associated with a significant change in the chemical composition, it is thus recommended not to continue the use of a unit process that is based on a particular (standard or default) chemical composition, but to use the more realistic one corresponding to the new composition. System engineering models like that of Matsuto mentioned above will be useful for this purpose.

From the point of view of solving the WIO model, the dependence of the input and output coefficients of waste treatment processes, that is, $A_{I,II}$, G_{II}, and R_{II} on waste's chemical properties introduces nonlinearity into the model. Suppose that a system engineering submodel of waste treatment of the type as above is available, which gives $A_{I,II}$, G_{II}, and R_{II} as a set of nonlinear functions, say F, of the chemical characteristics of the waste entering each of the treatment processes. Recalling that the chemical characteristics of the waste entering each treatment process depend on the way the vector of waste for treatment, w, is allocated to each process, the following general expression can be obtained

$$(A_{I,II}, G_{II}, R_{II}) = F(S, w) \tag{6.42}$$

With the engineering submodel F given this way, the nonlinear WIO model can be solved for x iteratively as illustrated in Figure 6.5. Starting from a given set of $A_{I,I}$, G_I, and R_I referring to the technology of production sectors, and final demand f_I, the first round values of production and waste generation, x_I^0 and w^0 can be obtained by solving the model for production sectors only (the first rectangular box). Application of F to w^0 and a given S then gives the matrices $A_{I,II}$, G_{II}, and R_{II} referring to the

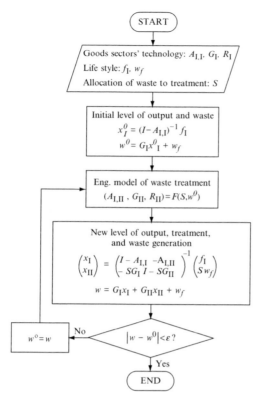

Fig. 6.5 Solving the WIO Model Integrating a System Engineering Model of Waste Management in an Iterative Fashion.

technology of waste treatment sectors (the second rectangular box). The resulting model encompassing both production and waste treatment sectors is then solved for production and waste treatment as well, yielding the net generation of waste, w (the third rectangular box), which may differ from its starting value w^0. If this is the case, the starting value is replaced by w in F, and a new set of $A_{I,II}$, G_{II}, and R_{II} is obtained. The process is iterated until the change in w becomes "small" by certain criteria.

A nonlinear WIO model incorporating an engineering submodel of waste treatment will be an extremely convenient one. Among others, it allows one to run simulations involving a significant change in the chemical composition of the waste entering each treatment process without having to prepare for each treatment a set of unit processes each corresponding to different sets of chemical compositions of the waste.

On the other hand, such a model is costly to develop, and the appropriate engineering model may not be readily available. In particular, one has to resort to more professional computational software like MATLAB, instead of readily available ones such as Excel.

Still, the above mentioned does not imply that the use of a nonlinear model is mandatory in WIO. Incineration is the process where the possible range of input and output coefficients is the widest, and hence its extent of nonlinearity is the largest. In fact, Matsuto [6] indicates that for other MSW treatment processes such as composting and gasification of garbage, the production of refuse derived fuels (RDF), shredding, and landfill the amounts of major inputs and outputs can be mostly represented as being proportional to the amount of waste. For most practical purposes, the linear model would provide useful approximation. Even for a simulation involving a significant change in the chemical composition of waste to be incinerated, the intrinsic nonlinearity of the incineration process can be coped with by preparing a set of unit processes corresponding to relevant chemical compositions of the waste.

Apart from the possibility of their being able to incorporate into WIO, engineering models as above are extremely useful in shedding light on the internal structure of processes, which are otherwise treated as a black box.

6.2.3 Effects of Changing the Allocation Pattern of Waste to Treatment Processes

In their first WIO study that incorporated the engineering model of waste treatment, Nakamura and Kondo [10] analyzed the effects of alternative sorting patterns of waste for different types of incinerators. The sorting patterns were represented by the allocation matrices S_0 and S_1 in Table 6.14. As for the types of incinerators, three types were identified that were representative as of year 1995:

1. Type A: the large-size full continuous type, with power generation from waste heat, and melting of ash
2. Type B: the medium-size full continuous type, with neither power generation from waste heat nor melting of ash
3. Type C: the small-size noncontinuous type, with neither power generation from waste heat nor melting of ash

The shares of these treatment capacity were 34%, 14%, and 51%, respectively. It was found that the introduction of a complete separation as in S_1 and the shift to Type A could reduce the amount of waste going to landfills by 9%, and the emission of GWP by 0.5–0.7% (see Table 8 of [10]).

A similar calculation was made by using the WIO table for the year 2000 and updated information about the distribution of incinerators. As was mentioned in Section 6.1.2.4, there has been a significant change in the meantime in the distribution of incinerator types in Japan, and the share of type A has increased, while that of type C has drastically declined.

The transition of the allocation pattern from S_0 to S_1 increases the share of waste plastics in the waste for incineration. This leads to an increase in the heat value and a decrees in the ash content of waste for incineration. Other things being equal, a decrease in the ash content of waste would lead to a reduction in the generation of residue per waste. As was discussed in Section 6.2, however, an increase in the share of plastics can work counter to the reduction in the generation of residue because of an increase in the HCl treatment of the flue gas.

Table 6.22 shows the values of major input-output coefficients of the incineration sector under S_0 and S_1 that were obtained by use of the engineering model discussed in Section 6.2. It is found that the amount of utilities produced from the heat of waste increased by 200%, while the amount of residue decreased by 23%. On the other side, the amount of GWP emission increased by 8% due to the increase in the fossil fuel content of waste.

Figure 6.6 (the bars in light gray denoted variable coefficients) shows the overall effects on the amount of landfill, incineration, and the GWP emission that were obtained from the WIO quantity model for a given final demand. It is found that

Table 6.22 The Major Inputs/Output Coefficients of the Incineration Sector Under Different Sorting Patterns.

	Units[a]	S_0	S_1
Chemicals (CHEM)	Million yen	0.0002	0.0002
Cement (CMNT)	Million yen	0.0003	0.0003
Machinery (MCHN)	Million yen	0.0196	0.0192
Utilities (UTIL)	Million yen	−0.0001	−0.0017
Services (SRVC)	Million yen	0.0003	0.0003
Ash (and slag)	10^6 kg	0.0752	0.0577
CO_2	10^6 kg	0.9610	1.0396

[a]per 10^6 kg of waste for incineration. See Table 6.20 for detailed definitions of the sectors and waste.

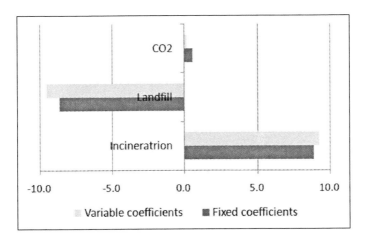

Fig. 6.6 Effects of the Shift of Allocation Patterns from S_0 to S_1. The Upper Bar Indicates the Case Where the Changes in the Input-Output Coefficients That Result from a Change in the Allocation Matrix Are Taken into Account by Using an Engineering Submodel of Waste Treatment. The Lower Bar Indicates the Case Where These Changes Are Not Considered, and the Same Set of Input-Output Coefficients Is Used When There Is a Change in the Allocation Matrix.

completing the separation of waste according to its chemical (not political) flammability could lead to an almost 10% reduction in the amount of waste for landfill, with almost negligible increase in the emission of GWP.

The bars in dark gray (denoted Fixed coefficients) in Figure 6.6 also show the results obtained under S_1 but without considering the associated changes in the input/output coefficients of the incineration sector (Table 6.22). It thus corresponds to the case where the same set of input coefficients matrix that is consistent with S_0 was wrongly used for the case of S_1. It is found that the use of fixed coefficients resulted in an overestimation of the emission of GWP, and an underestimation of the extent to which landfill is reduced and incineration is increased. This demonstrates the usefulness of incorporating an engineering submodel of waste treatment when there is going to be a substantial change in the chemical composition of waste, in particular for incineration.

6.3 The WIO Cost/Price Model

It was shown in Section 2.2.3 that corresponding to a Leontief quantity model, there exists a Leontief cost/price model. The same applies to WIO as well: corresponding to a WIO quantity mode, one can derive a WIO cost/price model. The presence of waste introduces specific features to the derivation of WIO cost/price models, to which we now turn. A numerical example is used for illustration. This section is based on [11].

6.3.1 The WIO Price Model with Waste Treatment

First consider the case where there is no recycling of waste materials. Writing p_I for an $n \times 1$ vector of the prices of goods & services, and p_{II} for a $k \times 1$ vector of the prices of waste treatment services, the unit cost equations will be given by

$$
\begin{aligned}
p_I &= p_I A_{I,I} + p_{II} S G_I^{\text{out}} + \pi_I \\
p_{II} &= p_I A_{I,II} + p_{II} S G_{II}^{\text{out}} + \pi_{II}
\end{aligned}
\tag{6.43}
$$

where π_I and π_{II} respectively refer to the row vectors of value added ratios of the goods & service sectors, and the waste treatment sectors, which are exogenously given. The second term on the right-hand side of each equation refers to the cost associated with the treatment of waste generated in production. Implicit in this formulation of the cost is the notion of the Polluter Pays Principle (PPP) or Extended Producer Responsibility (EPR) under which the cost of waste disposal is internalized into the cost of the product [14].

Solving for the prices yields

$$
\begin{pmatrix} p_I & p_{II} \end{pmatrix} = \begin{pmatrix} \pi_I & \pi_{II} \end{pmatrix} \begin{pmatrix} I - A_{I,I} & -A_{I,II} \\ -SG_I^{out} & I - SG_{II}^{out} \end{pmatrix}^{-1}
\tag{6.44}
$$

which gives the WIO version of the standard Leontief price model (2.168). This equation generalizes the Leontief EIO price model (5.38) to the case where there is no one-to-one correspondence between waste (pollutant) and its treatment (abatement).

6.3.2 The WIO Price Model with Waste Treatment and Recycling

6.3.2.1 Cost Items with Recycling

In the WIO price model mentioned above, the following cost items have been accounted for:

(a) The cost for the input of goods & services
(b) The cost for waste treatment and
(c) The cost for the input of primary factors

which respectively correspond to the first, the second, and the last terms on the right-hand of (6.43). Accounting for the recycling of waste calls for considering the following additional cost items in an explicit way:

(d) The cost for the input of waste materials and
(e) The revenue from the sale of waste materials

The item (d) refers to the expenditure that arises for the users of a waste material for its acquisition, and item (e) refers to the revenue that is passed back to its suppliers. It should be noted that mention of these items here does not mean that they are not considered in the standard IO table. They are certainly covered in the standard IO table as well, but in an implicit way, with no decomposition into components as above (see Section 5.2.3.1).

In the consideration of recycling, it is important to distinguish whether it deals with by-product (I) or by-product (II) (see Section 5.3.2.4). In the case of by-product (I), if its price were equal to its primary counterpart, the terms (d) and (e) would require no consideration because they would be subsumed in the usual cost term (a). In this case, for the users of the relevant input its origin (by-product or primary product) would be indistinguishable. For its suppliers, the term (e) is accounted for by the negative entry of the by-product as its input: the cost of production is reduced by the amount equal to the revenue from the sale of the by-product. If, however, the price of by-product (I) is not equal to its primary counterpart, a different procedure needs to be used, which explicitly takes separate account of the terms (d) and (e). This is the procedure to be applied to by-product (II), to which we now turn.

6.3.2.2 The Price of Waste for Recycling

Accounting for the cost terms (d) and (e) calls for an explicit consideration of the price of recycled waste items. As mentioned above, in the case of by-product (I), the price of its primary counterpart can be used as a reference price. This convention is not applicable to by-product (II), however, because, by definition, there is no endogenous sector where it is primarily produced.

Some waste items such as iron scrap, nonferrous metal scrap (copper and aluminum), and waste paper are characterized by the presence of well-established national markets. Figure 6.7 shows a time series of the prices of iron scrap and steel products in Japan. It is indicated that the price of scrap is characterized by high volatility, which is not shared with product prices. Furthermore, this aspect of the nature of scrap appears to hold universally (Figure 6.8).

In the Leontief EIO price model (5.38), it is implicitly assumed that the price of a recycled waste material (a pollutant) is equal to the price of its treatment (abatement) (see also Mathematical appendix of [5]). When it comes to practical application, this method suffers from two shortcomings. First, the highly volatile nature of the price of some scrap cannot be adequately dealt with by an IO based price model including the Leontief EIO model, where the prices of products are determined based on the cost of production, or the fundamentals, alone. Secondly, even for the case of scrap with less volatile price movements, the method of the Leontief EIO price model will not be directly applicable to the current case where there is

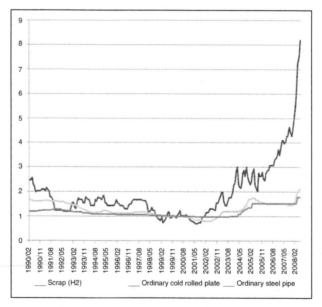

Fig. 6.7 Changes in Prices of Steel Products and Scrap: Index (January 2001 = 1) of the Average Price at the Tokyo Market. Source: Nippon Keizai Shimbun Inc. "Main Market Price and Iron Scrap", Steel Newspaper Inc. "Steel Newspaper", The Japan Iron and Steel Federation, "Steel Supply and Demand Statistics Monthly Report". H2 Refers to a Grade of Iron Scrap in Japan.

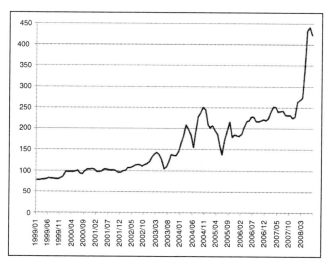

Fig. 6.8 Changes in the Price of Demolition Scrap: Index (2001=100) Calculated on the Basis of the Average Price in Euros for France, Germany, Italy, Spain, and the UK. Source: European Confederation of Iron and Steel Industries.

no one-to-one correspondence between waste and its treatment (see Section 5.3.1). Because a given waste item can be treated by multiple treatment methods, and a given waste treatment method can be applied to a multiple of waste items, it will not be possible to establish a unique price for a waste item based on (5.38).

It will now be clear that the consideration of recycling of waste (mostly by-product (II), but does not excluding by-product (I), see Section 5.3.2.4) in the WIO price model calls for some conceptual extension of the Leontief EIO price model. Based on the above discussion, two remarks can be made on the price of waste for recycling. First, the absence of the corresponding primary sector does not allow its endogenous determination in a straightforward fashion based on the IO price model. Secondly, the volatile nature of some scrap prices indicates inadequacy of applying an IO based model to their explanation. Because of this, except for when otherwise stated, the prices of recycled waste materials, which are denoted by an $m \times 1$ vector q^w, will henceforth be assumed to be exogenously given.

6.3.2.3 The Cost Equation with Recycling of Waste

Henceforth, it is assumed that the flow of waste generation and of waste inputs, W^{out} and W^{in}, are measured net of intra-sectoral transactions, that is:

$$(W^{\text{out}})_{ij} (W^{\text{in}})_{ij} = 0 \tag{6.45}$$

which implies that the case where both $(W^{\text{out}})_{ij}$ and $(W^{\text{in}})_{ij}$ take nonzero values simultaneously is excluded.

With the cost items (d) and (e) added, the cost equation for producing sector j can be given by

$$
p_j x_j = \underbrace{\sum_{i=1}^{n} p_i a_{ij} x_j}_{(a)} + \underbrace{\sum_{l=1}^{k} p_l \sum_{o=1}^{m} s_{lo}(g_{oj}^{in} x_j - R_{oj})}_{(b)} + \underbrace{V_j}_{(c)}
$$

$$
+ \underbrace{\sum_{o=1}^{m} q_o^w g_{oj}^{in} x_j}_{(d)} - \underbrace{\sum_{o=1}^{m} q_o^w R_{oj}}_{(e)} \qquad (6.46)
$$

where $R_{oj} \geq 0$ stands for the amount of waste o that is generated in sector j, and is used as input (recycled) in sectors other than j. A distinguishing feature of the WIO price model consists in the presence of the terms (b), (d), and (e) that explicitly represent the cost of waste treatment, and the sale as well as purchase of recovered waste materials. While these cost elements also occur in the traditional Leontief price model, their representation is implicit, and does not provide detailed information. When there is no recycling of waste materials, the terms (d) and (e) vanish, and the term (b) reduces to the treatment cost of wastes generated in sector j.

The sale of recovered waste materials is an important source of revenue for waste recyclers. A typical example is the disassembly of discarded automobiles, the major revenue source of which has been the sale of scrap metal to electric arc furnaces. There are, however, cases where the value of waste materials is negative, that is, the user is paid to accept the waste materials. For instance, several Japanese steel mills accepted waste plastics for fees and used them, after a series of pretreatments, as a reduction agent in blast furnaces, together with pulverized coal. Another example is the acceptance with fees of scrap tires and waste plastics by several Japanese cement manufacturers as supplementary heat sources in a cement kiln.

The price of waste can thus become negative. Based on its sign condition, three cases can be distinguished:

1. The waste has a positive value: $q_o^w > 0$.
2. It has zero value but can be accepted by other sectors as input with no payment: $q_o^w = 0$.
3. It has negative value and can be accepted only with payment: $q_o^w < 0$.

Henceforth, R_{oj} is called "sale of waste" regardless of whether the price of waste o is positive, zero, or negative.

The term (b) indicates that the amount of waste for treatment is reduced by the amount of R_{oj}. The sale of waste materials at positive prices can reduce the cost of production or treatment in two ways. First, it can reduce the cost directly by creating a new source of revenue other than the sale of primary product. The term (e) refers to this component. Secondly, it can reduce the waste treatment cost that would have been necessary if the waste materials were not sold but had to be treated with payment. The term (b) refers to this component. On the other hand, the sale of waste at negative prices reduces the production cost (increases the revenue) of the sectors that use the waste as input.

Rearranging the terms in (6.46), the contribution of the sale of waste materials to the cost can be shown in a more explicit way:

$$
p_j x_j = \underbrace{\sum_{i=1}^{n} p_i a_{ij} x_j}_{(a)} + \underbrace{\sum_{l=1}^{k} p_l \sum_{o=1}^{m} s_{lo} g_{oj}^{out} x_j}_{(f)} + \underbrace{\sum_{o=1}^{m} q_o^w g_{oj}^{in} x_j}_{(d)}
$$

$$
\underbrace{- \sum_{o=1}^{m} \left(q_o^w + \sum_{l=1}^{k} p_l s_{lo} \right) R_{oj} + V_j,}_{(g)} \tag{6.47}
$$

Here, the term (f) refers to the waste treatment cost that would have been necessary if no waste materials were sold. When waste is sold to other sectors, it can affect the cost via the term (g). The extent to which the cost can be reduced by the sale of waste depends on the sign condition of the expression inside the parentheses of (g). When $q_o^w > 0$, the sale of waste certainly reduces the cost of production. It is important to note that even if $q_o^w \leq 0$, the sale of waste could reduce the cost as long as the following condition is satisfied:

$$
q_o^w + \sum_{l=1}^{k} p_l s_{lo} > 0 \Longleftrightarrow |q_o^w| = -q_o^w < \sum_{l=1}^{k} p_l s_{lo}. \tag{6.48}
$$

This refers to the case where the sale of waste to other sectors at negative prices costs less than submitting it to waste treatment.

6.3.2.4 The Unit Cost Equation with Recycling

The term R_{oj} plays a vital role in the cost equation (6.47), but does not occur in the WIO quantity model (5.88). It is necessary to establish the relationship between R_{oj} and the elements occurring in the quantity model. For this purpose, denote by r_o the average rate of recycling of waste o that is defined as the ratio of the total amount of waste o used to the total amount generated:

$$
r_o = \sum_{l=1}^{n+k+n_f} (W^{in})_{ol} \bigg/ \sum_{l=1}^{n+k+n_f} (W^{out})_{ol} = \sum_{l=1}^{n+k+n_f} R_{ol} \bigg/ \sum_{l=1}^{n+k+n_f} (W^{out})_{ol}, \tag{6.49}
$$

where n_f refers to the number of final demand categories, and the second equality follows from (6.45), which impels $W_{oj}^{in} = 0$ when $R_{oj} > 0$.

We now assume that the rate of recycling of a given waste is the same for all of its users (which is reasonable when it deals with a homogeneous waste item):

$$
R_{oj} \big/ (W^{out})_{oj} = r_o, \quad ((W^{out})_{oj} > 0, o = 1, \cdots m, j = 1, \cdots, n+k+n_f).
$$

Recalling the definition of g_{kj}^{out}, it then follows:

$$R_{oj} = r_o(W^{\text{out}})_{oj} = r_o g_{oj}^{\text{out}} x_j.$$

Substituting this into (6.47) yields the following expression of the cost equation:

$$p_j x_j = \sum_{i=1}^{n} p_i a_{ij} x_j + \sum_{l=1}^{k} p_l \sum_{o=1}^{m} s_{lo} g_{oj}^{\text{out}} x_j$$

$$+ \sum_{o=1}^{m} q_o^w g_{oj}^{\text{in}} x_j - \sum_{o=1}^{m} \left(q_o^w + \sum_{l=1}^{k} p_l s_{lo} \right) r_o g_{oj}^{\text{out}} x_j + V_j.$$

Dividing both sides by x_j and rearranging the terms yields the following price equation (or unit cost function):

$$p_j = \sum_{i=1}^{n} p_i a_{ij} + \sum_{l=1}^{k} p_l \sum_{k=1}^{m} s_{lo} g_{oj}^{\text{out}} + \sum_{o=1}^{m} q_o^w g_{oj}^{\text{in}} - \sum_{o=1}^{m} \left(q_o^w + \sum_{l=1}^{k} p_l s_{lo} \right) r_o g_{oj}^{\text{out}} + \pi_j$$

$$= \sum_{i=1}^{n} p_i a_{ij} + \sum_{l=1}^{k} p_l \sum_{o=1}^{m} s_{lo} (1 - r_o) g_{oj}^{\text{out}} + \sum_{o=1}^{m} q_o^w (g_{oj}^{\text{in}} - r_o g_{oj}^{\text{out}}) + \pi_j, \qquad (6.50)$$

In (6.50), the first term on the right-hand side corresponds to the unit cost counterpart of (a) in (6.46) or (6.47), the second term refers to the treatment cost of waste that is generated per unit of production net of recycling, and the third term refers to the negative of the net revenue that results from the trade of waste with other sectors.

Recall from (6.45) that the nonzero element inside the parentheses on the third right-hand term of (6.50) can be either g_{oj}^{in} or $r_o g_{oj}^{\text{out}}$. The term $q_o^w g_{oj}^{\text{in}}$ refers to the unit input cost of waste o generated in other sectors. If $q_o^w < 0$, sector j provides a waste management service with regard to waste o. On the other hand, the term $q_o^w r_o g_{oj}^{\text{out}}$ refers to the unit revenue from the sale of waste o generated in sector j. If $q_o^w < 0$, the revenue is negative, and becomes the cost for sector j that is required for its waste o to be recycled in other sectors at the rate of r_o.

6.3.2.5 The WIO Price Model

Using obvious matrix notations, (6.50) can be rewritten as

$$(p_{\text{I}} \ p_{\text{II}}) = (p_{\text{I}} \ p_{\text{II}}) \begin{pmatrix} A_{\text{I,I}} & A_{\text{I,II}} \\ S(I - \hat{r})G_{\text{I}}^{\text{out}} & S(I - \hat{r})G_{\text{II}}^{\text{out}} \end{pmatrix}$$

$$+ q^w (G_{\text{I}}^{\text{in}} - \hat{r}G_{\text{I}}^{\text{out}} \quad G_{\text{II}}^{\text{in}} - \hat{r}G_{\text{II}}^{\text{out}}) + (\pi_{\text{I}} \ \pi_{\text{II}}), \qquad (6.51)$$

Solving for p_{I} and p_{II} would yield

$$\left(p_{\mathrm{I}} \; p_{\mathrm{II}}\right) = \left(q^w \left(G_{\mathrm{I}}^{\mathrm{in}} - \widehat{r}G_{\mathrm{I}}^{\mathrm{out}} \; G_{\mathrm{II}}^{\mathrm{in}} - \widehat{r}G_{\mathrm{II}}^{\mathrm{out}}\right) + \left(\pi_{\mathrm{I}} \; \pi_{\mathrm{II}}\right)\right)$$
$$\times \left(I - \left(\frac{A_{\mathrm{I},\mathrm{I}}}{S(I - \widehat{r})G_{\mathrm{I}}^{\mathrm{out}}} \; \frac{A_{\mathrm{I},\mathrm{II}}}{S(I - \widehat{r})G_{\mathrm{II}}^{\mathrm{out}}}\right)\right)^{-1} \qquad (6.52)$$

which is the WIO price model in the presence of recycling of waste.

Comparison of the inverse matrix in (6.52) with that in the quantity model (5.88) indicates that they become identical if and only if $G^{\mathrm{in}} = \widehat{r}G^{\mathrm{out}}$, which includes the case where there is no recycling of waste. When there is no recycling of waste, that is, $\widehat{r} = 0$, (6.52) reduces to (6.44).

Except for this special case, the system of price equations includes the rates of recycling. Because the rate of recycling for each waste item depends on the level of output of all the sectors and not only on the output of sector j (see (6.49)), it follows that the solution of the price model also depends on the level of output of all the sectors. In other words, the level of unit cost of a particular sector cannot be determined independent of the level of final demand, in spite of the fact that the technology is assumed to be subject to constant returns to scale. This can be seen as an example where the presence of joint products (recycling of waste) interferes with the nonsubstitution theorem (see Section 4.4.2).

6.3.2.6 A Special Case Where the Price of Waste Is the Negative of the Weighted Unit Costs of Its Treatment

A negative value of q_o^w implies that waste o can be accepted by its user against the payment of a fee. Consider now a special case where the inequality on the right-hand side of (6.48) is replaced by equality:

$$q_o^w + \sum_l p_l s_{lo} = 0 \qquad (6.53)$$

In this case, the price of waste o is exactly equal to the negative value of the average of its treatment costs weighted by the shares of treatment processes it was submitted to. So far as the monetary costs are concerned, for the supplier of this waste it makes no difference whether the waste is submitted to waste treatment or recycled as waste material. From the point of view of environmental concerns, however, it may be the case that one prefers recycling as useful material to disposal as waste. It therefore seems worthwhile to consider the implication of (6.53), the break-even case, for the WIO price model.

When (6.53) holds, the fourth term of the first equation of (6.50) vanishes, and becomes:

$$p_j = \sum_{i=1}^n p_i a_{ij} + \sum_{l=1}^k p_l \sum_{o=1}^m s_{lo} g_{oj}^{\mathrm{out}} - \sum_{o=1}^m \sum_{l=1}^k p_l s_{lo} g_{oj}^{\mathrm{in}} + \pi_j$$
$$= \sum_{i=1}^n p_i a_{ij} + \sum_{l=1}^k p_l \sum_{o=1}^m s_{lo}(g_{oj}^{\mathrm{out}} - g_{oj}^{\mathrm{in}}) + \pi_j \qquad (6.54)$$

Recalling that from (6.45) g_{oj}^{out} and g_{oj}^{in} cannot be nonzero at the same time, this equation should be read as

$$p_j = \sum_{i=1}^{n} p_i a_{ij} + \sum_{l=1}^{k} p_l \left(\sum_{g_{oj}^{out}>0} s_{lo} g_{oj}^{out} - \sum_{g_{oj}^{in}>0} s_{lo} g_{oj}^{in} \right) + \pi_j \qquad (6.55)$$

Writing in a matrix form, (6.54) becomes

$$\left(p_I \; p_{II} \right) = \left(p_I \; p_{II} \right) \begin{pmatrix} A_{I,I} & A_{I,II} \\ S(G_I^{out} - G_I^{in}) & S(G_{II}^{out} - G_{II}^{in}) \end{pmatrix} + \left(\pi_I \; \pi_{II} \right) \qquad (6.56)$$

with the solution for the prices given by

$$\left(p_I \; p_{II} \right) = \left(\pi_I \; \pi_{II} \right) \begin{pmatrix} I - A_{I,I} & -A_{I,II} \\ -S(G_I^{out} - G_I^{in}) & I - S(G_{II}^{out} - G_{II}^{in}) \end{pmatrix}^{-1}$$

$$= \left(\pi_I \; \pi_{II} \right) \begin{pmatrix} I - A_{I,I} & -A_{I,II} \\ -SG_I & I - SG_{II} \end{pmatrix}^{-1} \qquad (6.57)$$

which is the price model that is dual to the WIO quantity model (5.88). Note that this provides a closed solution for the prices, p_I and p_{II}.

In the case of a one-to-one correspondence between waste and waste treatment under which S becomes a unit matrix of order $m \times m$ with $m = k$, (6.57) reduces to the Leontief EIO price model (5.38). It follows that the Leontief EIO price model is a special case of the WIO price model when there is a one-to-one correspondence between waste and waste treatment, and the break-even condition (6.53) is also satisfied.

6.3.3 Numerical Example

For illustration, consider a hypothetical case given by Table 6.23, which consists of two producing sectors, two waste treatment sectors, one final demand sector, and three types of waste, of which waste 3 is the residue from waste incineration. Suppose that both waste 2 and 3 are waste for landfill, while waste 1 is for incineration, with the allocation matrix given by

$$S = \begin{pmatrix} 0 & 1 & 1 \\ 1 & 0 & 0 \end{pmatrix} \qquad (6.58)$$

6.3.3.1 The Case of No Recycling

We first consider the case where there is no recycling, that is, $G^{in} = (G_I^{in}, G_{II}^{in}) = 0$. Noting that

Table 6.23 A Numerical Example for the WIO-Price Model.

	Sector 1	Sector 2	Landfill	Incineration	Final demand
	$A_{\mathrm{I,I}}$			$A_{\mathrm{I,II}}$	f_{I}
Sector 1	0.15	0.4	0.01	0.2	30
Sector 2	0.4	0.1	0.1	0.1	70
	$G_{\mathrm{I}}^{\mathrm{out}}$			$G_{\mathrm{II}}^{\mathrm{out}}$	w_f
Waste 1	0.1	0	0	0	20
Waste 2	0	0.2	0	0	10
Waste 3	0	0	0	0.15	0
π	0.4	0.4	0.4	0.4	

$$SG^{\mathrm{out}} = \begin{pmatrix} 0 & 1 & 1 \\ 1 & 0 & 0 \end{pmatrix} \begin{pmatrix} 0.1 & 0.0 & 0.0 & 0.0 \\ 0.0 & 0.2 & 0.0 & 0.0 \\ 0.0 & 0.0 & 0.0 & 0.15 \end{pmatrix} = \begin{pmatrix} 0.0 & 0.2 & 0.0 & 0.15 \\ 0.1 & 0.0 & 0.0 & 0.0 \end{pmatrix}, \tag{6.59}$$

where $G^{\mathrm{out}} = (G_{\mathrm{I}}^{\mathrm{out}}, G_{\mathrm{II}}^{\mathrm{out}})$, from (6.44) it follows that

$$p = (p_{\mathrm{I}}, p_{\mathrm{II}}) = (0.4\ 0.4\ 0.4\ 0.4) \begin{pmatrix} 0.85 & -0.4 & -0.01 & -0.2 \\ -0.4 & -0.9 & -0.1 & -0.1 \\ 0.0 & -0.2 & 1.0 & -0.15 \\ -0.1 & 0.0 & 0.0 & 1.0 \end{pmatrix}^{-1}$$

$$= (1.0\ 1.0\ 0.5\ 0.8) \tag{6.60}$$

6.3.3.2 The Case of Recycling

Next consider the case where portions of waste 1 and 2 are respectively used as substitutes for product 2 and product 1, with $A_{\mathrm{I,I}}$ and $G_{\mathrm{I}}^{\mathrm{in}}$ given by

$$\left(\frac{A_{\mathrm{I,I}}}{G_{\mathrm{I}}^{\mathrm{in}}} \right) = \begin{pmatrix} 0.15 & 0.2 \\ 0.3 & 0 \\ 0 & 0.2 \\ 0.1 & 0 \\ 0 & 0 \end{pmatrix} \tag{6.61}$$

and the price of waste given by

$$q^w = (0.5\ 0.5\ -0.1) \tag{6.62}$$

Note that the sum of $(A_{\mathrm{I,I}})_{ij}$ and the corresponding inputs of waste coincides with the input coefficients matrix in Table 6.23.

In order to obtain the rate of recycling r, the WIO quantity model (5.88) needs to be solved. Noting

$$S(G^{\text{out}} - G^{\text{in}}) = \begin{pmatrix} -0.1 & 0.2 & 0 & 0.15 \\ 0.1 & -0.2 & 0 & -0.15 \end{pmatrix} \tag{6.63}$$

this becomes

$$x = \begin{pmatrix} 0.85 & -0.20 & -0.01 & -0.20 \\ -0.30 & 1.00 & -0.10 & -0.10 \\ 0.10 & -0.20 & 1.00 & -0.15 \\ -0.10 & 0.20 & 0.00 & 1.15 \end{pmatrix}^{-1} \begin{pmatrix} 30 \\ 70 \\ 10 \\ 20 \end{pmatrix} = \begin{pmatrix} 58 \\ 91 \\ 23 \\ 7 \end{pmatrix} \tag{6.64}$$

The amounts of waste generated and used are then given by

$$w^{\text{out}} = G^{\text{out}}x + f_w = \begin{pmatrix} 0.1 & 0 & 0 & 0 \\ 0 & 0.2 & 0 & 0 \\ 0 & 0 & 0 & 0.15 \end{pmatrix} \begin{pmatrix} 58 \\ 91 \\ 23 \\ 7 \end{pmatrix} + \begin{pmatrix} 20 \\ 10 \\ 0 \end{pmatrix} = \begin{pmatrix} 26 \\ 28 \\ 1 \end{pmatrix} \tag{6.65}$$

$$w^{\text{in}} = G^{\text{in}}x = \begin{pmatrix} 0 & 0.2 & 0 & 0.15 \\ 0.1 & 0 & 0 & 0 \\ 0 & 0 & 0 & 0 \end{pmatrix} \begin{pmatrix} 58 \\ 91 \\ 23 \\ 7 \end{pmatrix} = \begin{pmatrix} 19 \\ 6 \\ 0 \end{pmatrix} \tag{6.66}$$

and hence

$$r = w^{\text{in}}\widehat{w^{\text{out}}}^{-1} = \begin{pmatrix} 0.74 \\ 0.21 \\ 0.00 \end{pmatrix} \tag{6.67}$$

By use of the rate of recycling thus obtained, it follows

$$q^w \left(G_{\text{I}}^{\text{in}} - \hat{r}G_{\text{I}}^{\text{out}} \ \ G_{\text{II}}^{\text{in}} - \hat{r}G_{\text{II}}^{\text{out}} \right) + \left(\pi_{\text{I}} \ \pi_{\text{II}} \right) = \left(0.41 \ 0.48 \ 0.40 \ 0.48 \right) \tag{6.68}$$

and

$$S(I - \hat{r})G^{\text{out}} = \begin{pmatrix} 0.00 & 0.16 & 0.00 & 0.15 \\ 0.03 & 0.00 & 0.00 & 0.00 \end{pmatrix} \tag{6.69}$$

From (6.52), it then follows

$$p = \left(0.41 \ 0.48 \ 0.40 \ 0.48 \right) \begin{pmatrix} 0.85 & -0.20 & -0.01 & -0.20 \\ -0.30 & 1.00 & -0.10 & -0.10 \\ 0.00 & -0.16 & 1.00 & -0.15 \\ -0.03 & 0.00 & 0.00 & 1.00 \end{pmatrix}^{-1}$$

$$= \left(0.76 \ 0.71 \ 0.48 \ 0.77 \right) \tag{6.70}$$

Comparison with the case of no recycling (6.60) indicates that in the present case the recycling results in a reduction in the prices of products 1 and 2.

6.3.3.3 Effects of a Change in the Composition of Final Demand

Now consider the case where the composition of final demand changes

$$f_1 = \begin{pmatrix} 90 \\ 10 \end{pmatrix} \tag{6.71}$$

Similarly to (6.64), the level of production/activity now becomes

$$x = \begin{pmatrix} 0.85 & -0.20 & -0.01 & -0.20 \\ -0.30 & 1.00 & -0.10 & -0.10 \\ 0.10 & -0.20 & 1.00 & -0.15 \\ -0.10 & 0.20 & 0.00 & 1.15 \end{pmatrix}^{-1} \begin{pmatrix} 90 \\ 10 \\ 10 \\ 20 \end{pmatrix} = \begin{pmatrix} 122 \\ 50 \\ 11 \\ 19 \end{pmatrix} \tag{6.72}$$

with the rate of recycling given by (following (6.65) and (6.66))

$$r = \begin{pmatrix} 0.40 \\ 0.61 \\ 0.00 \end{pmatrix} \tag{6.73}$$

For this new set of r, it follows

$$q^w \left(G_I^{in} - \hat{r} G_I^{out} \quad G_{II}^{in} - \hat{r} G_{II}^{out} \right) + \left(\pi_I \quad \pi_{II} \right) = \left(0.43 \ 0.44 \ 0.40 \ 0.48 \right) \tag{6.74}$$

and

$$S(I - \hat{r})G^{out} = \begin{pmatrix} 0.00 \ 0.08 \ 0.00 \ 0.15 \\ 0.06 \ 0.00 \ 0.00 \ 0.00 \end{pmatrix} \tag{6.75}$$

Because of the change in r, these terms are numerically different from those in (6.68) and (6.69). Accordingly, the change in the composition of the final demand results in a set of prices that differs from (6.70):

$$p = \left(0.43 \ 0.44 \ 0.40 \ 0.48 \right) \begin{pmatrix} 0.85 & -0.20 & -0.01 & -0.20 \\ -0.30 & 1.00 & -0.10 & -0.10 \\ 0.00 & -0.08 & 1.00 & -0.15 \\ -0.06 & 0.00 & 0.00 & 1.00 \end{pmatrix}^{-1}$$

$$= \left(0.78 \ 0.63 \ 0.47 \ 0.77 \right) \tag{6.76}$$

which indicates that the change in the composition of final demand, f_1, leads to a noticeable reduction in the price of product 2.

6.3.3.4 The Case Where the Price of Waste Is Equal to the Negative of Its Weighted Treatment Costs

Finally, consider the break-even case considered in Section 6.3.2.6 where the price of waste is equal to the negative of the weighted unit treatment costs. From (6.57), the price is given by

$$p = \begin{pmatrix} 0.4\ 0.4\ 0.4\ 0.4 \end{pmatrix} \begin{pmatrix} 0.85 & -0.2 & -0.01 & -0.2 \\ -0.3 & 1 & -0.1 & -0.1 \\ 0.1 & -0.2 & 1 & -0.15 \\ -0.1 & 0.2 & 0 & 1.15 \end{pmatrix}^{-1}$$

$$= \begin{pmatrix} 0.66\ 0.51\ 0.46\ 0.57 \end{pmatrix} \tag{6.77}$$

It may be of interest that this case gives the lowest price level among the three cases considered.

References

1. Basel Convention on the Control of Transboundary Movements of Hazardous Wastes and Their Disposal. http://www.basel.int/. Cited 8 August 2008.
2. Kondo, Y., Takase, K., & Nakamura, S. (2002). On the compilation of 1995 WIO table for Japan. In S. Nakamura (Ed.), *Toward an economics of waste, in Japanese*. Tokyo: Waseda University Press.
3. Kondo, Y., & Nakamura, S. (2007). On LCA and LCC based on the WIO. In S. Nakamura (Ed.), *Life cycle input-output analysis, in Japanese*. Tokyo: Waseda University Press.
4. Laboratory of Solid Waste Disposal Engineering, Graduate School of Engineering, Hokkaido University. http://www.pref.hokkaido.lg.jp/NR/rdonlyres/43BAA3FE-AC81–46A3–9810–4763E92317BB/0/jittaichousa17.pdf, in Japanese. Cited 18 July 2008.
5. Leontief, W., & Ford, D. (1972). Air pollution and the economic structure: Empirical results of input-output computations. In A. Bróby A & A. C. Carter (Eds.), *Conference on input-output techniques*. Geneva, Switzerland: North-Holland.
6. Matsuto, T. (2005). *Municipal solid waste treatment system: analysis, planning and evaluation, in Japanese*. Tokyo: Gihoudou.
7. McDougall, F., White, P., Franke, M., & Hindel, P. (2001). *Integrated solid waste management: A life cycle inventory*. Oxford: Blackwell Science.
8. Nakamura, S. (1999). Input-output analysis of waste cycles. In *First International Symposium on Environmentally Conscious Design and Inverse Manufacturing, Proceedings*. Los Alamitos: IEEE Computer Society.
9. Nakamura, S. (1999). On the compilation of WIO table for Japan. *Waseda Journal of Political Science and Economics, 340*, 171–203 (in Japanese).
10. Nakamura, S., & Kondo, Y. (2002). Input-output analysis of waste management. *Journal of Industrial Ecology, 6*(1), 39–63.
11. Nakamura, S., & Kondo, Y. (2006). A waste input output life-cycle cost analysis of the recycling of end-of-life electrical home appliances. *Ecological Economics, 57*, 494–506.
12. Nakamura, S. *Waste input-output table*. http://www.f.waseda.jp/nakashin/WIO.html
13. Nansai, K., Moriguchi, Y., & Tohno, S. (2002). *Embodied Energy and Emission Intensity Data for Japan using input-output tables (3EID) -inventory data for LCA-*. Tsukuba, Japan: National Institute for Environmental Studies.
14. OECD Fact Sheet About EPR. http://www.oecd.org/document/53/0,3343,en_2649_34395_37284725_1_1_1_1,00.html
15. Sánchez-Chóliz, J., & Duarte, R. (2005). Water pollution in the Spanish economy: Analysis of sensitivity to production and environmental constraints. *Ecological Economics, 53*, 325–338.
16. Suh, S. (2006). Are services better for climate change? *Environmental Science and Technology, 40*(21), 6555–6560.
17. Tanaka, N. et al. (2003). *Basic knowledge of waste treatment engineering, in Japanese*. Tokyo: Gihoudou.

Chapter 7
Application of WIO to Industrial Ecology

Abstract This chapter is concerned with the application of WIO to major tools of IE such as LCA, MFA, LCC (life cycle costing), and its further extensions in various fields of IE. Introducing the use phase, which has remained unconsidered in the previous chapters, into the WIO quantity model yields the WIO-LCA model, which can be used for life cycle inventory analysis. In a similar fashion, its cost counterpart, the WIO-LCC model, is also obtained, which can be used for environmental LCC. A new method of MFA, WIO-MFA, is presented, which can consider the physical flows of many materials/substances simultaneously. This method can also be used to transform a monetary IO table into a physical table of the flow of materials. Extensions of WIO to regional models and a linear programming model are also dealt with.

7.1 Introduction

LCA (life cycle assessment) and MFA (materials flow analysis) are two major methodological tools of IE (Industrial Ecology). Closely related to LCA is its economic counterpart, environmental life cycle costing (LCC) [12], which refers to evaluating the economic cost of a product over its entire life cycle. The life cycle of a product consists of three phases: production (including R&D and distribution), use, and disposal. The production phase is considered by the standard IOA. The WIO has extended it to account for the EoL phase. By including the use phase, the WIO will be able to account for the whole life cycle of a product. This chapter is concerned with the application of WIO to these tools of IE, and its further extensions.

Section 7.2 introduces the use phase into WIO, and derives the WIO-LCA model which can be used for life cycle inventory analysis, and its cost counterpart, the WIO-LCC model which can be used for environmental LCC. Numerical examples are given to illustrate these methods, before showing examples of application to real data.

S. Nakamura, Y. Kondo, *Waste Input-Output Analysis*, Eco-Efficiency in Industry
and Science 26, © Springer Science+Business Media B.V. 2009

The WIO tables currently available are a hybrid of monetary IO tables and physical flows of waste. Section 7.3 deals with a method to convert a monetary IO table into a physical table by use of information on the material composition of products, which can be estimated by exploiting the triangular nature of the matrix of input coefficients, which was discussed in Section 3.7. This method, WIO-MFA, enables one to consider the physical flows of any number of materials simultaneously at the level of detail determined by the underlying IO table, and thus provides an alternative to the standard MFA and/or PIOT (physical input-output table). Some results of application to the flow of metals in Japan are shown.

Spatial aspects are of considerable importance in waste management because the locations of waste generation may considerably differ from the locations where they can be recycled. Section 7.4 deals with spatial (regional) extensions of WIO along the lines of regional IOA discussed in Section 3.1. Results of its implementation with reference to Japanese regional IO tables and to the IO table for Tokyo are shown.

In Section 3.5.2 the standard IOA was extended to a linear programming problem (LP) to allow for the possibility of substitution among alternative technologies. In a similar fashion, the WIO model can also be extended to an LP which allows for the presence of alternative technologies of both production and waste treatment. This is the subject of Section 7.5.

A brief survey of other applications of WIO in diverse fields of Industrial Ecology closes this chapter.

7.2 The Full Life Cycle WIO Model: Closing the Loop of The Product Life Cycle

According to the general ISO 14040 standard [14], LCA is defined as the compilation and evaluation of the inputs, outputs, and the environmental impacts of a product system throughout the life cycle. The entire LCA procedure is divided into four distinct phases (see [8] for further details on ISO 14040):

- Goal and scope definition
- Inventory analysis
- Impact assessment
- Interpretation

The phase with which this book is concerned is the inventory analysis which refers to the construction of product systems consisting of industrial production, household consumption, waste management, transport, and so forth.

This section deals with the use of WIO for life cycle inventory analysis. The functional unit to be considered is a unit of product that is used over T years, and then discarded to be subjected to end of life (EoL) processes.

7.2.1 The Use (and Discard) Process

Compared to the EoL phase, the incorporation of the use (and discard) phase into IOA is rather simple. In essence, it deals with the incorporation of an additional process, the use process, that refers to the inputs and outputs that are associated with the use of a unit of the product over its product life, and the discard process that refers to its discard after use.

Suppose that we are concerned with the life cycle of a unit of product j, which is used for T years, subsequently discarded as the EoL product (waste l), and subjected to EoL processes. Write by P_j^u an $(n+m+p) \times 1$ vector of the use process which consists of a vector of n inputs, a_j^u, a vector of m waste (including the EoL product), g_j^u, and a vector of p emissions, r_j^u:

$$
P_j^u = \begin{pmatrix} a_j^u \\ g_j^u \\ r_j^u \end{pmatrix} = \left(
\begin{array}{c}
\sum_{t=1}^{T} \alpha_{1j}^u(t) \\
\vdots \\
\sum_{t=1}^{T} \alpha_{j-1\,j}^u(t) \\
1 \\
\sum_{t=1}^{T} \alpha_{j+1\,j}^u(t) \\
\vdots \\
\sum_{t=1}^{T} \alpha_{nj}^u(t) \\
\hline
\sum_{t=1}^{T} \gamma_{1j}^u(t) \\
\vdots \\
\sum_{t=1}^{T} \gamma_{l-1\,j}^u(t) \\
0 \\
\sum_{t=1}^{T} \gamma_{l+1\,j}^u(t) \\
\vdots \\
\sum_{t=1}^{T} \gamma_{mj}^u(t) \\
\hline
\sum_{t=1}^{T} \zeta_{1j}^{u+e}(t) \\
\vdots \\
\sum_{t=1}^{T} \zeta_{pj}^u(t)
\end{array}
\right)
\begin{array}{l}
\left.\rule{0pt}{9em}\right\} \text{Inputs} \\[2em]
\left.\rule{0pt}{9em}\right\} \text{Waste generation} \\[2em]
\left.\rule{0pt}{4em}\right\} \text{Emissions}
\end{array}
\tag{7.1}
$$

where $\alpha_{ij}^u(t)$, $\gamma_{oj}^u(t)$, and $\zeta_{sj}^u(t)$ respectively refer to the input of i, the generation of waste o, and the emission of s per year in the tth year of the use phase. Note that the value for the jth input is unity because this process refers to the use of a unit of product j, and that the generation of waste l is zero because it refers to the discard of product j, which is not accounted for by P_j^u.

The discard of product j as waste l is represented by the following $n + m + p$ vector of the discard process:

$$
P_j^e = \begin{pmatrix} a_j^e \\ g_j^e \\ r_j^e \end{pmatrix} = \begin{pmatrix} 0_n \\ 0_{l-1} \\ 1 \\ 0_{m-l} \\ 0_p \end{pmatrix}, \tag{7.2}
$$

where 0_h refers to an $h \times 1$ vector of zeros.

In (7.1) the coefficients are made dependent on t to account for possible changes in the amounts of inputs and outputs over time in the use phase: the efficiency of the product may decline over time, which results in an increase in the use of certain inputs, say electricity, and/or in the need for repair, which may increase the input of repairs, the amounts of waste generation (repaired parts etc.) and emissions. In a simple case, where there is no change in the amounts of inputs and outputs over time, α_{ij}^u, γ_{oj}^u, and ζ_{sj}^u become constants, and their amounts over T years become

$$
\sum_{t=1}^{T} \alpha_{ij}^u(t) = T\,\alpha_{ij}^u(1), \quad \sum_{t=1}^{T} \gamma_{oj}^u(t) = T\,\gamma_{oj}^u(1), \quad \sum_{t=1}^{T} \zeta_{sj}^u(t) = T\,\zeta_{sj}^u(1), \tag{7.3}
$$

where $i \neq j$ and $o \neq l$. In their IOA-based LCA of a mid size passenger car, Hendrickson et al. [9] assume that the vehicle fuel economy (as well as the emissions) remains constant throughout the vehicle's lifetime. The assumption of fixed coefficients at the use phase over the product life is a standard one in process-based LCA (see for example the LCA of air conditioners by Yokota et al. [45]).

7.2.2 Incorporating the Use (and Discard) Phase

The use (and discard) process can be incorporated into WIO in several different ways. Which one is to be used will depend on the purpose of analysis, and the ease of computation and data management.

7.2.2.1 The Use and Discard Process in Place of Final Demand

The functional unit in the present case refers to the use over T years and the subsequent discard of product j. The vectors (7.1) and (7.2) represent the amounts of inputs and outputs that are directly associated with the functional unit.

Life cycle inventory analysis is concerned with the estimation of the amounts of environmental burden that are directly and indirectly generated to realize the functional unit for a given production system. This estimation can be facilitated by a WIO with the final demand replaced by the first $n + m$ elements of (7.1) and (7.2), the balance equation of which is given by

$$\begin{pmatrix} A_{I,I} & A_{I,II} \\ G_I & G_{II} \end{pmatrix} \begin{pmatrix} x_I \\ x_{II} \end{pmatrix} + \begin{pmatrix} a_j^u + a_j^e \\ g_j^u + g_j^e \end{pmatrix} = \begin{pmatrix} x_I \\ w \end{pmatrix} \tag{7.4}$$

Following the discussion in Section 5.3, for a given allocation matrix S, this can be transformed to a system of simultaneous equations

$$\begin{pmatrix} A_{I,I} & A_{I,II} \\ SG_I & SG_{II} \end{pmatrix} \begin{pmatrix} x_I \\ x_{II} \end{pmatrix} + \begin{pmatrix} a_j^u + a_j^e \\ S(g_j^u + g_j^e) \end{pmatrix} = \begin{pmatrix} x_I \\ x_{II} \end{pmatrix} \tag{7.5}$$

where $x_{II} = Sw$, which can be solved for x as

$$\begin{pmatrix} x_I \\ x_{II} \end{pmatrix} = \begin{pmatrix} I - A_{I,I} & -A_{I,II} \\ -SG_I & I - SG_{II} \end{pmatrix}^{-1} \begin{pmatrix} a_j^u + a_j^e \\ S(g_j^u + g_j^e) \end{pmatrix} \tag{7.6}$$

with the total environmental burden (in terms of the $p \times 1$ vector of emissions) given by

$$\varepsilon = (R_I \ R_{II}) \begin{pmatrix} I - A_{I,I} & -A_{I,II} \\ -SG_I & I - SG_{II} \end{pmatrix}^{-1} \begin{pmatrix} a_j^u + a_j^e \\ S(g_j^u + g_j^e) \end{pmatrix} + r_j^u \tag{7.7}$$

Takase et al. [39] used this type of model for a WIO-based analysis of sustainable consumption.

An advantage of this formulation is that it enables one to obtain the expression for ε that corresponds to each life cycle stage in a simple fashion. Write i_j for an $n \times 1$ unit vector, all the elements of which are zero except for the jth element which is equal to unity. The final demand vector on the right-hand side can be decomposed into the three life stages as

$$\begin{pmatrix} a_j^u + a_j^e \\ S(g_j^u + g_j^e) \end{pmatrix} = \underbrace{\begin{pmatrix} i_j \\ 0_m \end{pmatrix}}_{\text{Production}} + \underbrace{\begin{pmatrix} \breve{a}_j^u \\ Sg_j^u \end{pmatrix}}_{\text{Use}} + \underbrace{\begin{pmatrix} a_j^e \\ Sg_j^e \end{pmatrix}}_{\text{End of Life}}, \tag{7.8}$$

where $\breve{a}_j^u$ refers to a_j^u with the jth element set equal to zero. The total life cycle burdens can then be decomposed into the three life cycle phases as follows

$$\varepsilon^{\text{Production}} = (R_I \ R_{II}) \begin{pmatrix} I - A_{I,I} & -A_{I,II} \\ -SG_I & I - SG_{II} \end{pmatrix}^{-1} \begin{pmatrix} i_j \\ 0_m \end{pmatrix} \tag{7.9}$$

$$\varepsilon^{\text{Use}} = (R_I \ R_{II}) \begin{pmatrix} I - A_{I,I} & -A_{I,II} \\ -SG_I & I - SG_{II} \end{pmatrix}^{-1} \begin{pmatrix} \breve{a}_j^u \\ Sg_j^u \end{pmatrix} + r_I^u \tag{7.10}$$

$$\varepsilon^{\text{End of life}} = (R_I \ R_{II}) \begin{pmatrix} I - A_{I,I} & -A_{I,II} \\ -SG_I & I - SG_{II} \end{pmatrix}^{-1} \begin{pmatrix} a_j^e \\ Sg_j^e \end{pmatrix} \tag{7.11}$$

7.2.2.2 Incorporation into the Production Process

An alternative to the above way is to incorporate the use (and discard) process into the production process [27]. Augmented by the use process, the unit process of production, P_j, becomes

$$
P_j^{p+u+e} = \begin{pmatrix} a_j + \check{a}_j^{\mathrm{u}} + a_j^{\mathrm{e}} \\ g_j + g_j^{\mathrm{u}} + g_j^{\mathrm{e}} \\ r_j + r_j^{\mathrm{u}} + r_j^{\mathrm{e}} \end{pmatrix} = \begin{pmatrix} \check{a}_j^{p+u+e} \\ g_j^{p+u+e} \\ r_j^{p+u+e} \end{pmatrix}
\tag{7.12}
$$

With the use (and discard) phase incorporated into the production phase, the demand for product j becomes the only nonzero element of the final demand:

$$
\begin{pmatrix} f_{\mathrm{I}} \\ w_f \\ r_f \end{pmatrix} = \begin{pmatrix} i_j \\ 0_m \\ 0_p \end{pmatrix},
\tag{7.13}
$$

where r_f refers to the emission from the final demand.

Write $\check{A}_{\mathrm{I,I}}$ and $\check{G}_{\mathrm{I}}$ for $A_{\mathrm{I,I}}$ and G_{I} with the jth columns respectively replaced by $\check{a}_j^{p+u+e}$ and $\check{g}_j^{p+u+e}$. The system of balance equations for goods and waste (7.4) then becomes

$$
\begin{pmatrix} \check{A}_{\mathrm{I,I}} & A_{\mathrm{I,II}} \\ \check{G}_{\mathrm{I}} & G_{\mathrm{II}} \end{pmatrix} \begin{pmatrix} x_{\mathrm{I}} \\ x_{\mathrm{II}} \end{pmatrix} + \begin{pmatrix} i_j \\ 0_m \end{pmatrix} = \begin{pmatrix} x_{\mathrm{I}} \\ w \end{pmatrix}
\tag{7.14}
$$

For a given allocation matrix S, the solution is given by (note that $S0_m = 0_k$)

$$
\begin{pmatrix} x_{\mathrm{I}} \\ x_{\mathrm{II}} \end{pmatrix} = \left(I - \begin{pmatrix} \check{A}_{\mathrm{I,I}} & A_{\mathrm{I,II}} \\ S\check{G}_{\mathrm{I}} & SG_{\mathrm{II}} \end{pmatrix} \right)^{-1} \begin{pmatrix} i_j \\ 0_k \end{pmatrix},
\tag{7.15}
$$

and the associated environmental burdens by

$$
\begin{pmatrix} \check{R}_{\mathrm{I}} & R_{\mathrm{II}} \end{pmatrix} \left(I - \begin{pmatrix} \check{A}_{\mathrm{I,I}} & A_{\mathrm{I,II}} \\ S\check{G}_{\mathrm{I}} & SG_{\mathrm{II}} \end{pmatrix} \right)^{-1} \begin{pmatrix} i_j \\ 0_k \end{pmatrix}
\tag{7.16}
$$

where $\check{R}_{\mathrm{I}}$ stands for R_{I} with its jth column replaced by r_j^{p+u+e}. This equation can be called the WIO-LCA model. This model was used for an LCA of air conditioners by [27], and of washing machines by [33].

The two ways mentioned above are by no means exclusive ones for incorporating the use process into WIO. Another possible way is to expand the inputs and outputs coefficients matrix by introducing a new column vector referring to the use and discard process, and a new row vector referring to the input from the use and discard process:

$$
\begin{pmatrix} A_{\mathrm{I,I}} & A_{\mathrm{I,II}} & a_j^{u+e} \\ G_{\mathrm{I}} & G_{\mathrm{II}} & g_j^{u+e} \\ 0_n^{\top} & 0_m^{\top} & 0 \end{pmatrix},
\tag{7.17}
$$

where

$$a_j^{u+e} = a_j^u + a_j^e \tag{7.18}$$
$$g_j^{u+e} = g_j^u + g_j^e \tag{7.19}$$

All the elements of the last row will be zero, because there will be no intermediate demand for the use and discard process: the input of the use process refers to the functional unit, and occurs as the final demand.

An advantage of the previous two ways over this one is that they do not involve any extension of the order of IO matrices, which can be quite tedious in actual computation. For practitioners of a process-based LCA, who are familiar with this type of matrix expansion, this way may appear more straightforward.

Of the three ways of incorporating the use process, the second one is characterized by its integration of the use and discard processes into the production process. This feature is convenient for considering the life cycle costs based on the idea of Polluter-Pays Principle or Extended Polluter Responsibility (see Section 6.3.1), that is, life cycle costing, to which we now turn.

7.2.3 LCC: The Cost and Price Model

For a product with favorable environmental performance to be able to realize its desirable effects in the economy, it needs to be widely used. Otherwise, its potential to reduce environmental loads remains mostly unexploited. An important factor for it to be widely used will be that it is economically affordable. This calls for the LCA of a product to be accompanied with a complementary evaluation of its affordability, namely its cost aspects [31]. The aspect of cost here should be the one that encompasses the cost associated with the whole life cycle of a product, that is, the life cycle cost [12]. The life cycle cost is in general not visible, because the market price of a product does not usually reflect the cost in the use and end-of-life (EoL) phases, but the costs associated with the production (and distribution) phase only. The life cycle cost therefore needs to be estimated just as one needs an LCA to evaluate the environmental impacts of a product. Environmental life cycle costing (LCC) is a tool that is designed to meet this requirement [17] (for the latest development, see [12]).

7.2.3.1 The WIO-LCC Model

The WIO price model (6.52) gives a unit cost of products that incorporates both production and EoL costs. Introducing the use process as discussed above, the WIO price model will be able to account for the costs associated with all the stages of a product life cycle, that is, the life cycle cost.

Implemented with reference to the present case as given by (7.14), the square matrix of order $n + k$ on the right-hand side of (6.51), which is henceforth denoted by Ψ, becomes (recall that the matrix G refers to of net waste generation, with $G = G^{\text{out}} - G^{\text{in}}$)

$$\Psi = \begin{pmatrix} \check{A}_{\text{I,I}} & A_{\text{I,II}} \\ S(I - \hat{r})\check{G}_{\text{I}}^{\text{out}} & S(I - \hat{r})G_{\text{II}}^{\text{out}} \end{pmatrix} \tag{7.20}$$

Without loss of generality, let $j = 1$, and rewrite Ψ as

$$\Psi = \begin{pmatrix} \check{a}_1^{p+u+e} & \\ S(I - \hat{r})g_1^{p+u+e \text{ out}} & \bar{\Psi} \end{pmatrix}, \tag{7.21}$$

where $\bar{\Psi}$ refers to Ψ with the first column elements excluded.

Writing p_1^{p+u+e} for the life cycle unit cost of product 1, and $\bar{p}$ for the vector of the remaining $n + k - 1$ prices, the cost balance equation (6.51) becomes

$$\left(p_1^{p+u+e} \quad \bar{p} \right) = \left(p_1^{p+u+e} \quad \bar{p} \right) \Psi + \left(\omega_1 \quad \bar{\omega} \right) \tag{7.22}$$

where

$$\omega_1 = q^w (g_1^{p+u+e \text{ in}} - \hat{r} g_1^{p+u+e \text{ out}}) + \pi_1 \tag{7.23}$$

$$\bar{\omega} = \left(q^w (\bar{G}_{\text{I}}^{\text{in}} - \hat{r} \bar{G}_{\text{I}}^{\text{out}}) + \bar{\pi}_{\text{I}} \quad q^w (G_{\text{II}}^{\text{in}} - \hat{r} G_{\text{II}}^{\text{out}}) + \pi_{\text{II}} \right) \tag{7.24}$$

Here, $\bar{G}_{\text{I (II)}}^{\text{in (out)}}$ refers to $G_{\text{I (II)}}^{\text{in (out)}}$ with the first column elements excluded, and $\bar{\pi}_{\text{I}}$ to π_{I} with the first element, π_1, excluded.

Solving (7.22) yields

$$\left(p_1^{p+u+e} \quad \bar{p} \right) = \left(\omega_1 \quad \bar{\omega} \right) \left(I - \Psi \right)^{-1} \tag{7.25}$$

Recalling (7.21–7.23) implies the following expression for the life cycle cost

$$p_1^{p+u+e} = \left(p_1^{p+u+e} \quad \bar{p} \right) \begin{pmatrix} \check{a}_1^{p+u+e} \\ S(I - \hat{r})g_1^{p+u+e \text{ out}} \end{pmatrix}$$
$$+ q^w (g_1^{p+u+e \text{ in}} - \hat{r} g_1^{p+u+e \text{ out}}) + \pi_1 \tag{7.26}$$

Writing p_1^p, p_1^u, and p_1^e, for the components of p_1^{p+u+e} corresponding to production, use, and discard phases

$$p_1^{p+u+e} = p_1^p + p_1^u + p_1^e \tag{7.27}$$

one obtains by use of (7.12)

$$p_1^p = \left(p_1^{p+u+e} \quad \bar{p} \right) \begin{pmatrix} a_1^p \\ S(I - \hat{r})g_1^{p \text{ out}} \end{pmatrix} + q^w (g_1^{p \text{ in}} - \hat{r} g_1^{p \text{ out}}) + \pi_1 \tag{7.28}$$

$$p_1^u = \left(p_1^{p+u+e} \quad \bar{p} \right) \begin{pmatrix} \check{a}_1^u \\ S(I - \hat{r})g_1^{u \text{ out}} \end{pmatrix} + q^w (g_1^{u \text{ in}} - \hat{r} g_1^{u \text{ out}}) \tag{7.29}$$

$$p_1^e = \left(p_1^{p+u+e} \quad \bar{p} \right) \begin{pmatrix} a_1^e \\ S(I - \hat{r})g_1^{e \text{ out}} \end{pmatrix} + q^w (g_1^{e \text{ in}} - \hat{r} g_1^{e \text{ out}}) \tag{7.30}$$

When product 1 is not used as an input in the production of any other product, which will be the case when it deals with a final product, the prices of other products, $\bar{p}$, would not be affected by the pricing of product 1 based on its life cycle cost. If, furthermore, product 1 is not used as an input in the production of any product including product 1 itself, that is, $a_{11}^p = 0$, p_1^p given by (7.28) would be identical with the solution of the standard WIO-Price model where the price of a product is determined based on the production cost alone without considering the use and EoL costs. The fact that an LCC is usually performed for durable final products such as aircraft, buildings, and appliances implies that these conditions will be satisfied.

7.2.3.2 External Costs

The basic idea of environmental LCC with which we are concerned here is characterized by the fact that only those items with internal and internalized costs are to be considered, but not those that are not borne by any of the actors in the life cycle during the relevant time period ([33]). In other words, the LCC in this section does not account for external costs such as the costs associated with the loss of biodiversity to the extent that they are not internalized.

There are, however, exceptions, which apply to "those externalities that are shown, based on preliminary or prior analyses, to introduce significant (potential) costs in the future due to internalization via regulatory measures (e.g., anticipated CO2 taxes, renewable energy subsidies)" [32]. Accounting for a possible internalization of such items in (7.25) is straightforward: they can be introduced as additional items of ω's.

As an example of the internalization of this type of externality, consider the case where a CO$_2$ tax of τ per kg of CO$_2$ on fuel consumption is introduced. Writing r_1. for a $1 \times (n+k)$ vector, the jth element of which refers to the emission of CO$_2$ (kg) per unit of product j, its effects on costs are given by

$$\left(\Delta p_1^{p+u+e} \ \Delta \bar{p}\right) = \left(\tau r_{11} \ \tau \times \bar{r}_1.\right)\left(I - \Psi\right)^{-1}, \tag{7.31}$$

where r_{11} refers to the first element of r_1, and $\bar{r}_1$. to its remaining elements. Following (7.28–7.30), Δp_1^{p+u+e} can also be decomposed into its life cycle components:

$$\Delta p_1^p = \left(\Delta p_1^{p+u+e} \ \Delta \bar{p}\right)\left(\frac{a_1^p}{S(I - \hat{r})g_1^p \ \text{out}}\right) + \tau r_{11}^p \tag{7.32}$$

$$\Delta p_1^u = \left(\Delta p_1^{p+u+e} \ \Delta \bar{p}\right)\left(\frac{\breve{a}_1^u}{S(I - \hat{r})g_1^u \ \text{out}}\right) + \tau r_{11}^u \tag{7.33}$$

$$\Delta p_1^e = \left(\Delta p_1^{p+u+e} \ \Delta \bar{p}\right)\left(\frac{a_1^e}{S(I - \hat{r})g_1^e \ \text{out}}\right) + \tau r_{11}^e, \tag{7.34}$$

where r^p, r^u, and r^e respectively refer to the element of r_1 at the production, use, and EoL phases.

7.2.3.3 Discounting

Unlike LCA, discounting (of future costs relative to the present one) could be a matter of concern in LCC, in particular, when it deals with the real cost of borrowing [12]. Writing ρ for the rate of discounting, the discounted life cycle cost of product j, say, $\dot{p}_j^{p+u+e}$, is given by

$$\dot{p}_j^{p+u+e} = p_j^p + \sum_{t=1}^{T} \frac{p_j^u(t)}{(1+\rho)^t} + \frac{p_j^e}{(1+\rho)^T} \tag{7.35}$$

where $p_j^u(t)$ refers to the use cost at time t, which is from (7.3) given by

$$p_j^u(t) = \frac{p_j^u}{T} \tag{7.36}$$

7.2.4 Numerical Examples

For illustration of the above methodologies, consider a simple hypothetical example consisting of four production processes, two waste treatment processes, three types of waste, and one type of emission given by Table 7.1. The appliance is assumed to consist solely of metal, and weigh 25 kg per unit. Note that the shredding process transforms a unit of EoL appliance that weighs 25 kg into 20 kg of metal scrap and 5 kg of shredding residue, and that the production of metals requires 0.2 kg of metal scrap per 1 kg of its production. As for the allocation matrix S of order 2×3, the following is assumed

$$S = \begin{pmatrix} 1 & 0 & 0 \\ 0 & 1 & 1 \end{pmatrix}, \tag{7.37}$$

Consider the case where the functional unit consists of the use of the appliance for 10 years, and its subsequent submission to the shredding process, with the use process (including the disposal of the EoL appliance) given by

Table 7.1 A Numerical Example for WIO-LCA: Unit Processes.

	Units	Appliance	Metals	Electricity	Others	Shredding	Landfill
Appliance	unit	0	0	0	0	0	0
Metals	kg	25	0.2	0	0	0	0
Electricity	kWh	130	60	0	3	100	1
Others	kg	5	0.8	0.01	0	0	0
EoL product	unit	0	0	0	0	0	0
Metal scrap	kg	0	−0.2	0	0	20	0
Residuals	kg	0	0	0.03	0	5	0
CO_2	kg	0	0.4	0.8	0.2	2.5	0.05

$$p_{\text{appliance}}^{u+e} = \begin{pmatrix} a_1^{u+e} \\ g_1^{u+e} \\ r_1^{u+e} \end{pmatrix} = \begin{pmatrix} \dfrac{\begin{matrix} 1 \text{ unit} \\ 0 \\ 900 \text{ kWh/year} \times 10 \text{ years} \\ 0 \end{matrix}}{\begin{matrix} 1 \text{unit} \\ 0 \\ 0 \\ 0 \end{matrix}} \end{pmatrix} \qquad (7.38)$$

For the sake of simplicity, it is assumed that (7.3) holds.

7.2.4.1 WIO-LCA

We now turn to the application of WIO-LCA, and start with the case of Section 7.2.2.1 where the use and discard process occur in place of final demand.

The use and discard process in place of final demand

In this case, the final demand vector becomes equal to P^{u+e}:

$$\begin{pmatrix} f_{\text{appliance}} \\ f_{\text{metals}} \\ f_{\text{electricity}} \\ f_{\text{others}} \\ w_{f \text{ EoL product}} \\ w_{f \text{ metal scrap}} \\ w_{f \text{ residues}} \\ r_{f \text{ CO}_2} \end{pmatrix} = \begin{pmatrix} 1 \\ 0 \\ 9000 \\ 0 \\ 1 \\ 0 \\ 0 \\ 0 \end{pmatrix}, \qquad (7.39)$$

and

$$\begin{pmatrix} f_1 \\ S w^f \end{pmatrix} = \begin{pmatrix} 1 \\ 0 \\ 9000 \\ 0 \\ 1 \\ 0 \end{pmatrix} \qquad (7.40)$$

The solution (7.6) then becomes

$$
\begin{pmatrix} x_{\text{appliance}} \\ x_{\text{metals}} \\ x_{\text{electricity}} \\ x_{\text{others}} \\ x_{\text{shredding}} \\ x_{\text{landfill}} \end{pmatrix} = \begin{pmatrix} 1 & 0 & 0 & 0 & 0 & 0 \\ -25 & 0.8 & 0 & 0 & 0 & 0 \\ -130 & -60 & 1 & -3 & -100 & -1 \\ -5 & -0.8 & -0.01 & 1 & 0 & 0 \\ 0 & 0 & 0 & 0 & 1 & 0 \\ 0 & 0.2 & -0.03 & 0 & -25 & 1 \end{pmatrix}^{-1} \begin{pmatrix} 1 \\ 0 \\ 9000 \\ 0 \\ 1 \\ 0 \end{pmatrix}
$$

$$
= \begin{pmatrix} 1 \\ 31 \\ 11930 \\ 149 \\ 1 \\ 377 \end{pmatrix} \tag{7.41}
$$

with

$$
CO_2 = \begin{pmatrix} 0 & 0.4 & 0.8 & 0.2 & 2.5 & 0.05 \end{pmatrix} \begin{pmatrix} 1 \\ 31 \\ 11930 \\ 149 \\ 1 \\ 377 \end{pmatrix} + 0.0 = 9607, \tag{7.42}
$$

where the last term before the second equality refers to the direct emission from the use phase, which is zero.

Next, consider decomposing the emission of CO_2 into the three life cycle phases along the lines of (7.9–7.11). Noting that in the present case the decomposition (7.8) becomes

$$
\begin{pmatrix} 1 \\ 0 \\ 9000 \\ 0 \\ 1 \\ 0 \end{pmatrix} = \underbrace{\begin{pmatrix} 1 \\ 0 \\ 0 \\ 0 \\ 0 \\ 0 \end{pmatrix}}_{\text{Production}} + \underbrace{\begin{pmatrix} 0 \\ 0 \\ 9000 \\ 0 \\ 0 \\ 0 \end{pmatrix}}_{\text{Use}} + \underbrace{\begin{pmatrix} 0 \\ 0 \\ 0 \\ 0 \\ 1 \\ 0 \end{pmatrix}}_{\text{EoL}}, \tag{7.43}
$$

the total level of production/activity induced by the whole life cycle, (7.41), can be decomposed into its life cycle components as follows

$$
\begin{pmatrix} 1 \\ 31 \\ 11930 \\ 149 \\ 1 \\ 377 \end{pmatrix} = \underbrace{\begin{pmatrix} 1 \\ 31 \\ 2222 \\ 52 \\ 0 \\ 60 \end{pmatrix}}_{\text{Production}} + \underbrace{\begin{pmatrix} 0 \\ 0 \\ 9574 \\ 96 \\ 0 \\ 287 \end{pmatrix}}_{\text{Use}} + \underbrace{\begin{pmatrix} 0 \\ 0 \\ 133 \\ 1 \\ 1 \\ 29 \end{pmatrix}}_{\text{EoL}} \tag{7.44}
$$

Applying this decomposition to (7.42), the total emission of CO_2 can then be decomposed into the three life cycle phases as follows

$$9607 = \underbrace{1804}_{\text{Production}} + \underbrace{7693}_{\text{Use}} + \underbrace{111}_{\text{EoL}}, \tag{7.45}$$

which indicates that about 80% of emission occurs in the use phase, and that the emission associated with the EoL phase is of minor importance.

Incorporation into the Production Process

Next, consider the case where P^u and P^e are incorporated into the production process as in (7.12):

$$P^{p+u+e} = \begin{pmatrix} 0 \\ 25 \\ 9130 \\ 5 \\ 1 \\ 0 \\ 0 \\ 0 \end{pmatrix} \tag{7.46}$$

The coefficients matrix that occurs on the left-hand side of (7.14) is then given by

$$\begin{pmatrix} \check{A}_{\mathrm{I,I}} & A_{\mathrm{I,II}} \\ \check{G}_{\mathrm{I}} & G_{\mathrm{II}} \end{pmatrix} = \begin{pmatrix} 0 & 0 & 0 & 0 & 0 & 0 \\ 25 & 0.2 & 0 & 0 & 0 & 0 \\ 9130 & 60 & 0 & 3 & 100 & 1 \\ 5 & 0.8 & 0.01 & 0 & 0 & 0 \\ 1 & 0 & 0 & 0 & 0 & 0 \\ 0 & -0.2 & 0 & 0 & 20 & 0 \\ 0 & 0 & 0.03 & 0 & 5 & 0 \end{pmatrix} \tag{7.47}$$

The use of the allocation matrix (7.37) transforms the above nonsquare coefficients matrix into the following square matrix

$$\begin{pmatrix} \check{A}_{\mathrm{I,I}} & A_{\mathrm{I,II}} \\ S\check{G}_{\mathrm{I}} & SG_{\mathrm{II}} \end{pmatrix} = \begin{pmatrix} 0 & 0 & 0 & 0 & 0 & 0 \\ 25 & 0.2 & 0 & 0 & 0 & 0 \\ 9130 & 60 & 0 & 3 & 100 & 1 \\ 5 & 0.8 & 0.01 & 0 & 0 & 0 \\ 1 & 0 & 0 & 0 & 0 & 0 \\ 0 & -0.2 & 0.03 & 0 & 25 & 0 \end{pmatrix} \tag{7.48}$$

The solution (7.15) then becomes

$$
\begin{pmatrix} x_{\text{appliance}} \\ x_{\text{metals}} \\ x_{\text{electricity}} \\ x_{\text{others}} \\ x_{\text{shredding}} \\ x_{\text{landfill}} \end{pmatrix} = \begin{pmatrix} 1 & 0 & 0 & 0 & 0 & 0 \\ -25 & 0.8 & 0 & 0 & 0 & 0 \\ -9130 & -60 & 1 & -3 & -100 & -1 \\ -5 & -0.8 & -0.01 & 1 & 0 & 0 \\ -1 & 0 & 0 & 0 & 1 & 0 \\ 0 & 0.2 & -0.03 & 0 & -25 & 1 \end{pmatrix}^{-1} \begin{pmatrix} 1 \\ 0 \\ 0 \\ 0 \\ 0 \\ 0 \end{pmatrix} = \begin{pmatrix} 1 \\ 31 \\ 11930 \\ 149 \\ 1 \\ 377 \end{pmatrix},
$$

$$(7.49)$$

which gives the same result as (7.41).

Extending the Technology (Coefficients) Matrix

Finally, consider the case where the use (and discard) process is introduced into the technology matrix as a new sector. The matrix is then given by (7.17)

$$
\begin{pmatrix} A_{\text{I,I}} & A_{\text{I,II}} & a_j^{u+e} \\ G_{\text{I}} & G_{\text{II}} & g_j^{u+e} \\ 0_n & 0_m & 0 \end{pmatrix} = \begin{pmatrix} 0 & 0 & 0 & 0 & 0 & 0 & 1 \\ 25 & 0.2 & 0 & 0 & 0 & 0 & 0 \\ 130 & 60 & 0 & 3 & 100 & 1 & 9000 \\ 5 & 0.8 & 0.01 & 0 & 0 & 0 & 0 \\ 0 & 0 & 0 & 0 & 0 & 0 & 1 \\ 0 & -0.2 & 0.03 & 0 & 25 & 0 & 0 \\ 0 & 0 & 0 & 0 & 0 & 0 & 0 \\ 0 & 0 & 0 & 0 & 0 & 0 & 0 \end{pmatrix},
$$

$$(7.50)$$

the last column of which is given by the first seven elements of the vector (7.39). Solving for x yields

$$
\begin{pmatrix} x_{\text{appliance}} \\ x_{\text{metals}} \\ x_{\text{electricity}} \\ x_{\text{others}} \\ x_{\text{shredding}} \\ x_{\text{landfill}} \\ x_{\text{use}} \end{pmatrix} = \begin{pmatrix} 1 & 0 & 0 & 0 & 0 & 0 & -1 \\ -25 & 0.8 & 0 & 0 & 0 & 0 & 0 \\ -130 & -60 & 1 & -3 & -100 & -1 & -9000 \\ -5 & -0.8 & -0.01 & 1 & 0 & 0 & 0 \\ 0 & 0 & 0 & 0 & 1 & 0 & -25 \\ 0 & 0.2 & -0.03 & 0 & -25 & 1 & 0 \\ 0 & 0 & 0 & 0 & 0 & 0 & 1 \end{pmatrix}^{-1} \begin{pmatrix} 0 \\ 0 \\ 0 \\ 0 \\ 0 \\ 0 \\ 1 \end{pmatrix}
$$

$$
= \begin{pmatrix} 1 \\ 31 \\ 11930 \\ 149 \\ 1 \\ 377 \\ 1 \end{pmatrix}
$$

$$(7.51)$$

7.2.4.2 WIO-LCC

Suppose that value-added ratios, π, and the price of waste, q^w are given by (the monetary unit can be the euro)

$$\pi = \begin{pmatrix} 400 & 0.2 & 0.2 & 0.8 & 1.25 & 0.01 \end{pmatrix} \tag{7.52}$$

$$q^w = \begin{pmatrix} -15 & 1 & -0.5 \end{pmatrix}, \tag{7.53}$$

which indicates that the EoL appliance is shredded for a fee of 15 monetary units per unit, the residues are landfilled for a fee of 0.5 monetary units per 1 kg, while scrap metal is sold at 1 monetary unit per kg. The relatively large value of π for the appliance is due to the fact that it refers to a unit of appliance, while for the other sectors they are in terms of 1 kg or 1 kWh.

Following (6.65) and (6.66), for the level of activity obtained above, the rate of recycling r becomes

$$r = w^{\text{in}}(\widehat{w}^{\text{out}})^{-1} = \begin{pmatrix} 0 \\ 6.25 \\ 0 \end{pmatrix} \begin{pmatrix} 1 & 0 & 0 \\ 0 & 20 & 0 \\ 0 & 0 & 377 \end{pmatrix}^{-1} = \begin{pmatrix} 0.00 \\ 0.31 \\ 0.00 \end{pmatrix} \tag{7.54}$$

Note that the recycling rate of an EoL appliance is zero, because it is allocated to the shredding process, a treatment sector, while the metal scrap recovered from the shredding process is partially recycled. From (7.23) and (7.24), ω's become

$$\omega = \begin{pmatrix} -15 & 1 & -0.5 \end{pmatrix} \left\{ \begin{pmatrix} 0 & 0 & 0 & 0 & 0 & 0 \\ 0 & 0.2 & 0 & 0 & 0 & 0 \\ 0 & 0 & 0 & 0 & 0 & 0 \end{pmatrix} - \begin{pmatrix} 1.00 & 0.00 & 0.00 \\ 0.00 & 0.69 & 0.00 \\ 0.00 & 0.00 & 1.00 \end{pmatrix} \right.$$

$$\left. \begin{pmatrix} 1 & 0 & 0 & 0 & 0 & 0 \\ 0 & 0 & 0 & 0 & 20 & 0 \\ 0 & 0 & 0.03 & 0 & 5 & 0 \end{pmatrix} \right\} + \begin{pmatrix} 400 & 0.2 & 0.2 & 0.8 & 1.25 & 0.01 \end{pmatrix}$$

$$= \begin{pmatrix} 415 & 0.4 & 0.215 & 0.8 & -10 & 0.01 \end{pmatrix} \tag{7.55}$$

Noting

$$S(I-\widehat{r})\check{G}_{\text{I}}^{\text{out}} = \begin{pmatrix} 1 & 0 & 0 \\ 0 & 1 & 1 \end{pmatrix} \begin{pmatrix} 1 & 0 & 0 \\ 0 & 0.69 & 0 \\ 0 & 0 & 1 \end{pmatrix} \begin{pmatrix} 1 & 0 & 0 & 0 \\ 0 & 0 & 0 & 0 \\ 0 & 0 & 0.03 & 0 \end{pmatrix} \begin{pmatrix} 1 & 0 & 0 & 0 \\ 0 & 0 & 0.03 & 0 \end{pmatrix},$$

$$S(I-\widehat{r})G_{\text{II}}^{\text{out}} = \begin{pmatrix} 1 & 0 & 0 \\ 0 & 1 & 1 \end{pmatrix} \begin{pmatrix} 1 & 0 & 0 \\ 0 & 0.69 & 0 \\ 0 & 0 & 1 \end{pmatrix} \begin{pmatrix} 0 & 0 \\ 20 & 0 \\ 5 & 0 \end{pmatrix} \begin{pmatrix} 0 & 0 \\ 18.75 & 0 \end{pmatrix},$$

the matrix Ψ is given by

$$
\Psi = \begin{pmatrix}
0 & 0 & 0 & 0 & 0 & 0 \\
25 & 0.2 & 0 & 0 & 0 & 0 \\
9130 & 60 & 0 & 3 & 100 & 1 \\
5 & 0.8 & 0.01 & 0 & 0 & 0 \\
1 & 0 & 0 & 0 & 0 & 0 \\
0 & 0 & 0.03 & 0 & 18.75 & 0
\end{pmatrix},
\tag{7.56}
$$

the first column elements of which can be decomposed into individual life cycle phases as follows

$$
\begin{pmatrix} a_1^p \\ S g_1^{p\ \text{out}} \end{pmatrix} = \begin{pmatrix} 0 \\ 25 \\ 130 \\ 5 \\ 0 \\ 0 \end{pmatrix}, \quad \begin{pmatrix} \breve{a}_1^u \\ S g_1^{u\ \text{out}} \end{pmatrix} = \begin{pmatrix} 0 \\ 0 \\ 9000 \\ 0 \\ 0 \\ 0 \end{pmatrix}, \quad \begin{pmatrix} a_1^e \\ S g_1^{e\ \text{out}} \end{pmatrix} = \begin{pmatrix} 0 \\ 0 \\ 0 \\ 0 \\ 1 \\ 0 \end{pmatrix}
\tag{7.57}
$$

From (7.25) we obtain

$$
\left(p_1^{p+u+e} \; \bar{p} \right) = (3105.55 \; 19.83 \; 0.24 \; 1.51 \; 18.40 \; 0.25)
\tag{7.58}
$$

The amount of life cycle cost turns out to be 3,106 monetary units, which can be further divided into the individual life cycles by use of (7.28–7.30):

$$
\left(p^p \; p^u \; p^u \right) = (934.17 \; 2137.98 \; 33.40)
\tag{7.59}
$$

It turns out that the life cycle cost is more than three times the purchase (production) cost of the appliance. The cost at the use phase constitutes 68% of the total life cycle cost, followed by the production cost of 30%.

7.2.5 Applications to LCA and LCC

We now turn to two examples where the WIO-LCA and WIO-LCC models were simultaneously applied to real data. They refer to [27] on air conditioners of different price and efficiency with the same cooling and heating capacity, and to [33] on a washing machine. In both studies, the Japanese WIOT for year 2000 with 396 endogenous sectors was used as the major body of data. The data on emissions were taken from the 3EID database [30].

In Japan, four kinds of End of Life (EoL) home appliances are to be subjected to an intensive recycling process by the home appliance recycling law (see Section 5.3.2.5). This applies to both air conditioners and washing machines. Accordingly, in both studies the EoL phase was represented by an intensive recycling process, where EoL products are disassembled into iron scrap, copper scrap, aluminum scrap, waste plastics, and shredding residues, and (in the case of air conditioners) coolant such as CFCs are recovered and properly treated (see [18]). A prototype of the EoL process of an air conditioner (the EoL process for a washing machine is almost identical to this except that coolant is not involved) is given by

$$
p^{\mathrm{EoL}} = \begin{pmatrix} \dfrac{a}{g^{\mathrm{out}}} \\ \dfrac{}{g^{\mathrm{in}}} \\ \dfrac{}{r} \end{pmatrix} = \left(\begin{array}{c} \text{utilities} \\ \text{spare parts} \\ \vdots \\ \hline \vdots \\ \text{iron scrap} \\ \text{copper scrap} \\ \text{aluminum scrap} \\ \text{waste plastics} \\ \text{shredding residues} \\ \text{residues of coolant treatment} \\ \hline 0 \\ \vdots \\ 0 \\ \hline CO_2 \\ SO_x \\ NO_x \\ \vdots \end{array} \right) \tag{7.60}
$$

7.2.5.1 Air Conditioners with Different Prices and Efficiencies

Figure 7.1 shows the scatter diagram of the purchase price and the electricity consumption per year for 20 models of air conditioners (2.5 kW type) that were available in the Japanese market in the winter of 2002. The diagram indicates the presence of a clear negative correlation between the purchase price and the efficiency. The cheapest model costs half of the most expensive one, but its efficiency is likewise half that of the most expensive.

The main concern of [27] was to perform an LCA and environmental LCC of air conditioner models with a marked difference in the purchase price and efficiency by use of WIO, that is, the cheapest model with the lowest efficiency (the low-end model) and the most expensive model with the highest efficiency (the high-end model). Electricity is the only significant input in the use phase of an air conditioner: it is the only nonzero element in the use process, and is at the same time a measure of efficiency at the use phase. The functional unit was an air conditioner of the 2.5 kW type that is used for 10 years, and then subjected to the EoL process as given by (7.60).

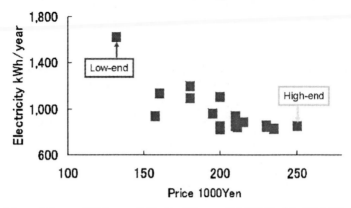

Fig. 7.1 Price and Energy Efficiency of Air Conditioners: 2.5 kW Models, 2002 Winter, Japan. Source [27].

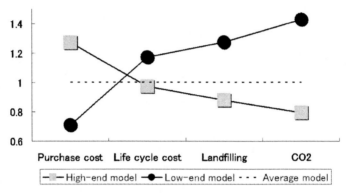

Fig. 7.2 Cost and Environmental Load of Different Air Conditioners Types: Relative Values with the Levels of Cost, Landfilling, and GWP (GWP100 in CO_2 eq) Set Unity for the Average Model (Discount Rate = 0). Source [27].

The Japanese IOT identifies the manufacturing of air conditioners as an independent sector. However, accounting for the differences in prices among different appliance models calls for some modification of the original input coefficients, which refer to the average values of all the air conditioner models that were produced in that year. Due to lack of availability at that time of the material composition of appliances with different efficiencies, the differences in the prices were accounted for by the differences in the amount of inputs of services and R&D: appliances with different prices (efficiency) were assumed to differ in their amount of inputs of services and R&D.

The results of both LCA and LCC are summarized in Figure 7.2. The high-end model performs the best in terms of both GWP and landfill, while the low-end model performs the worst. As for life cycle costs, the high-end model turns out to be superior to (lower than) the low-end model, and is not inferior to (higher than) the average model. While the low-end model has the lowest purchase price, inclusion of the cost in the use and EoL phases makes it the most expensive one. The large amount of waste for landfill of the low-end model is attributable to the generation of fly ash from electricity generation: the higher electricity consumption due to lower efficiency contributes to a larger amount of fly ash generated over the use phase (fly ash is assumed to be landfilled here, although in reality most of it is recycled).

The above results were based on no discounting of future costs. Consideration of discounting based on (7.35) revealed that the cost advantage of the high-end model over the models with lower environmental performance was sensitive to the rate of discount. The cost advantage of the high-end model over the average model disappeared at a discount rate of 5%. The high-end model became more expensive than the average model at a discount rate of 10%, and more expensive than the low-end model at 15%.

7.2.5.2 The Case of a Washing Machine

In [33], the same methodology was applied to a washing machine under a representative Japanese washing pattern, the basic data of which are reproduced in Table 7.2. The functional unit consisted of the use of a washing machine for 9 years with a washing cycle of 1.4 times a day (this value is from [22]), and the subsequent disposal in an EoL process. While repeated washing may shorten the lives of clothes, these effects were kept outside the system boundary, and not considered.

The use phase of a washing machine is more complicated than that of an air conditioner because it involves the inputs of water and detergents in addition to electricity. The use of water implies the discharge of an equal amount of water to sewage systems. On the other hand, the EoL phase of a washing machine is less complicated due to its rather simple material composition and the absence of environmentally sensitive items such as coolant, as occurs in the case of an air conditioner.

The results indicate that the use phase represents the largest share (60–70%) in both the life cycle cost and in the emissions of CO_2, NO_x, and SO_x. The use cost was mostly attributed to water and sewage, while about half the CO_2 emission at the use phase was attributed to the use of electricity, with the rest shared by the use of water/sewage and detergents. Saving of water emerged as the most important factor for reducing the life cycle cost, whereas saving of electricity was found to be the most important factor for reducing GWP (see [33] for further details).

Table 7.2 Data for the Washing Machine Example.

	Price (euro)	Use over 9 years
Washing machine	704/unit	1 unit (29 kg)
Detergents	2.11/kg	160 kg
Electricity	0.16/kWh	600 kWh
Water	$0.73/m^3$	$691 \ m^3$
Sewage	$0.54/m^3$	$691 \ m^3$

Source [33], Table 3.4.

7.3 Application of WIO to MFA

Material Flow Analysis (MFA) is widely used, among others, to identify and trace the flows of materials/substances among different sectors of the economy over space and time (see [1] for a recent review). This section is concerned with the application of WIO to MFA along the lines of [26] and [28].

7.3.1 Two Major Methods of MFA

In regard to the number of materials or substances that are simultaneously accounted for, MFA can be separated into two groups:

1. [MFA/SFA] One type of well defined/detailed material/substance, such as metal elements, at a time [7, 21, 36, 37].
2. [Bulk MFA] A broad range of rather aggregated materials such as metals, fossil fuels, industrial minerals and so forth [23, 34, 43].

Behind this grouping are the differences in primary objectives of the relevant analysis. The first group is primarily concerned with specific environmental problems related to certain types of impacts per unit flow of substances, materials, or products, while the prime concern of the second group is in problems of environmental concerns related to the throughput of firms, sectors, or regions (see [1] for details).

From the methodological point of view, the second group of MFA, bulk MFA, is closely connected to the concept of the PIOT (physical input-output table), in particular when it focuses on the throughput among sectors constituting a whole economy [4, 5, 38]. However, the rather highly aggregated nature of currently available PIOTs such as the German PIOT (see Section 5.2.3.3) does not allow one to trace the flows of individual substances such as lead or cadmium in the economy, which are usually required in MFA/SFA.

On the other hand, the first group of MFA, MFA/SFA, has the shortcoming that it is not able to address issues related to the fact that in reality many materials or substances occur not separately, but in combination. For instance, the flows of nickel and chromium are not independent of the flow of iron, because significant portions of nickel and chromium are used for producing stainless steel, an alloy of iron, nickel, and chromium. Combined use of different types of materials/substances implies the high likelihood of their combined occurrence at the EoL phase as well. This may have important implications for issues of material recovery from EoL products, or recycling of waste materials in general [16]. For instance, the occurrence of copper in iron scrap is known to reduce its resource value as an iron source [3, 13]. On the other side, the growing use of electronics in automobiles has increased its copper content, and hence increased the possibility of iron scrap from EoL vehicles being mixed up with copper in its disassembling or shredding processes. Simultaneous consideration of materials is thus an important issue to be addressed by Industrial Ecology.

The absence of analytical models and of standardized accounting frameworks can also be mentioned as a shortcoming of MFA/SFA [1, 4]. On the other hand, as we have seen in previous chapters, IOA is endowed with well-established accounting frameworks, and highly developed analytical models. This may suggest the great advantage of resorting to a PIOT as a conceptual framework of MFA [4, 25, 34]. However, even apart from their rather aggregated nature and low degree of resolution, as we have seen in Section 5.2.3.3, the applicability of PIOTs such as the German PIOT to analytical purposes appears rather limited. Besides, a PIOT is usually compiled independent of available MIOTs (monetary input-output tables), a fact which often represents prohibitive costs for its compilation [5].

The rest of this section deals with a new approach based on WIO to MFA, WIO-MFA, which was developed to cope with these problems [26, 28]. Among others, this approach provides a convenient and economical way to transform a MIOT into a PIOT.

7.3.2 WIO-MFA: Methodology

Currently available WIO tables such as those mentioned in Section 6.1.2 consist of two parts, a part referring to the flows of goods and services, and a part referring to the flows of waste. The second part is in physical units, while the first part is in monetary units (because it was taken from a monetary table, except for the columns referring to waste treatment). To make WIO tables compatible with MFA, the first part in monetary units needs to be converted to physical units.

7.3.2.1 The Mass Filter and Yield Matrices

MFA is concerned with the flows of materials in a mass unit, say, kg. On the other hand, monetary IO tables refer to the flows of both goods and services, and hence include the flows without mass, to which most services (except for those involving physical flows such as repair and maintenance) and energy inputs belong. The mass of a product corresponds to the sum of the masses of its physical components, which in turn consist of the inputs with mass which were used for their production.

Physical inputs can be divided into primary input and ancillary input. Primary input and ancillary input differ from each other in that whereas the former can constitute the mass of a product, the latter cannot. An example of ancillary input is limestone used in blast furnaces for iron production: it ends up as slag, without entering the product, pig iron. An example of primary input is a metal, say copper, used in producing a metal alloy, say brass. A given physical input can become both

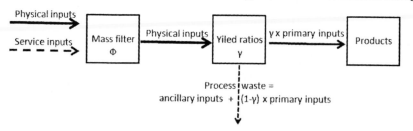

Fig. 7.3 The Flows of Inputs and the Input Composition of a Product.

primary and ancillary input depending on the way it is used: acid is primary input when used in the production of chemicals, but is ancillary input when used in a cleaning process for steel products.

While a physical product consists of primary inputs, not all primary inputs enter the product. It is usual that some of them end up as process waste without entering the product. In other words, the yield ratio of a primary input is in general less than one. An ancillary input can be represented as a physical input, the yield ratio of which is zero.

It follows that the mass of a product consists of the sum of the mass of its primary inputs adjusted for the yield ratios (Figure 7.3). Let ϕ_i be an index that takes unity when input i is physical (has mass) and zero otherwise, and let $\gamma_{ij} \in [0, 1]$ be the yield ratio of input i used in the production of product j. For a given input coefficient $a_{ij} = (A_{I,I})_{ij}$, the portion that becomes a component of product j, say $\widetilde{a}_{ij}$, is then given by

$$\widetilde{a}_{ij} = \gamma_{ij}\,\phi_i\,a_{ij} \tag{7.61}$$

Write Φ for an $n \times n$ diagonal matrix, the ith diagonal element of which is ϕ_i, and call it the mass filter [26]. Furthermore, call the $n \times n$ matrix $\Gamma = (\gamma_{ij})$ the yield matrix. The portions of the input coefficients matrix $A_{I,I}$ that enter products are then given by

$$\widetilde{A}_{I,I} = \Gamma \odot (\Phi A_{I,I}) \tag{7.62}$$

where $\odot$ refers to the Hadamard product (the element-wise product of two matrices).

The coefficients matrices that are relevant to the material components of products are $A_{I,I}$ and G_I^{in} that refer to the inputs for products. On the other hand, the matrices $A_{I,II}$ and G_{II}^{in} are not relevant for the material components of products, because they refer to the inputs for waste treatment.

Analogous to (7.62), the portions of the waste input coefficients matrix G_I^{in} that enter products are given by

$$\widetilde{G}_I^{in} = \Gamma_w \odot G_I^{in}, \tag{7.63}$$

where Γ_w refers to the $m \times n$ matrix of the yield of waste materials. The mass filter does not occur in this equation because by definition waste has mass.

7.3.2.2 Resources, Materials, and Products: A Formal Definition Based on Degrees of Fabrication

Simultaneous consideration of the flow of many materials calls for one to pay proper attention to avoid double counting, which occurs when items of different degrees of fabrication (say, metals and metal alloys) are counted as "materials" (the problem of double counting is addressed in 5.21 and 5.23 of [5]). This implies the need for a formal definition of "materials", which appears lacking in the literature on MFA. Closely related to a formal definition of "materials" are formal definitions of "resources" and "products", because "materials" are made of "resources", and "products" are made of "materials".

Partition the set of inputs with mass into mutually exclusive and exhaustive sets of "resources", R, "materials", M, and "products", P, and write

$$\begin{pmatrix} \widetilde{A}_{I,I} \\ \widetilde{G}_I^{in} \end{pmatrix} = \begin{pmatrix} \widetilde{A}_{PP} & \widetilde{A}_{PM} & \widetilde{A}_{PR} \\ \widetilde{A}_{MP} & \widetilde{A}_{MM} & \widetilde{A}_{MR} \\ \widetilde{A}_{RP} & \widetilde{A}_{RM} & \widetilde{A}_{RR} \\ \widetilde{G}_P^{in} & \widetilde{G}_M^{in} & \widetilde{G}_R^{in} \end{pmatrix} \tag{7.64}$$

The definition of "materials", "resources", and "products" is then reduced to the way in which the set of inputs is partitioned into the subsets consisting of R, M, and P. In [26], this partition is defined as follows:

1. [R] Resources are not produced but given from outside the system under consideration.
2. [M] Materials are made of resources.
3. [P] Products are made of products, materials, and/or waste.

This definition is based on the degree of fabrication, with "resources" at the lowest level of fabrication, and "products" at the highest level of fabrication. Note that in this definition "resources" have to be first transformed into "materials" to enter "products": no "resources" can enter "products" directly. Note also that no "materials" are made out of "materials". This is necessary to avoid the double counting of materials (see [28] for a formal proof on this point).

With this definition imposed, (7.64) becomes

$$\begin{pmatrix} \widetilde{A}_{I,I} \\ \widetilde{G}_I^{in} \end{pmatrix} = \begin{pmatrix} \widetilde{A}_{PP} & O & O \\ \widetilde{A}_{MP} & O & O \\ O & \widetilde{A}_{RM} & O \\ \widetilde{G}_P^{in} & O & O \end{pmatrix} \tag{7.65}$$

Note that the imposition of the above definition transforms the matrix of input coefficients into a triangular one, which was discussed in Section 3.7. In other words, for a given $A_{I,I}$, any partition of inputs that gives rise to the triangular structure (7.65) will be a valid definition of "resources", "materials", and "products" in the sense of the above definition.

The above definition of "resources" is a relative one in the sense that it depends on which level of the fabrication of inputs is chosen as "materials", that is, as the objects the flow of which is to be accounted for by MFA. Accordingly, the term "resources" in our definition is a general one that refers to inputs to the fabrication stage which immediately precedes that of "materials", and hence can diverge considerably from its conventional meanings. For example, if metal alloys are to be considered as "materials", not metal ores but metals would be counted as "resources" (see [28] for further details on this point). This feature suffices to merit calling the above definition *formal*.

In the rest of this section, the terms resources, materials, and products are used in the sense of the above definition.

In the above definition, waste is counted as alternative materials. Comment is due on this point. In most cases, waste can be counted this way. Examples are iron scrap used in EAF (electric arc furnaces) to produce steel bars for construction purposes, aluminum scrap used in die casting, waste paper used in paper mills, glass cullet used to produce glass bottles, waste wood used to produce particle board, waste plastics used to produce daily products, and so forth. Common to these examples is the fact that in its use waste (by-product II) is distinguished from its virgin counterparts because of difference in quality. In general, waste materials are inferior in quality to their virgin counterparts owing to contamination with impurities, foreign objects, and the decay of components (in the case of paper, plastics, and woods).

There are exceptional cases, however, where materials of the same quality as their virgin counterparts can be obtained from waste. Important examples to which this applies is copper, lead, and zinc scrap subjected to electrolysis: copper, lead, and zinc obtained from scrap subjected to electrolytic processes are indistinguishable in quality to those obtained from ores. In these exceptional cases, waste should be counted not as alternative to materials, but as alternative to resources (see [28] for further details).

7.3.2.3 The Materials Composition Matrix

Consider the following Leontief inverse matrix obtained from (7.65):

$$(I - \tilde{A}_{I,I})^{-1} = \begin{pmatrix} (I - \tilde{A}_{PP})^{-1} & O & O \\ \tilde{A}_{MP}(I - \tilde{A}_{PP})^{-1} & I & O \\ \tilde{A}_{RM}\tilde{A}_{MP}(I - \tilde{A}_{PP})^{-1} & \tilde{A}_{RM} & I \end{pmatrix} \qquad (7.66)$$

The matrix $\tilde{A}_{MP}(I - \tilde{A}_{PP})^{-1}$ gives the materials composition of products, with its ith row and jth column element representing the amount of material i that is contained in a unit of product j [26]. When products are made of materials alone, that is, $\tilde{A}_{PP} = 0$, this reduces to $\tilde{A}_{MP}$. The term $(I - \tilde{A}_{PP})^{-1}$ is necessary when the production of products requires the inputs of products of lower degrees of fabrication.

In analogous manner, the composition of waste materials in products will be given by (see [28])

$$\widetilde{G}_P^{in}(I - \widetilde{A}_{PP})^{-1} \tag{7.67}$$

Put together, these matrices can be termed the matrix of material composition, C:

$$C = \begin{pmatrix} \widetilde{A}_{MP} \\ \widetilde{G}_P^{in} \end{pmatrix}(I - \widetilde{A}_{PP})^{-1} \tag{7.68}$$

Write c_j for the sum of the elements of the jth column of C. When all the materials are measured in kg, c_j gives the weight of product j in kg per unit. If product j is also measured in kg, c_j will be unity.

The above result implies that by use of C (or its column sum c) a MIOT can easily be converted to a PIOT. When the materials are measured in kg, the flow table of products in monetary units, X_{PP}^*, can be converted to a physical flow by

$$\hat{c}X_{PP}^* \tag{7.69}$$

Furthermore, if one is interested in the physical flow of a particular material, say material k, that is associated with the flow X_{PP}^*, it can be obtained by

$$\widehat{(C)_k}.X_{PP}^* \tag{7.70}$$

where $\widehat{(C)_k}.$ is a diagonal matrix with C_{kj} as its jth diagonal element. In other words, by use of (7.70) one can recover the physical flow of any material under consideration from a MIOT: by use of the material composition matrix C, a MIOT can easily be converted to a flow table of materials or substances of interests.

7.3.3 Application of WIO-MFA to Metals

Nakamura and Nakajima [26] applied the above methodology to the flow of base metals in the Japanese economy. Eleven types of metals consisting of pig iron, ferroalloys, copper, lead, zinc, aluminum, and their scraps (except for ferroalloys) were considered as materials (M). The Japanese IO table for the year 2000 was used as a major data source after having extended/modified it by use of detailed and mostly physical information on the production and supply of metals and related products. The resulting IO table consisted of 416 inputs, which include the 11 types of metals measured in 10^3 kg and 10 resources (ores, stones including limestone, coal, petroleum and natural gas), and 407 endogenous sectors, of which 297 refer to sectors producing physical products (P) and the rest to services or energy sectors.

The Metal Composition of Some Products

Table 7.3 shows the estimated C for some products. For the products measured in kg, the sum is equal to unity, whereas for the products measured in monetary units (the six products at the bottom of the table), the sum gives the weights (10^3 kg) per one million Japanese yen (at the time of writing the rate of exchange was about 100 yen per dollar). Comparison of iron and steel products shows the presence of substantial variations in the use of iron scrap among them. While iron scrap occupies 90% of the material composition of ordinary steel bars, it is less than 10% for coated steel. Special steel contains larger amounts of ferroalloys than ordinary steel. So far as the metal components are concerned, electric wires and cables consist almost exclusively of copper and aluminum. Boilers, turbines, and metal machine tools are estimated to weigh around 500–600 kg per one million yen on average, while engines weigh around 830 kg per one million yen. Compared to five other products at the bottom of Table 7.3, industrial robots are characterized by a remarkably higher share of copper, which indicates the use of a larger amount of electronics.

Testing for the Model

In order to check for the reliability of the WIO-MFA model, the estimated metal weight of a passenger car was compared with real data. It was found that the estimated weight compares well (with a difference of less than 5%) with the real weight of a representative car type. Furthermore, the estimated metal composition reproduced the real composition of representative car types from both years 1997 and 2001 fairly well (Table 7.4).

Tracing the Input Origins of Materials

When product j consists of a large number of parts and combined materials such as an appliance or car, C_{ij} is the sum of material i that is contained in all these parts and combined materials. For instance, if j is a passenger car, and i is aluminum, C_{ij} refers to the sum of aluminum that is contained, among others, in the engine, body, electric parts, motors, and electric cables. It may then be of interest to search for the input origin of a certain material, for instance, the portion of copper contained in a car that originates from electrical components.

Nakamura et al. [28] showed that a slight modification of C enables one to trace the input origins of materials to preceding stages of fabrication of arbitrary order. For instance, the material composition of products that are used as direct inputs (that is, the inputs at the last stage of fabrication) to product j, say a passenger car, will be given by

$$C\,(\widetilde{A}_{PP})_{.j} \tag{7.71}$$

where $(\widetilde{A}_{PP})_{.j}$ refers to the diagonalized matrix of the jth column elements of $\widetilde{A}_{PP}$. The kth row and mth column element of this matrix refers to the amount of material

7.3 Application of WIO to MFA

Table 7.3 An Example of the Material Composition Matrix C_{MP} for Selected Products.

	Pig iron	Ferroalloys	Copper	Lead	Zinc	Aluminum	Iron scrap	Copper scrap	Lead scrap	Zinc scrap	Aluminum scrap	Total
Crude steel (converters) ordinary steel	0.928	0.010	0.000	0.000	0.000	0.000	0.061	0.000	0.000	0.000	0.000	1.00
Crude steel (converters) special steel	0.938	0.020	0.000	0.000	0.000	0.000	0.040	0.000	0.000	0.000	0.000	1.00
Crude steel (electric furnaces) ordinary steel	0.033	0.008	0.000	0.000	0.000	0.001	0.959	0.000	0.000	0.000	0.000	1.00
Crude steel (electric furnaces) special steel	0.066	0.130	0.000	0.000	0.000	0.001	0.803	0.000	0.000	0.000	0.000	1.00
Ordinary steel shapes	0.333	0.009	0.000	0.000	0.000	0.001	0.658	0.000	0.000	0.000	0.000	1.00
Ordinary steel sheets and plates	0.880	0.010	0.000	0.000	0.000	0.000	0.109	0.000	0.000	0.000	0.000	1.00
Ordinary steel strip	0.899	0.010	0.000	0.000	0.000	0.000	0.090	0.000	0.000	0.000	0.000	1.00
Ordinary steel bar	0.059	0.008	0.000	0.000	0.000	0.001	0.933	0.000	0.000	0.000	0.000	1.00
Steel pipes and tubes	0.761	0.021	0.000	0.000	0.001	0.000	0.216	0.000	0.000	0.000	0.000	1.00
Cold-finished steel	0.870	0.014	0.000	0.000	0.003	0.000	0.113	0.000	0.000	0.000	0.000	1.00
Coated steel	0.868	0.009	0.000	0.000	0.024	0.001	0.096	0.000	0.000	0.001	0.000	1.00
Electric wires and cables	0.001	0.000	0.734	0.002	0.000	0.063	0.000	0.192	0.000	0.000	0.004	1.00
Rolled and drawn aluminum	0.001	0.000	0.000	0.000	0.000	0.642	0.000	0.000	0.000	0.000	0.357	1.00
Boilers	0.362	0.010	0.006	0.000	0.002	0.005	0.134	0.006	0.000	0.000	0.001	0.53
Turbines	0.365	0.012	0.007	0.000	0.002	0.005	0.157	0.007	0.000	0.000	0.001	0.56
Engines	0.492	0.012	0.004	0.000	0.003	0.014	0.296	0.003	0.000	0.000	0.003	0.83
Industrial robots	0.280	0.010	0.010	0.000	0.002	0.007	0.130	0.004	0.000	0.000	0.002	0.44
Metal machine tools	0.361	0.014	0.002	0.000	0.002	0.007	0.190	0.002	0.000	0.000	0.001	0.58
Metal processing machinery	0.155	0.005	0.002	0.000	0.001	0.005	0.056	0.001	0.000	0.000	0.001	0.23

For the products with Total equal to unity, the figures refer to the amount of materials (kg) per 1 kg of each product. For the products with Total different from unity, the figures refer to the amount of materials (kg) per 1,000 Japanese yen of each product. Source: own computation. See [26] for details of the data used.

Table 7.4 The Metal Composition of a Passenger Car.

	Iron	Aluminum	Copper	Lead	Zinc
JAMA 1997	0.88	0.096	0.015	0.006	0.003
JAMA 2001	0.904	0.078	0.01	0.006	0.001
WIO-MFA	0.877	0.085	0.023	0.008	0.007

JAMA refers to the values due to Japan Automobile Manufacturers Association. Source: [26].

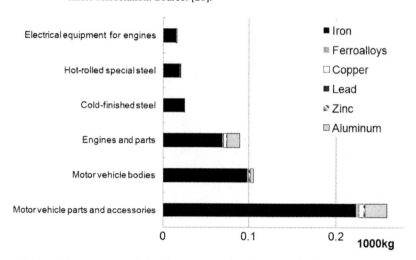

Fig. 7.4 Material Composition of the Metal Products that Constitute One Million Yen of a Passenger Car.

k that enters product j in the form of product m. For illustration, Figure 7.4 shows the origins of metals occurring in a passenger car traced to the direct inputs. It is shown that most aluminum and copper enter a passenger car in the forms of "parts & accessories" and "engine and parts".

A car engine consists of, among others, cast and forged metals, electric parts, bearings, bolts, and electric cables. One may be interested in breaking the amount of a certain material, say aluminum, that occurs in an engine into these components. Extending (7.71), it is possible to trace the input origins of materials to lower stages of fabrication [28]. The kth row and rth column element of the following matrix gives the amount of material k (say, aluminum) that occurs in product i (say, an engine) used for product j (say, a passenger car) in the form of product r (say, aluminum castings):

$$C\widehat{D_{ij}} \tag{7.72}$$

where

$$D_{ij} = (\widetilde{A}_{PP})_{\cdot i}(\widetilde{A}_{PP})_{ij} \tag{7.73}$$

Figure 7.5 shows the results of (7.72) applied to an automobile engine. It is indicated that the aluminum component of an automobile engine can mostly be

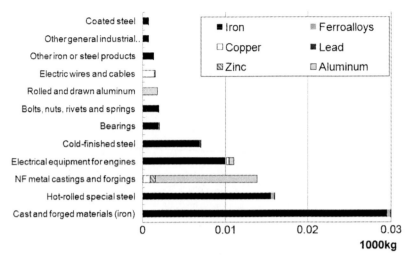

Fig. 7.5 Material Compositions of Metal Products that Constitute "Engines" Per Million Yen of a Passenger Car.

attributed to "nonferrous casting & forgings". Similarly, its copper component can be traced to "electric wires" and "nonferrous metal castings and forgings". Effective disassembling of these parts would be required if aluminum and copper are to be recovered from the engine of an EoL vehicle.

Application to Precious Metals

Nakamura et al. [29] used the WIO-MFA to evaluate the effects on the flow of metals of the introduction of lead-free solders (SAC solder), which are based on silver instead of lead, in place of traditional lead-based solders. For this purpose, the IO data mentioned above were further extended to incorporate 427 endogenous sectors including a solder sector by use of, among others, detailed information on the operations of major Japanese copper, lead, and zinc smelters. Besides the metals mentioned above, silver, gold, and tin were also added as "materials". It was found that if all the solders were replaced by SAC solders, passenger motor cars would emerge as the largest single user of metallic silver in Japan, and that most of the silver occurs in "parts and accessories" and "engines and parts".

7.4 Regional WIO Models

In Section 3.1 we discussed regional extensions of IOA. In analogous manner, regional extensions of WIO can also be considered. Industries such as cement, steel, and nonferrous metals are known to play an important role in waste recycling. Their plants, however, tend to be regionally unevenly distributed. It will usually be the case that their locations do not coincide with the regional patterns in which waste for recycling is being generated. Efficient recycling of waste materials thus calls for consideration of regional interdependence in addition to technological inter-industry relationships, including possible exploitation of regional industrial symbiosis or the formation of eco-industrial parks involving diverse industries [2]. On the other hand, the need for further improvement in the efficiency of waste treatment is likely to foster the introduction of larger treatment facilities, which in turn result in the increased need for collecting waste from wider areas (which appears to be the case in Japan at least). Consideration of these issues calls for a regional extension of the WIO model, which is the subject matter of this section.

7.4.1 Interregional WIO Model

Based on an idea similar to that of the two-region closed models discussed in Section 3.1.2, Kagawa et al. [15] proposed a regional extension of the WIO model and applied it to Japanese regional IOTs for nine regions.

7.4.1.1 Isard Type Model

Consider first the extension based on the Isard model, which, for the case of no waste flows and of two regions (a and b), is given by

$$\begin{pmatrix} x^a \\ x^b \end{pmatrix} = \left\{ I - \begin{pmatrix} A^{aa} & A^{ab} \\ A^{ba} & A^{bb} \end{pmatrix} \right\}^{-1} \begin{pmatrix} f^{aa} + f^{ab} \\ f^{ba} + f^{bb} \end{pmatrix} \tag{3.19}$$

where the first suffix on each vector and matrices refers to the region of origin, and the second suffix refers to the region of use.

The corresponding WIO model with regional extension will be obtained by extending x, f, and A to incorporate waste and waste treatment as follows

$$x^r = \begin{pmatrix} x_I^r \\ x_{II}^r \end{pmatrix} \quad (r = a, b), \tag{7.74}$$

$$f^{rs} = \begin{pmatrix} f_I^{rs} \\ S^r w_f^{rs} \end{pmatrix} \quad (r, s = a, b), \tag{7.75}$$

$$A^{rs} = \begin{pmatrix} A_{I,I}^{rs} & A_{I,II}^{rs} \\ S^r G_I^{rs} & S^r G_{II}^{rs} \end{pmatrix} \quad (r, s = a, b), \tag{7.76}$$

where the ith element of w_f^{rs}, $(w_f^{rs})_i$, refers to the amount of waste i generated by the final demand of region s and treated or recycled in region r, the ith row and jth column element of $A_{I,II}^{rs}$, $(A_{I,II}^{rs})_{ij}$, refers to the input coefficient of good i produced in region r into waste treatment sector j in region s, $(G_I^{rs})_{ij}$ refers to the amount of waste i that is generated per unit activity of production sector j in region s, and treated or recycled in region r, $(G_{II}^{rs})_{ij}$ refers to the amount of waste i that is generated per unit of activity of waste treatment sector j in region s, and treated or recycled in region r, and S^r refers to the allocation matrix that applies to region r.

Associated with the matrix G^{rs} is the transport of waste from region s to region r. In other words, the matrix G^{rs} refers to the input by regions s of waste management services produced in region r. Substitution of these equations into (3.19) gives the following WIO counterpart.

$$
\begin{pmatrix} x_I^a \\ x_{II}^a \\ x_I^b \\ x_{II}^b \end{pmatrix} = \left\{ I - \begin{pmatrix} A_{I,I}^{aa} & A_{I,II}^{aa} & A_{I,I}^{ab} & A_{I,II}^{ab} \\ S^a G_I^{aa} & S^a G_{II}^{aa} & S^a G_I^{ab} & S^a G_{II}^{ab} \\ A_{I,I}^{ba} & A_{I,II}^{ba} & A_{I,I}^{bb} & A_{I,II}^{bb} \\ S^b G_I^{ba} & S^b G_{II}^{ba} & S^b G_I^{bb} & S^b G_{II}^{bb} \end{pmatrix} \right\}^{-1} \begin{pmatrix} f_I^{aa} + f_I^{ab} \\ S^a w_f^{aa} + S^a w_f^{ab} \\ f_I^{ba} + f_I^{bb} \\ S^b w_f^{ba} + S^b w_f^{bb} \end{pmatrix}
$$

$$
= \begin{pmatrix} B_{I,I}^{aa} & B_{I,II}^{aa} & B_{I,I}^{ab} & B_{I,II}^{ab} \\ H_I^{aa} & H_{II}^{aa} & H_I^{ab} & H_{II}^{ab} \\ B_{I,I}^{ba} & B_{I,II}^{ba} & B_{I,I}^{bb} & B_{I,II}^{bb} \\ H_I^{ba} & H_{II}^{ba} & H_I^{bb} & H_{II}^{bb} \end{pmatrix} \begin{pmatrix} f_I^{aa} + f_I^{ab} \\ S^a w_f^{aa} + S^a w_f^{ab} \\ f_I^{ba} + f_I^{bb} \\ S^b w_f^{ba} + S^b w_f^{bb} \end{pmatrix} \tag{7.77}
$$

The H matrices and the matrices with the suffix "I,II" are unique to the WIO model. They do not occur in the standard regional model which does not involve waste flows. For instance, $B_{I,II}^{ab}$ gives the effects on the production activities in region a of the treatment or recycling of waste in region b, H_{II}^{ab} gives the effects on the waste treatment activities in region a of the treatment or recycling of waste in region b, and H_I^{ba} gives the effects on the waste treatment activities in region b of the production activities in region a.

7.4.1.2 Chenery-Moses Type Model

Characteristic of the Isard model is the use of detailed information on the inter-regional movements of goods and waste, or, to be more specific, the origins of inputs and the destinations of waste. Depending on the purpose of analysis and/or the availability of data, one may opt for using Chenery-Moses type models, which are less demanding in terms of the information needed for their implementation.

In Chenery-Moses type models, information on the origins of inputs and the destinations of waste is no longer required: regionally decomposed matrices or vectors such as A^{ab}, A^{ba}, G^{ab}, G^{ba}, and f^{ba} are not required, but are estimated from the matrices and vectors of competitive types A^a, A^b, G^a, G^b, and f^a, which make no distinction about the origins of inputs or the destination of waste. This estimation is done by use of "share coefficients of imports" and "trade coefficients" of waste.

Denote by d^a a vector, with $(d^a)_i$ referring to the total demand for input i in region a including imports (from region b), and by m^a a vector, with $(m^a)_i$ referring to the import of i in region a (from region b). In a manner analogous to that in Section 3.1.2.2, the coefficients referring to the shares of competitive imports can then be given by

$$\mu^a = m^a(\widehat{d^a})^{-1}, \quad \mu^b = m^b(\widehat{d^b})^{-1} \tag{7.78}$$

Furthermore, define the "trade coefficients" of waste between the two regions as follows:

$$\eta^a = h^a(\hat{w}^a)^{-1}, \quad \eta^b = h^b(\hat{w}^b)^{-1}, \tag{7.79}$$

where the ith element of w^a, $(w^a)_i$, refers to the amount of waste i generated in region a, and $(h^a)_i$ refers to the portion of $(w^a)_i$ that was treated or recycled in region b. The coefficient $(\eta^a)_i$ refers to the share of waste i that was generated in region a, and treated or recycled in region b. In other words, $(\eta^a)_i$ shows the extent to which region a depends on waste management service in region b for the treatment of waste i it generated.

Assuming the constancy of these coefficients, the following Chenery-Moses version of the WIO model will be obtained

$$
\begin{pmatrix} x_{\mathrm{I}}^a \\ x_{\mathrm{II}}^a \\ x_{\mathrm{I}}^b \\ x_{\mathrm{II}}^b \end{pmatrix} = \left\{ I - \begin{pmatrix} (I-\hat{\mu}^a)A_{\mathrm{I,I}}^a & (I-\hat{\mu}^a)A_{\mathrm{I,II}}^a & \hat{\mu}^b A_{\mathrm{I,I}}^b & \hat{\mu}^b A_{\mathrm{I,II}}^b \\ S^a(I-\hat{\eta}^a)G_{\mathrm{I}}^a & S^a(I-\hat{\eta}^a)G_{\mathrm{II}}^a & S^a\hat{\eta}^b G_{\mathrm{I}}^b & S^a\hat{\eta}^b G_{\mathrm{II}}^b \\ \hat{\mu}^a A_{\mathrm{I,I}}^a & \hat{\mu}^a A_{\mathrm{I,II}}^a & (I-\hat{\mu}^b)A_{\mathrm{I,I}}^b & (I-\hat{\mu}^b)A_{\mathrm{I,II}}^b \\ S^b\hat{\eta}^a G_{\mathrm{I}}^a & S^b\hat{\eta}^a G_{\mathrm{II}}^a & S^b(I-\hat{\eta}^b)G_{\mathrm{I}}^b & S^b(I-\hat{\eta}^b)G_{\mathrm{II}}^b \end{pmatrix} \right\}^{-1}
$$

$$
\times \begin{pmatrix} (I-\hat{\mu}^a)f_{\mathrm{I}}^a + \hat{\mu}^b f_{\mathrm{I}}^b \\ S^a(I-\hat{\eta}^a)w_f^a + S^a\hat{\eta}^b w_f^b \\ \hat{\mu}^a f_{\mathrm{I}}^a + (I-\hat{\mu}^b)f_{\mathrm{I}}^b \\ S^b\hat{\eta}^a w_f^a + S^b(I-\hat{\eta}^b)w_f^b \end{pmatrix} \tag{7.80}
$$

7.4.1.3 Application to Japanese Regional IO Tables

Kagawa et al. [15] compiled regional WIO tables of the Isard type for nine regions of Japan (Figure 7.6) based on regional IO tables for the year 1995 and survey data on the regional flows of industrial waste, and analyzed interregional movements of industrial waste by use of (7.80).

Figure 7.7 shows the effects on the level of waste incineration in Chugoku and Shikoku regions of household consumption in the Kanto region (the region with the largest population and economic size), which lies hundreds of kilometers away from the two regions. The results indicate that household consumption of the Kanto region has a significant impact on the level of waste incineration in these rather distant regions not only via its purchase of the products produced in these regions (with the associated generation of waste for incineration) but via the indirect effects of the purchase of the products produced in the Kanto region as well (production in the Kanto region requires inputs from these regions).

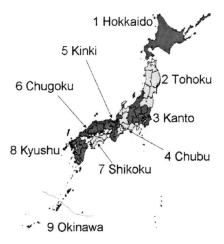

Fig. 7.6 Nine Regions of Japan. Source: Geographical Survey Institute (GSI).

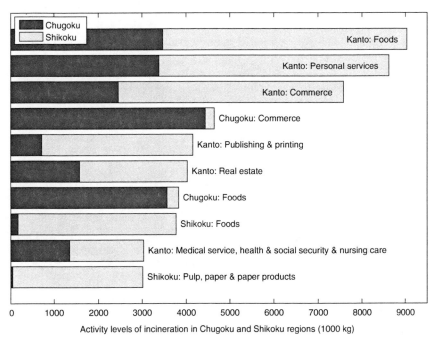

Fig. 7.7 The Effects of Household Consumption in the Kanto Region on the Level of Waste Incineration in the Chugoku and Shikoku Regions. 'Kanto: Foods' Refers to the Amount of Waste Incineration in Chugoku and Shikoku that was Induced by the Final Demand for Foods Produced in Kanto, While 'Chugoku: Foods' Refers to the Amount of Waste Incineration in Chugoku and Shikoku that was Induced by the Final Demand of Kanto for Foods Produced in Chugoku. Source [15].

Table 7.5 The Basic Structure of the Interregional WIO Table for Tokyo by Tsukui.

	Industry sectors		Waste treatment		Final demand		
	Tokyo	Others	Tokyo	Others	Tokyo	Others	Sum
Goods, Tokyo	$X_{I,I}^{TT}$	$X_{I,I}^{TR}$	$X_{I,II}^{TT}$	$X_{I,II}^{TR}$	F_I^{TT}	F_I^{TR}	x_I^T
Goods, other regions	$X_{I,I}^{RT}$	$X_{I,II}^{RR}$	$X_{I,II}^{RT}$	$X_{I,II}^{RR}$	F_I^{RT}	F_I^{RR}	x_I^R
Waste, Tokyo	W_I^{TT}	W_I^{TR}	W_{II}^{TT}	W_{II}^{TR}	W_f^{TT}	W_f^{TR}	w_f^T
Waste, other regions	W_I^{RT}	W_I^{RR}	X_{II}^{RT}	W_{II}^{RR}	W_f^{RT}	W_f^{RR}	w_f^R
Value added	V_I^T	V_I^R	V_{II}^T	V_{II}^R			V
Environmental load	E_I^T	E_I^R	E_{II}^T	E_{II}^R	E_f^R	E_f^R	E

T refers to Tokyo, and R to the rest of Japan. See Section 7.4.1.1 for further notations.

7.4.2 Regional WIO Table for Tokyo

While the focus of Kagawa et al. [15] was on the regional flows of industrial waste, Tsukui [40, 41] represent an interesting application of regional WIO which addressed both MSW (municipal solid waste) and industrial waste.

She compiled a WIO table for Tokyo of the Isard type by use of the IO table for the year 1995 of Tokyo, the 1995 WIO table for Japan, and additional information on waste flows and treatment in Tokyo (the Tokyo WIO table is provided on the Internet [42]). Table 7.5 shows the basic structure of the WIO table for Tokyo, the notations of which correspond to those of (7.77).

By use of a model similar to (7.77) implemented with reference to the Tokyo WIO, she analyzed the effects on production activities, GWP, and waste generation in both Tokyo and the rest of Japan of the introduction of a particular type of garbage disposal appliance in the households of Tokyo. It deals with a sort of small scale composting equipment driven by electricity, which can significantly reduce the weight and volume of garbage. It was found that in the event of its large scale introduction in Tokyo, the emission of GWP in Tokyo would decrease, but the emission from the rest of Japan would increase. The overall effects on GWP turned out to be a net increase in its emission except for when the rate of diffusion of the appliance was below 10% of the total households in Tokyo. Furthermore, it was found that while this measure would contribute to a significant reduction in the amount of waste generated in Tokyo, no corresponding results were obtained for the rest of Japan. The large degree of waste reduction in Tokyo was attributable to the reduction in incineration residues.

7.5 The Choice of Technology: WIO-LP

In our discussions of WIO so far, it has been assumed that for each product, there is only one production process available, just as is the case with the standard IOA. Recall that in Section 3.5.2 the standard IOA was extended to a linear programming problem (LP) by replacing the square technology matrix of the standard IOA by a rectangular matrix which incorporates alternative production processes, thus allowing for the possibility of substitution among alternative technologies.

In a similar fashion, the WIO model can also be extended to an LP which allows for the presence of alternative technologies of both production and waste treatment. Issues of substitution among alternative technologies are highly relevant to the recycling of waste materials. For instance, some limited portions of coke used for the production of pig iron in blast furnaces can be replaced by waste plastics. That is, pig iron can be produced with and without waste plastics. Similarly, paper can be produced with and without waste paper, and cement can be produced with and without slag. As these examples indicate, accounting for recycling of waste materials as substitutes for virgin materials will call for considering the presence of alternative production processes which differ from each other with regard to the use of waste materials.

With this background, Kondo and Nakamura [19] proposed a waste input-output linear programming model (the WIO-LP model), which extends the WIO model with a square technology matrix to an LP-based model with a rectangular technology matrix where the number of processes exceeds the number of products, to which we now turn.

7.5.1 Waste Input-Output Linear Programming Model

When there is a one-to-one correspondence between products and production processes, the balance equations for goods and waste are given by (5.87)

$$
\begin{pmatrix} A_{I,I} & A_{I,II} \\ SG_I & SG_{II} \end{pmatrix} \begin{pmatrix} x_I \\ x_{II} \end{pmatrix} + \begin{pmatrix} f_I \\ Sw_f \end{pmatrix} = \begin{pmatrix} x_I \\ x_{II} \end{pmatrix} \tag{5.87}
$$

Let the number of goods, waste, and waste treatment be n, m, and k. Then $A_{I,I}$ is of order $n \times n$, G_I is of order $m \times n$, and x_I is of order $n \times 1$.

In order to allow for the possibility of substitution among alternative technologies, let there be l unit processes (technologies) with $n \leq l$. The balance equations then become

$$
Dx_I = A_{I,I}x_I + A_{I,II}x_{II} + f_I, \tag{7.81}
$$

$$
w = G_I x_I + G_{II} x_{II} + w_f, \tag{7.82}
$$

$$
x_{II} = Sw \tag{7.83}
$$

where D is an $n \times l$ matrix, the elements of which are either 0 or 1. The matrix D plays the same role as D^+ in (3.128), with $d_{ij} = (D)_{ij}$ defined as

$$
d_{ij} = \begin{cases} 1 & \text{(when good } i \text{ is produced by process } j) \\ 0 & \text{(otherwise)} \end{cases} \tag{7.84}
$$

In the case where no alternative processes exist, $D = I_n$, and (7.81) reduces to the standard balance equation for x_I. Note that $A_{I,I}$ is now of order $n \times l$, G_I is of order $m \times l$, and x_I is of order $l \times 1$.

Introducing an objective function to the set of constraints (7.81–7.83) yields the following LP model, the WIO-LP model:

$$
\begin{aligned}
\text{minimize} \quad & c_I x_I + c_{II} x_{II} \\
\text{subject to} \quad & D x_I = A_{I,I} x_I + A_{I,II} x_{II} + f_I \\
& w = G_I x_I + G_{II} x_{II} + w_f \\
& x_{II} = S w \\
& x_I \geq 0_l, \ x_{II} \geq 0_k, \ w \geq 0_m \\
\text{with respect to} \quad & x_I, x_{II}, w
\end{aligned}
\tag{7.85}
$$

The restrictions in the fifth line are nonnegativity conditions, which have to be satisfied in order for the solutions to have real meanings.

The parameters (weights) of the objective function (c_I, c_{II}) can be chosen depending on the problem of concern. For instance, they can be the emission coefficients of GWP, the input coefficients of scarce materials, or of labor.

7.5.2 Making Allocation Matrices Variable

In the above LP, the allocation matrix S is given from outside and fixed (see Section 5.3.2.5 for the conditions under which S can be regarded as a constant matrix). However, one may be interested in searching for optimal ways of allocating waste to alternative waste treatment processes instead of adhering to a given S matrix. In fact, searching for optimal combinations of waste treatment processes in a way which is consistent with the flows of goods and waste will be an important component of integrated waste management policies (see [24] for the concept of integrated waste management).

It was obtained by [19] that a generalized version of (7.85) that allows for variable allocation matrices is given by the following LP (a straightforward extension of (7.85) where the elements of S occur as decision variables makes the problem nonlinear because of the presence of the term Sw, the solution of which may be hard to obtain in actual computations: see [19] for details):

$$
\begin{aligned}
\text{minimize} \quad & c_I x_I + c_{II} x_{II} \\
\text{subject to} \quad & D x_I = A_{I,I} x_I + A_{I,II} x_{II} + f_I \\
& w = G_I x_I + G_{II} x_{II} + w_f \\
& x_{II} = Z 1_m \\
& w = Z^\top 1_k \\
& x_I \geq 0_l, \ x_{II} \geq 0_k, \ w \geq 0_m, \ Z \geq 0_{k,m} \\
\text{with respect to} \quad & x_I, x_{II}, w, Z
\end{aligned}
\tag{7.86}
$$

where Z is an $k \times m$ matrix with $z_{ij} = (Z)_{ij}$ referring to the amount of waste j that is treated by waste treatment process i. The sum of the ith row elements of Z gives the amount of waste treated by waste treatment process i (the fourth equation in (7.86)), while the sum of its jth column elements gives the total amount of waste j for treatment (the fifth equation in (7.86)). Once the solution of (7.86) has been obtained, the "optimal" allocation matrix will be obtained as

$$S = Z\hat{w}^{-1} \tag{7.87}$$

Any number of additional constraints can be introduced into (7.86) to make the model more realistic. For instance, the presence of upper limits of productive or waste treatment capacities can be accounted for by introducing the constraints referring to the upper bounds of relevant activity levels. In a country or region where the space for a landfill is limited, its available amount can also be considered as an additional constraint.

7.5.3 Application to the Case Involving Alternative Waste Recycling and Treatment Technologies

Kondo and Nakamura [19] implemented the WIO-LP model (7.86) with reference to the Japanese WIO table for the year 1995 with 80 goods-producing sectors, 4 waste treatment sectors, and 42 waste types. The emissions of GWP (CO_2 originating from fossil fuels and lime stone, CH_4 from landfills, the release of CFCs contained in air conditioners and refrigerators) and the requirements for landfills were considered as possible objective functions. As for alternative production processes, the following were considered:

- The injection of waste plastics into blast furnaces (substitution of coke)
- The substitution of virgin materials by waste materials in the production of metals and glass:

    - EAF (electric arc furnace) steel in place of converter steel to increase the use of iron scrap
    - Copper scrap in the production of rolled and drawn copper in place of virgin input
    - Aluminum scrap in the production of rolled and drawn aluminum in place of virgin input
    - Glass cullet in place of silica stone in the production of glass products

Furthermore, additional waste treatment processes were considered, including the gasification of garbage, intensive recycling of electric appliances, and advanced large scale incineration processes.

The results indicate that the way of allocating waste to individual treatment processes is sensitive to the choice of the objective function and the constraints. For instance, when the emission of GWP was minimized, landfilling was a preferable treatment option of waste plastics to incineration, whereas when the consumption of landfills was minimized, the reverse results were obtained. It is interesting to note that the use of waste plastics in blast furnaces for iron production was found preferable to both incineration and landfill. Furthermore, it was found that, with other things being equal, minimization of the emission of GWP tends to increase the use of landfills, while minimization of landfills tends to increase the emission of GWP, which indicates the possible presence of a trade-off relationship between GWP and landfill consumption (see [19] for further details).

7.6 Other Applications of WIO

This section gives a brief survey of other applications of WIO that were conducted in diverse fields of Industrial Ecology.

Yamamoto et al. [44] applied WIO to evaluate the effects on landfill requirements, the emission of GWP, and resource productivity of the introduction of an advanced steel production technology (Scrap Melting Process, SMP) of a closed-loop type, which replace iron production based on blast furnaces. Furthermore, the effects were also considered of the formation of an eco-industrial park formed by steel mills based on the above new technology, and the end-of-life vehicle (ELV) disassembling and recycling facilities that provide iron scrap and other waste materials (such as waste tires as a substitute for fossil fuels) to the steel mills. The analysis was carried out based on the Japanese WIO table for the year 1995, after having extended it by incorporating processes referring to the SMP technology and advanced disassembling of ELV, which was not present in the original WIO table. It was found that these measures would result in a reduction in the emission of GWP, the use of natural resources, and the requirement of landfills.

The material flows of parts and waste that are associated with ELV was the main focus of Fuse et al. [6], whose concern was to evaluate the environmental effects of alternative recycling strategies of ELV in Japan. For this purpose, they developed an extended form of WIO (Automobile Recycling IO table) which explicitly focuses on the treatment of ELV, and the recycling of recovered waste materials as well, a schematic form of which is given by Table 7.6. As for possible alternatives of ELV treatment, among other things, intensive disassembling and the reuse of recovered parts, recovery of high quality scrap (with little contamination with other substances such as copper) and its use in the production of converter steel (not EAF steel), and conventional shredding with the recycling of shredding residues (thermal recovery) were considered. The first two measures were found to be preferable to the third one in terms of both the emission of GWP and the requirements for landfills.

Takase et al. [39] represents an application of WIO to issues of sustainable consumption, where, besides the emission of GWP, the requirement for landfills was also considered as a factor in environmental burden. A combined use was made of the WIO table for the year 2000, and detailed data on consumption patterns of Japanese households as well as associated waste generation at the consumption

Table 7.6 A Schematic form of Automobile Recycling IO Table [6].

	Industry sectors	Waste treatment sectors		Final demand
	$1 \cdots n$	ELV	Others	
Goods	$X_{I,I}$	$X_{I,V}$	$X_{I,II}$	F_I
ELV: waste materials	$W_{M,I}$	$W_{M,E}$	0	M_F
ELV: residues	$W_{R,I}$	$W_{R,E}$	0	R_F
Other waste	$W_{O,I}$	0	$W_{O,II}$	O_F

$W_{M,E}$ refers to the parts and materials recovered from ELV, $W_{M,I}$ to the portion used in industry sectors, and M_F to the portion absorbed by the final demand (export). $W_{R.}$ and $W_{O.}$ refer to the corresponding flows of ELV residues and other waste. Note that $W_{O,I}$ contain positive elements, while $W_{M,I}$ and $W_{R,I}$ do not, because ELV related waste materials and sresidues are generated by the ELV treatment sectors only.

stage. It was found that the purchase of agricultural products, stone and clay products, and cement contributed to reducing the requirements for landfills. On the other hand, the purchase of beverages and tobacco was found to make up about 20% of the total requirements for landfills induced by household consumption. The issues of "rebound effects" ([10, 11]), which are known to be a factor of importance to be accounted for in an LCA of consumption patterns, were also addressed (see also [20]).

Finally, Settani [35] represents an application of WIO to Management Accounting, where the algebra of WIO and WIO price models were used as an analytical way of dealing with end-of-pipe environmental costs at the enterprise level.

References

1. Bringezu, S., & Moriguchi, Y. (2001). Material flow analysis. In R. Ayres & L. Ayres (Eds.), *A handbook of industrial ecology*. Elgar, UK: Cheltenham.
2. Chertow, M., & Lombardi, D. (2005). Quantifying economic and environmental benefits of collocated firms. *Environmental Science & Technology, 39*(17), 6535–6541.
3. Daigo, I., Fujimaki, D., Matsuno, Y., & Adachi, A. (2004). Development of a Dynamic Model for Assessing Environmental Impact Associated with Cyclic Use of Steel. In: Proceedings, The 6th international conference on EcoBalance, Tsukuba October 2004: 25–27.
4. Daniels, P. (2002). Approaches for quantifying the metabolism of physical economies: A comparative survey, Part II: Review of individual approaches. *Journal of Industrial Ecology, 6*(1), 65–88.
5. Eurostat (2001). *Economy-wide material flow accounts and derived indicators: A methodological guide*. Luxembourg: Office for Official Publications of the European Communities.
6. Fuse, M., Kashima, S., & Yagita, H. (2005). Input-output analysis of automobile recycling. *Journal of Life Cycle Assessment Japan, 2*(1), 65–72 (in Japanese with English summary).
7. Graedel, T.E., van Beers, D., Bertram, M., Fuse, K., Gordon, R.B., Gritsinin, A., Kapur, A., Klee, R., Lifset, R., Memon, L., Rechberger, H., Spatari, S., & Vexler, D. (2004). The multilevel cycle of anthropogenic copper. *Environmental Science & Technology, 38*, 1253–1261.
8. Guinee, J. et al. (Eds.). (2002). *Handbook on life cycle assessment: Operational guide to the ISO standards*. Dordrecht: Kluwer.
9. Hendrickson, C., Lave, L., & Matthews, S. (2006). Environmental life cycle assessment of goods and services, an input-output approach. Washington, DC: Resources for the Future.

10. Hertwich, E.G. (2005). Consumption and the rebound effect: An industrial ecology perspective. *Journal of Industrial Ecology, 9*(1–2), 85–98.

11. Hertwich, E.G. (2005). Life cycle approaches to sustainable consumption: A critical review. *Environmental Science and Technology, 39*(13), 4673–4684.

12. Hunkeler, D., Lichtenvort, K., & Rebitzer, G. (Eds.). (2008). *Environmental life cycle costing.* Pensacola, FL: SETAC Press.

13. Igarashi, Y., Daigo, I., Matsuno, Y., & Adachi, Y. (2007). Estimation of quality change in domestic steel production affected by steel scrap exports. *ISIJ International, 47*(5), 753–757.

14. ISO (1997). *Environmental management. Life cycle assessment. Principles and framework.* Geneva: ISO.

15. Kagawa, S., Nakamura, S., Inamura, H., & Yamada, M. (2007). Measuring spatial repercussion effects of regional waste management. *Resources, Conservation and Recycling, 51,* 141–174.

16. Karlsson, S. (2006). Dematerialization of the metals turnover: Some reasons and prospects. In A. von Gleich, R. U. Ayres, S. Gössling-Reisemann (Eds.), *Sustainable metals management securing our future - steps towards a closed loop economy* (pp. 237–247). Dordrecht: Springer.

17. Klöpffer, W. (2003). Life-cycle based methods for sustainable product development. *International Journal of Life Cycle Assessment, 8*(3), 157–159.

18. Kondo, Y., & Nakamura, S. (2004). Evaluating alternative life-cycle strategies for electrical appliances by the waste input-output model. *International Journal of Life Cycle Assessment, 9*(4), 236–246.

19. Kondo, Y., & Nakamura, S. (2005). Waste input-output linear programming model with its application to eco-efficiency analysis. *Economic Systems Research, 17*(4), 393–408.

20. Kondo, Y., & Takase, K. (2007). An integrated model for evaluating environmental impact of consumer's behavior: Consumption 'technologies' and the waste input-output model. In S. Takada & Y. Umeda (Eds.), *Advances in life cycle engineering for sustainable manufacturing business.* London: Springer.

21. Lanzano, T., Bertram, M., De Palo, M., Wagner, C., Zyla, K., & Graedel, T. E. (2006). The contemporary European silver cycle. *Resources, Conservation and Recycling, 46,* 27–43.

22. Matsuno, Y., Tahara, K., & Inaba, A. (1996). Life cycle inventories of washing machines. *Journal of the Japan Institute of Energy, 75–12,* 1050–1055 (in Japanese with English summary).

23. Matthews, E., Amann, C., Bringezu, S., Fischer-Kowalski, M., Huttler, W., Kleijn, R., Moriguchi, Y., Ottke, C., Rodenburg, E., Rogich, D., Schandl, H., Schutz, H., van der Voet, E., & Weisz, H. (2000). *The weight of nations.* Washington, DC: World Resource Institute.

24. McDougall, F., White, P., Franke, M., & Hindel, P. (2001). *Integrated solid waste management: A life cycle inventory.* Oxford: Blackwell Science.

25. Moriguchi, Y. (2001). Material flow analysis and industrial ecology studies in Japan. In R. Ayres & L. Ayres (Eds.), *A handbook of industrial ecology* (pp. 301–310). Cheltenham, UK: Elgar.

26. Nakamura, S., & Nakajima, K. (2005). Waste input-output material flow analysis of metals in the Japanese economy. *Materials Transactions, 46,* 2550–2553.

27. Nakamura, S., & Kondo, Y. (2006). Hybrid LCC of appliances with different energy efficiency. *International Journal of Life Cycle Assessment, 11*(5), 305–314.

28. Nakamura, S., Nakajima, K., Kondo, Y., & Nagasaka, T. (2006). The waste input-output approach to materials flow analysis concepts and application to base metals. *Journal of Industrial Ecology, 11*(4), 50–63.

29. Nakamura, S., Murakami, S., Nakajima, K., & Nagasaka, T. (2008). A hybrid input-output approach to metal production and its application to the introduction of lead-free solders. *Environmental Science & Technology, 42*(10), 3843–3848.

30. Nansai, K., Moriguchi, Y., & Tohno, S. (2002). *Embodied energy and emission intensity data for Japan using input-output tables (3EID) -inventory data for LCA-.* Tsukuba, Japan: National Institute for Environmental Studies. http://www-cger.nies.go.jp/publication/D031/index.html

31. Rebitzer, G. (2002). Integrating life cycle costing and life cycle assessment for managing costs and environmental impacts in supply chain analysis and LCA. In S. Seuring & M. Goldbach (Eds.), *Cost management in supply chains* (pp. 128–146). Heidelberg: Physica-Verlag.
32. Rebitzer, G., & Hunkeler, D. (2003). Life cycle costing in LCM: Ambitions, Discussing a Framework. *International Journal of Life Cycle Assessment, 8*(5), 253–256.
33. Rebitzer, N., & Nakamura, S. (2008). Environmental life cycle costing. In D. Hunkeler, K. Lichtenvort, & G. Rebitzer (Eds.), *Environmental life cycle costing* (pp. 35–57). Pensacola, FL: SETAC Press.
34. Schandl, H., & Schulz, N. (2002). Changes in the United Kingdom's natural relations in terms of society's metabolism and land-use from 1850 to the present day. *Ecological Economics, 41*, 203–221.
35. Settanni, E., & Heijungs, R. (2008). *Feedback loops and closed-loop recycling as a driver for dynamics.* Seville, Spain: International Input Output Meeting on Managing the Environment.
36. Spatari, S., Bertram, M., Fuse, K., Shelov, E., & Graedel, T. E. (2002). The contemporary European zinc cycle: One-year stocks and flows. *Resources, Conservation, and Recycling, 39*, 137–160.
37. Spatari, S., Bertram, M., Fuge, D., Fuse, K., Graedel, T. E., & Rechberger, H. (2002). The contemporary European copper cycle: One-year stocks and flows. *Ecological Economics, 42*, 27–42.
38. Strassert, G. (2001). Physical input-output accounting. In R. Ayres & L. Ayres (Eds.), *A handbook of industrial ecology.* Cheltenham, UK: Elgar.
39. Takase, K., Kondo, Y., & Washizu, A. (2005). An analysis of sustainable consumption by the waste input-output model. *Journal of Industrial Ecology, 9*, 201–219.
40. Tsukui, M. (2007). LCA of garbage disposal for household use by interregional waste input-output analysis. *Journal of Life Cycle Assessment Japan, 13*(4), 212–220 (in Japanese with English summary).
41. Tsukui, M. (2007). Analysis of structure of waste emission in Tokyo by interregional waste input-output table. *11th International Waste Management and Landfill Symposium (Sardinia),* 1 October 2007, CD Rom.
42. Makiko Tsukui's HP http://www.tiu.ac.jp/~makiko/English/e-index. htm. Cited 1 September 2008.
43. Weisz, H., Krausmann, F., Amann, C., Eisenmenger, N., Erb, K. H., Hubacek, K., & Fischer-Kowalski, M. (2006). The physical economy of the European Union: Cross-country comparison and determinants of material consumption. *Ecological Economics, 58*, 676–698.
44. Yamamoto, Y., Yoshida, N., Morioka, T., & Moriguchi, Y. (2007). Material flow analysis of technological change and eco-industrial development in steel production Industry using waste input-output model. *Transactions of the Japan Society of Civil Engineers G, 63*(4), 304–312.
45. Yokota, K., Matsuno, Y., Yamashita, M., & Adachi, Y. (2003). Integration of life cycle assessment and population balance model for assessing environmental impacts of product population in a social scale. *International Journal of Life Cycle Assessment, 8*(3), 129–136.

Index